LINEAR ANALYSIS
OF FRAMEWORKS

T. R. GRAVES SMITH, M.A., Ph.D., C.Eng., M.I.C.E.

Senior Lecturer, Department of Civil Engineering
University of Southampton

ELLIS HORWOOD LIMITED
Publishers · Chichester

Halsted Press: a division of
JOHN WILEY & SONS
New York · Chichester · Brisbane · Ontario

First published in 1983 by
ELLIS HORWOOD LIMITED
Market Cross House, Cooper Street, Chichester, West Sussex, PO19 1EB, England

The publisher's colophon is reproduced from James Gillison's drawing of the ancient Market Cross, Chichester

Distributors

Australia, New Zealand, South-east Asia:
Jacaranda-Wiley Ltd., Jacaranda Press
JOHN WILEY & SONS INC.,
GPO Box 859, Brisbane, Queensland 40001, Australia

Canada:
JOHN WILEY & SONS CANADA LIMITED
22 Worcester Road, Rexdale, Ontario, Canada

Europe, Africa:
JOHN WILEY & SONS LIMITED
Baffins Lane, Chichester, West Sussex, England

North and South America and the rest of the world:
Halsted Press: a division of
JOHN WILEY & SONS
605 Third Avenue, New York, NY 10016, USA

© 1983 T. R. Graves Smith/Ellis Horwood Ltd.

British Library Cataloguing in Publication Data
Graves Smith, T. R.
Linear analysis of frameworks. –
(Ellis Horwood series in civil engineering)
1. Structural frames 2. Structures. Theory of.
I. Title
624.1'773 TA658.2

Library of Congress Card No. 83-4332

ISBN 0–85312–613–5 (Ellis Horwood Limited Publishers – Library Edn.)
ISBN 0–85312–614–3 (Ellis Horwood Limited Publishers – Student End.)
ISBN 0–470–27449–2 (Halsted Press)

Typeset in Press Roman by Ellis Horwood Ltd.
Printed in Great Britain by Butler & Tanner, Frome, Somerset.

Table of Contents

Preface

Structural analysis is a highly complex subject based on simple physical and mathematical concepts. Differing viewpoints are possible on how it should be presented, and I should like at once to acknowledge the many excellent texts by fellow workers in this field. Nevertheless, over the years that I have addressed the task of teaching structural analysis in a finite number of lectures to undergraduates, I have always missed the presence of a satisfactory teaching text. By this I mean a text that adequately covers all the basic concepts of the subject *and* its complexity, yet manages to be easily assimilable by a young intelligent reader encountering the subject for the first time. The complexity of structural analysis in its detailed applications must be adequately considered in a teaching course, whether by formal lecturing or by directed reading, for otherwise a thorough grasp of the subject is impossible to acquire. Amongst this complexity, perhaps the most significant is the programming problem. Since structural analysis programs are now widely available and have become routine design office tools, it should at least be the aim of undergraduate courses on this subject to cover the basics of the numerical analysis of structures. The ideal teaching text will therefore include a description of practical computation.

In writing this book, I have attempted to provide such a text. I have restricted the discussion to linear framework analysis to allow a thorough presentation of the basic aspects of the theory to be made in a book of reasonable length. The two objectives of coverage and assimilability are then primarily achieved by the selection and organisation of the material. Thus taking as a starting point the body of knowledge of a student reader who has undertaken a typical first year course on the strength of materials, the subject of structural analysis divides into a natural progression of ideas and methods. I have followed this progression from the physical idea of static equilibrium needed for analysing statically determinate frameworks, through determining deflections and flexibility, to the flexibility method for analysing statically indeterminate frameworks. The emphasis then progressively shifts from physical to numerical ideas in considering next the physical aspects of the stiffness method, then its geometrical and numerical aspects, and finally the programming problem. Each of the above subject areas becomes progressively more complicated when related to structures of increasing complexity. In illustrating this, I have made extensive use of that second tool for assimilability – the worked example, trying in particular both to integrate the subject of structural analysis and to demonstrate its complexity. Thus I have chosen, as examples, structures that are more ambitious than are usually considered in textbooks, ranging from a pin-jointed frame panel, through a continuous beam bridge, a pitched-roof portal frame and a curved arch, to a cranked space frame. The various areas of structural analysis are all then illustrated with these few structures. This should enable the reader to make a simple comparison between the two main analytical methods mentioned above: the flexibility and the stiffness methods.

In the first introductory chapter, I have defined the problem of structural analysis, introduced notation, and for conciseness have merely summarised the results of the engineering theory of beams and related one-dimensional theories. These theories are normally covered in first-year undergraduate courses, and I have omitted their derivation with some regret while recognising that a knowledge of the theories, though useful, is not *essential* for understanding structural analysis.

The material in the second chapter, concerning the analysis of statically determinate structures, is also normally covered in first year undergraduate courses, but since this *is* essential for understanding structural analysis, I have given it extensive treatment. I have particularly emphasised the construction of bending moment diagrams, because this problem frequently causes difficulty for undergraduates. The analysis of statically determinate pin-jointed frameworks traditionally receives considerable emphasis in elementary courses on structural analysis, because of its simplicity and its power in illustrating equilibrium concepts. However, when considering where to place pin-jointed frameworks in a chapter which is essentially introducing various types of structure to the reader, I have found that the logical hierarchy of structures runs from the simply supported beam, through rigid-jointed plane frameworks to space frameworks, and *then* to structures with the added complexity of internal articulation, such as compound beams and finally pin-jointed frameworks.

Chapter 3 covers the unit force method for finding the deflections of statically determinate frameworks. This method uses the equation of virtual work and here I have taken the unusual step of starting with a general statement of the equation and simplifying it for particular types of framework. This is because I do not find convincing the usual alternative procedure of starting from a restricted derivation of the virtual work equation for pin-jointed frameworks, and trying to generalise from this. The proof of the general equation of virtual work is beyond the scope of a course on structural analysis; indeed it is of little relevance. I have briefly outlined it in an appendix, but have asked the reader simply to regard the virtual work equation as a useful analytical tool. After the discussion in Chapter 2, the pin-jointed framework should be familiar to the reader. I have thus begun the discussion of the unit force method with a pin-jointed framework example, and then progressed through simple beams and rigid-jointed plane frameworks to space frameworks. In the transition up to rigid-jointed frameworks I have included two examples, respectively showing the combined effects of shear and bending, and axial and bending stress resultants. This is to avoid a common weakness of textbooks on this subject, that of separately considering axial effects in pin-jointed frameworks, and bending effects in building frames. I feel that this early separation of stress resultants in a presentation, can make the subsequent analysis of combined effects difficult to understand.

I have introduced the flexibility method in Chapter 4, by an example showing the obvious use of compatibility to solve a singly-redundant pin-jointed framework. The method is then presented in terms of obtaining the flexibility matrix for the released primary structure and solving the compatibility equations for unknown release forces. This is the presentation of the method in its classical form, and I have illustrated its use in solving fairly complex structures by hand. I have indicated how the flexibility method can be useful for the numerical analysis of continuous beams, but otherwise have made no attempt to present the method as a general numerical tool. In particular, I have omitted the numerically oriented formulation of the method in terms of individual member flexibilities, since this does seem to be an unnecessary sophistication for a first undergraduate text. The numerical application of the flexibility method is after all, a specialist area of structural analysis.

The discussion of the stiffness method again naturally divides into two parts. The first, contained in Chapter 5, is concerned with the physical aspects of the method and once more using the pin-jointed framework as a first example, I have demonstrated how the global stiffness matrix can be assembled from the member stiffness matrices and transformation matrices. Examples of rigid-jointed plane frameworks then illustrate the classical method of the slope deflection equations, but in a form that is compatible with the general stiffness method presented towards the end of the chapter. I have included a concise discussion of the method of moment distribution, as an example of solving the equilibrium equations by hand. Moment distribution will always be instructive of the physical behaviour of frameworks, and I suppose it is potentially useful for that almost mythical figure, the engineer in the jungle without access to a computer. I have not, however, included the use of moment distribution for sway frames, since it is difficult to imagine that such a calculation will ever again be undertaken. The second part of the discussion of the stiffness method is concerned with geometry and computation. This I have placed in Chapter 6. The problem of geometry concerns the calculation of the member transformation matrices from simple structural data. This is easy enough for pin-jointed frameworks and rigid-jointed plane frameworks, but is highly complex for the arbitrarily oriented members in space frameworks. I have discussed this as the last problem in the book. Chapter 6 is illustrated by two FORTRAN computer programs which contain all the basic subroutines needed for stiffness analysis. These programs for analysing respectively pin-jointed plane frameworks and rigid-jointed plane frameworks, are described in considerable detail, and the variables and arrays are labelled in a way that is consistent with the notation of the text. At the expense of some slight loss of efficiency, the programs are written in a form that can be easily read and understood. They generally illustrate the programming of a fairly complex numerical problem. I have also made several suggestions to start the reader on his own exploration of the programming of structural analysis.

In the same way as I suspect, many of my predecessors writing on structural analysis, I have been particularly concerned with the problem of notation. In describing such a complex subject, any final notation must inevitably be a compromise. In order to familiarise the reader with a *numerically* oriented notation I have distinguished the joints by numbers and the members by letters; this being a departure from the well-established classical notation of lettering the joints, (joints A and B say), and distinguishing members by the joints to which they are attached, (member AB say). Lettering the members then enables subscripts to be used to distinguish properties associated with particular members. This can be helpful for clarity, but I have been conscious that an undergrowth of subscripts is a deterrent to a reader exploring this subject for the first time. Therefore I have introduced devices for reducing the subscripts and have only used them (a) where they really are helpful for clarity and (b) where it would be misleading not to use them. The joint numbering system though, does lead to one awkwardness. Where I have wished to refer to specific points in a framework, I have distinguished them by the usual capital letter. However, when these points coincide with joints it has seemed sensible to use the joint numbers. This then leads to the possibility of statements such as, 'let us consider the section of the beam between points 1 and A'. One decision facing any writer on framework analysis is how to orient the bending moment diagram. Certainly for reinforced concrete frameworks, it is very helpful to draw the bending moment diagram on the tension side of members, the diagram then giving an immediate indication of where the main reinforcement should be placed. However, this orientation implies that the bending moment diagram in a beam should be drawn positive downwards, that is, in the opposite direction to the y coordinate. This is not mathematically attractive, which is no doubt why many writers have used the opposite convention of drawing the bending moment diagram on the compression side of members. Nevertheless, I do feel that the engineering significance of the first alternative is too important to be disregarded, and this is the alternative I have followed. Concerning matrix notation, the organisation of the material in this book allows the notation to be introduced gradually, although I have in fact assumed that the reader will be reasonably familiar with the use of matrices.

In writing this preface, I have referred throughout to the problems of undergraduate learning and teaching. I should like to add that having specifically tried to write an easily assimilable book, I hope that it will be equally useful to more senior engineers who may wish to renew their study of structural analysis, particularly with reference to the numerical programming problem.

Finally I should like to thank my colleagues at Southampton University for interesting discussions over the years, and my students for being such patient recipients of my own gradually maturing viewpoint on structural analysis. I should like to thank Professor Roy Butterfield for his encouragement during my period of writing, and I am indebted to Mrs Glynis Cooper for patiently translating a manuscript black with alterations into a pristine typescript. Lastly, I wish to record my pleasure in working with Ellis Horwood and his colleagues, and my appreciation of their friendly advice and help in producing this book.

T. R. Graves Smith
Southampton, 1983

The computer program **RJSFA** referred to in Chapter 6 is available on tape from

ELLIS HORWOOD LIMITED
Market Cross House, Cooper Street, Chichester, West Sussex, PO19 1EB, England

Although this program has been carefully tested, users should satisfy themselves that its criteria are completely valid and safe when applied to practical problems.
Neither the author nor the publishers can accept any responsibility for any liability whatsoever that might arise through its use.

To my wife Daphne

Notation

SYMBOLS

A: Cross-sectional area of a member.

b: Breadth of a section.

C: Number of constraints to produce a rigid-jointed framework.

c: Distance of the centroid of a section to a specified axis.

$c;\mathbf{c}$:   Direction cosine of a member; direction cosine matrix.

d: Depth of a section.

$d;\mathbf{d}$:   Displacement of a framework, degree of freedom; column matrix of displacements.

E: Young's modulus.

$f;\mathbf{f}$:   Flexibility coefficient; flexibility matrix of a framework.

G: Shear modulus; first moment of area of a section about a specified axis.

g: Gravitational acceleration.

I: Second moment of area of a section about a specified axis; rotational inertia of a body about an axis through the centre of mass.

J: Torsional constant of a section; number of joints in a framework.

k: Shape factor for virtual work due to shear.

$k;\mathbf{k}$:   Stiffness coefficient; stiffness matrix of a framework.

l: Length of a member.

M: Bending moment; moment of a force about an axis; mass of a particle; number of members in a framework.

$m;\mathbf{m}$:   Bending moment due to a unit force; column matrix of bending moments due to a set of unit forces.

N: Axial force; degree of redundancy of a framework; number of forces acting on a framework or body.

$n;\mathbf{n}$:   Axial force due to a unit force; column matrix of axial forces due to a set of unit forces.

$P,\mathbf{P}$:   Concentrated force on a framework; column matrix of concentrated forces.

p: Distributed force on a framework.

$Q;\mathbf{Q}$:   Intermediate concentrated force on a member; column matrix of intermediate concentrated forces.

Q: Number of cuts to produce cantilevers from a rigid-jointed framework.

$q;\mathbf{q}$:   Intermediate distributed force on a member; column matrix of intermediate distributed forces.

$R;\mathbf{R}$:   Reaction on a framework; column matrix of reactions.

R: Number of reactions on a framework.

r: Radial coordinate; radius of a section or of a curved member.

$r;\mathbf{r}$:   Reaction due to a unit force; column matrix of reactions due to a set of unit forces.

$\mathbf{r}$:   Matrix relating member displacements at one end of a member to the global displacements at that end.

S: Shear force; number of support settlements; surface area of a body.

$\mathbf{S}$:   Column matrix of stress resultants.

s: Shear force due to a unit force.

$\mathbf{s}$:   Column matrix of stress resultants due to a unit force.

T: Torque.

t: Torque due to a unit force; tension coefficient; wall thickness.

$\mathbf{t};\mathbf{t}'$:   Transformation matrix of a member; condensed transformation matrix.

$u;\mathbf{u}$:   Relative displacement at the release of a primary structure; column matrix of relative displacements.

V: Volume of a body.

W: Work done on a framework or body (W_{I} — internal work, W_{E} — external work).

x,y,z: Cartesian coordinates.

$X;\mathbf{X}$:   Release force on a primary structure: column matrix of release forces.

α: Temperature coefficient of expansion.

γ: Engineering shear strain; rotational orientation of a member in a space framework.

δ: Distribution factor.

Δ: Member extension (Δ_{E} — elastic extension, Δ_{Θ} — temperature extension, Δ_{C} — construction error).

$\Delta\theta$:   Temperature change.

ϵ: Direct strain.

θ: Curvilinear coordinate.

ν: Poisson's ratio.

ρ: Density.

σ: Direct stress.

τ: Shear stress.

$\boldsymbol{\Phi}$:   Matrix of flexibility properties of a member.

ψ: Curvilinear coordinate; (EI/l) for a member.

SUPERSCRIPTS

* :   Properties of an equilibrium force system; virtual work.

(g):   Member matrix transformed relative to the global coordinate system.

SUBSCRIPTS

e: Property of an element e of a cross-section.

F: Fixing force.

i, j etc: Indices defining particular forces, reactions, unit forces, or displacements, or properties related to them; indices in indicial notation.

m: Property of a member m in a framework.

P: Property arising from the external forces or actions on a primary structure; property arising from the member forces on a member.

Q: Property arising from the intermediate forces on a member; equivalent global force.

s: Support settlement or related property.

X: Property arising from the release forces on a primary structure.

x, y, z :   Component in the x, y or z coordinate directions; property related to the x, y or z coordinate axes.

MATHEMATICAL OPERATORS

In addition to the usual mathematical operators, the following special operators are used in this book:

$\{\ \}_m$:   All symbols inside the curly brackets are assigned the subscript m.

$\sum\limits_m$:   Summation for all the members in a framework.

$\oint(\ )\,dx$: $\displaystyle{}_m\int_0^{l_m}(\ )_m\,dx_m$

$\sum\limits_x \rightarrow$:   The equation produced by resolving in the x direction.

$\sum\limits_z \text{Ⓐ}$:   The equation produced by taking moments about a z axis passing through the point A.

NOTE ON THE SYSTÈME INTERNATIONAL (SI) UNITS

The basic SI units used in this book are the metre, m, the kilogram, kg, and the second, sec. Other units are the radian, rad, and the degree Celsius (centigrade) °C. The following prefixes are employed:

m (milli, 10^{-3}); k (kilo, 10^3); M (mega, 10^6); G (giga, 10^9).

The unit of force is the newton, N, (a force of one newton giving a kilogram mass an acceleration of 1.0 m/sec^2). Stresses are expressed in the explicit units, N/m^2. Note that in structural problems, stresses normally have magnitudes of the order of MN/m^2. Such magnitudes may alternatively be expressed as N/mm^2.

The following table gives the conversion of typical SI units to appropriate British/US units.

Property	SI units	British/US units
length	1.0 mm	39.37×10^{-3} in
		(= 1/25.4 in)
	1.0 m	3.28 ft
area	1.0 mm^2	1.550×10^{-3} in^2
	1.0 m^2	10.76 ft^2
second moment	1.0 mm^4	2.403×10^{-6} in^4
of area	1.0 m^4	115.9 ft^4
mass	1.0 kg	2.205 lb
density	1.0 kg/m^3	62.4×10^{-3} lb/ft^3
force	1.0 N	0.225 lbf
	1.0 kN	0.225 kip
stress	1.0 MN/m^2	0.145 kip/in^2
elastic modulus	1.0 GN/m^2	145.0 kip/in^2
moment	1.0 Nm	0.738 lbf ft
	1.0 kNm	0.738 kip ft
temperature change	1.0 °C	1.8 °F

CHAPTER 1

Introduction

1.1 INTRODUCTION

This book is concerned with the linear analysis of frameworks. It describes the two basic methods of analysis, the flexibility method and the stiffness method, and presents the former as a method suitable for rapid hand analysis and preliminary design, and the latter as a method suitable for computer programming. Programs for solving frameworks are now widely available for microcomputers, and one of the objects of this book is to give the reader a thorough understanding of how such programs are developed.

The analysis of frameworks is an important problem in civil and structural engineering simply because frameworks are the basis of so many engineering structures. Thus their uses range from simple supports for signs to the enormous and complicated structures required for oil rigs. Between these extremes, they are employed in building frames of all kinds (for example, multistorey office blocks, factories and warehouses), in bridge trusses, pylons, radio and television masts, derricks, tower cranes, grandstands for sports arenas, rocket launching gantries and aircraft hangars. Furthermore, other types of structure, such as slab and box bridges can also be analysed by treating them as frameworks in the form of continuous beams and grillages. Programs for analysing such frameworks are therefore an indispensable tool in bridge design offices.

The stiffness method for framework analysis contains all the important features of the finite element method for solving plate and shell structures. Thus, as well as being important in its own right, a study of framework analysis also provides an essential introduction to modern numerical techniques for solving engineering structures in widely differing fields, such as aircraft, satellite and rocket structures, shell roofs, ship-hulls, boilers, pressure vessels and liquid storage tanks.

Before beginning a detailed account of the linear analysis of frameworks, it is worth briefly considering the place of linear analysis in modern structural design. By 'linear' we mean that the analysis is restricted to structures which are completely elastic, and whose deflections can be treated as being infinitesimally small compared with their geometry. This analysis therefore excludes those very effects such as plastic yielding and buckling, which determine the **ultimate limit state** or collapse of the structure. Current design codes are based on the limit state design philosophy [1.1], [1.2] which requires the engineer to ensure that the design loading on a structure (including a partial load factor) is less than or equal to the collapse loading (divided by a partial strength factor). The direct assessment of collapse loading by the plastic analysis of collapse mechanisms, is possible for a few important types of structure such as building frames and slabs [1.3]. For the majority of other types of structure, the assessment of collapse loading by say a non-linear elasto-plastic analysis is completely uneconomic for the designer. For these structures, limit state design involves calculating the internal forces in particular elements of the structures due to the factored design loading, and ensuring that these forces are less than the strengths of the *elements* (again divided by the strength factor). The design codes supply information from which the strengths of elements can be calculated, and it thus remains to analyse the structures for the internal forces. It is at this stage that linear analysis is used in the design process, since for the majority of routine structures, non-linear analysis is again uneconomic. Linear analysis is also appropriate for one further design function, that of checking the deflections of structures for their **serviceability limit states** [1.4], the structures then being subject to reduced serviceability loading.

Referring to the non-linear elasto-plastic analysis mentioned above, we can also note that the majority of computer programs that have been developed to account for plasticity and large deflections in structures, are based on the stiffness method. Therefore a thorough knowledge of the method applied to linear problems, is certainly essential for an understanding of this advanced area of structural analysis.

In this chapter we shall describe in general the objectives of structural analysis, and define frameworks, loading and structural response. We shall conclude the chapter with a review of those basic aspects of the theory of elasticity that are needed in the subsequent text. Of necessity, the chapter is written concisely, and much of the material reviewed is contained within the traditional engineering field of study called the Strength of Materials. References for further reading will be made to texts that cover aspects of the theory particularly well. In general, the reader may find it helpful to consult a classic text by Case and Chilver [1.5] and a more recent text by Edmunds [1.6].

1.2 STRUCTURAL ANALYSIS

Structural analysis can be defined as the process of finding the **response** of a **structure** to **actions**. Each of the terms in this definition will now be briefly discussed.

The basis function of a *structure* is to transmit forces from one point to another. Thus a bridge deck for example, transmits wheel loading from the deck to the abutments; a roof structure transmits wind loading from the roof to the walls of a building and thence to the foundations; a piston in an engine transmits compression forces from the cylinder to the crank; and a wing of an aircraft transmits aerodynamic lift from the wing to the fuselage.

Actions on a structure influence its behaviour, the most important actions being the applied external forces. These forces are also called the **loading**, and engineers distinguish between various types of loading as follows. Thus **dead loading** refers to the constant forces corresponding to the self-weight of the structure; **superimposed dead loading** refers to constant forces applied after the structure is completed such as the weight of the road surfacing on a bridge deck; and **live loading** refers to forces that change with time and can be due for example, to wind, traffic, or earthquakes. The other main actions whose effects must be considered in structural analysis are **temperature changes, support settlements** and **prestress**, the latter being either deliberate prestress, or prestress due to construction errors.

The *response* of a structure is considered in two forms; the **internal forces** generated within the structure, and the **deflections**. As we have noted in the previous section, the internal forces have to be calculated as part of the limit state design process. The deflections on the other hand, are particularly important in deflection-sensitive structures, such as for example, telecommunication towers where the transmitting and receiving parabolas have to remain closely aligned under wind loading. As we have also noted in Section 1.1, linear structural analysis is carried out assuming that the material in the structure remains elastic and the deflections of the structure are small compared with its geometry. Then, the structural response is linear and directly proportional to the actions, and the responses to several actions can be linearly superimposed. Thus if two forces P_1 and P_2 acting separately on a linear structure produce two deflections d_1 and d_2 at a point, then P_1 and P_2 acting together produce the deflection $(d_1 + d_2)$ at the point. This property is called **superposition**.

1.3 FRAMEWORKS

Frameworks are two- or three-dimensional structures composed of **one-dimensional elements**.

A *one-dimensional element* is long and thin, so that all its properties can be reasonably defined by a single axial coordinate. Thus in Fig. 1.1(a), which illustrates the general case of a curved tapering element, the axis of the element is defined as the x coordinate line passing through the centroids† of all the cross-sections. We are then able to define the cross-sectional area by a one-dimensional formula such as

$$A = 0.005 + 0.0025x_1^2 \text{ m}^2 \ . \tag{1.1}$$

---

† Detailed definitions of the centroid and first and second moments of area of a cross-section are given in Appendix A1.1, pp. 52–57.

It might help in understanding this concept to note that if we wished to define the thickness t of a variable thickness slab as in Fig. 1.1(b) we should have to use the two dimensions, x_1, y_1 as shown. The slab is therefore a **two-dimensional element**.

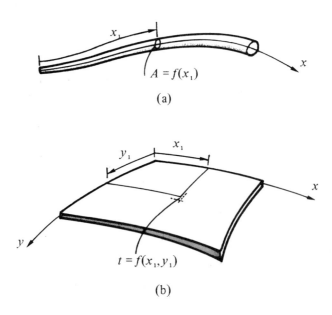

(a)

(b)

Fig. 1.1

Obviously there is a limitation on the breadth of an element that can be considered as one-dimensional, otherwise significant errors arise from the approximate one-dimensional theories used to describe its behaviour. The limitation depends on whether overall or local structural response is sought, but certainly an element whose greatest breadth is of the order of one-tenth of its length, will qualify as one-dimensional. Such an element in a framework is also referred to as a **member**, this being the usual engineering term. It will be used throughout the present text. Members have also been given different names to denote their function. Thus in pin-jointed frameworks, a **strut** is in compression and a **tie** is in tension. In rigid-jointed frameworks, a **beam** carries loading at right angles to its axis and is usually horizontal and loaded by vertical gravity loads. Its main structural action is in bending. A **column** on the other hand, carries axial loading, is usually vertical, and its main structural action is in compression.

    Two-dimensional frameworks are called **plane frameworks** and are frameworks in which all the members lie in a single plane and are loaded by forces in this plane as in Fig. 1.2(a). A **grillage** is a two-dimensional framework with the

forces acting at right angles to its plane as in Fig. 1.2(b). Frameworks where
either the members and/or the forces occupy several planes are called **space
frameworks** as in Fig. 1.2(c).

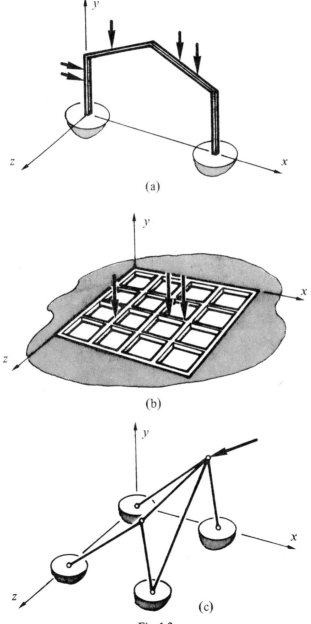

(a)

(b)

(c)

Fig. 1.2

The members of frameworks meet at **joints**. (The term **nodes** is also used in mathematical discussions.) They can either be **rigid** or **pinned**. At a rigid joint the members are welded or rivetted together as in Fig. 1.3(a). At a pinned joint, the members can rotate freely with respect to one another. In space frameworks this condition is difficult to realise practically and requires a complex articulated arrangement as in Fig. 1.3(b). In plane frameworks, pinned joints can be formed by arranging the members to rotate about a common axle at right angles to the plane of the framework as in Fig. 1.3(c). It is then occasionally helpful for clarity to refer to such joints as being **hinged**. Pinned joints will be represented in figures as in Fig. 1.3(d). Since the members of a framework are of finite cross-section, the question arises as to what point in space should be regarded as the location of the joint. The convention is to treat the framework as an assembly of **line elements** coinciding with the axes of the members, and the joints as being located at the intersection of these elements, such as point 1 in Fig. 1.3(a). In the stiffness method of analysis, joints are also given the mathematical function of defining the ends of members. This distinction is important for two cases: for the free end of a cantilever as in Fig. 1.4(a), and for the supported end of a member as in Fig. 1.4(b). In the analysis, these ends of *single* members are treated as joints.

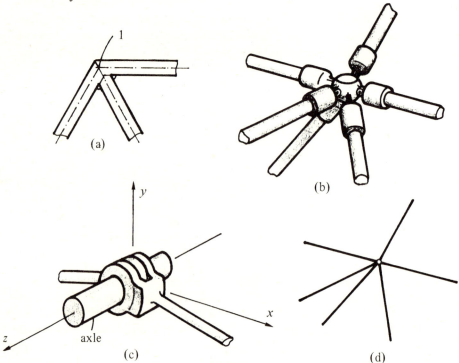

Fig. 1.3

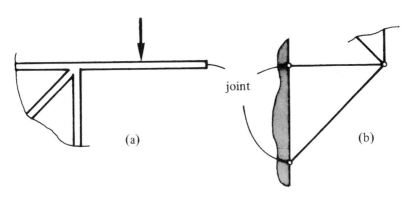

Fig. 1.4

Frameworks need to be described by a system of notation. For computer applications this is achieved by assigning a number to each joint and a number to each member. This however is not convenient for a written presentation, and we shall compromise in this text by *numbering* the joints and *lettering* the members, each in an arbitrary order, using the sequences 1, 2, 3 etc., and a, b, c, etc. The portal frame in Fig. 1.5 might well then be given the joint numbering and member lettering shown. When helpful for clarity, we shall add an appropriate letter subscript to parameters which are associated with a particular member. Thus the length of member b for example will be written l_b.

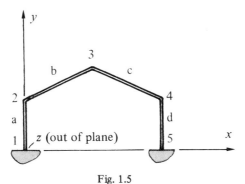

Fig. 1.5

Complete frameworks are described by **global coordinate systems**. These are composed of orthogonal coordinates x, y and z, of arbitrary orientation and with an arbitrary origin. Coordinate systems are always right-handed, with the x, y and z directions following the thumb and first and second fingers of the right hand when extended orthogonally as in Fig. 1.6. For plane frameworks we shall for convenience, always arrange the x axis horizontally and the y axis vertically, with the x–y plane coinciding with the plane of the frame. In figures, the z axis then comes out of the paper, as in Fig. 1.5.

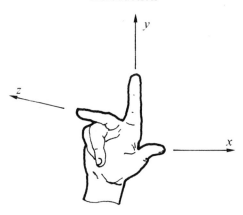

Fig. 1.6

Separate **local** or **member coordinate systems** are used to describe the members. Thus for a particular member m, the x_m coordinate coincides with the axis of the member, running from the end with the lower joint number to the end with the higher. The y_m and z_m coordinates then lie in cross-sectional plane of the member. They are always chosen to coincide with the **principal axes of the cross-section**, being those axes about which the second moment of area of the cross-section is a maximum or a minimum. In the symmetrical I-section shown in Fig. 1.7, for example, the major principal axis is orthogonal to the web, while the minor principal axis is parallel to the web. For structural reasons, it is usual to construct plane frameworks with the major principal axes of the members at right angles to the plane. This direction is chosen for the member z_m coordinates. The member coordinate systems are then completely defined, with the y_m coordinates lying in the plane of the framework. The member coordinate systems for the portal frame with the joint numbering of Fig. 1.5, for example, are arranged as in Fig. 1.8.

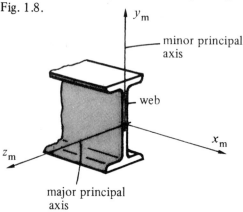

Fig. 1.7

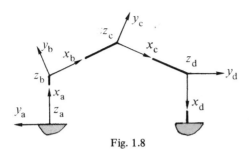

Fig. 1.8

1.4 LOADING

The loading on structures is represented by **forces**. We remind the reader that a force is an entity defined entirely in terms of its effect on the acceleration of particles. Thus a unit force applied in a particular direction to a particle of unit mass gives the particle a unit acceleration in that direction. Possessing both magnitude and direction, forces are vector quantities. Their units in the SI system are newtons (kgm/s^2).

In analysing frameworks, forces are considered in three forms as follows.

(i) **Concentrated direct forces** act on the members at specific points or on the joints. In practice the points of action can be considered to lie on the axes of the members. The adjective 'direct' distinguishes the forces from concentrated couples described below. Concentrated direct forces are, for example, applied to roof beams of building frames by subsidiary structural members called **purlins** which themselves support the roof cladding as in Fig. 1.9(a). They will be

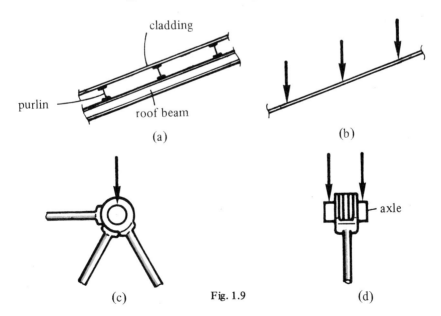

Fig. 1.9

depicted in figures by the bold arrows shown in Fig. 1.9(b). Forces on hinged joints in plane frameworks, act through the axles of the joints as shown in Figs. 1.9(c) and (d).

(ii) **Distributed forces** act on particular lengths of members and their basic units are newtons per metre. The self-weight of a beam for example, acts vertically downwards over the whole length of the beam and is equal to $\rho A g \mathrm{N/m}$, where ρ is the density of the beam material $(\mathrm{kg/m^3})$†, A is the cross-sectional area $(\mathrm{m^2})$, and g is the acceleration due to gravity $(\mathrm{m/s^2})$. Another example might be wind loading acting on a member in a particular direction. In this text, distributed forces will be depicted in figures in the manner shown in Fig. 1.10, indicating the lengths over which they act.

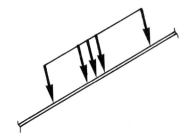

Fig. 1.10

(iii) **Concentrated couples** are hypothetical force *systems* again acting on the members at specific points or on rigid joints. Thus a concentrated couple can be thought of as being equivalent to two equal and opposite direct forces acting on the ends of a crank welded to the member at a point as in Fig. 1.11(a). If then the member were a free body, the couple would induce a rotational acceleration about an axis at right angles to the plane containing the two forces [1.7]. The magnitude of the couple is equal to the magnitude of the forces times the perpendicular distance between them, its basic units therefore being newton-metres. A couple will be depicted in figures by a bold circular arrow about a particular axis, indicating the direction of the corresponding angular acceleration, as in Fig. 1.11(b). Although it is unlikely that concentrated couples would actually be applied in the above form as part of the external loading on a framework, it will become apparent that they play an important theoretical role in structural analysis. Thus we shall see, for example, in Chapter 3, that they are applied as hypothetical external loading on a framework in order to find the rotations of the members at particular points.

---

† Values for ρ for various structural materials are listed in Table 1.2, p. 44.

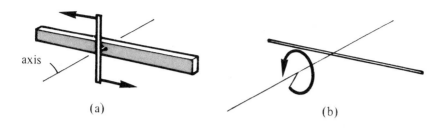

Fig. 1.11

Concerning the magnitudes of the forces usually encountered in framework analysis, since a newton is about equivalent to the self-weight of an apple, these forces are normally of the order of kilonewtons (kN) or kilonewton-metres (kNm).

A concentrated direct force being a vector, is classically denoted by a symbol in bold type such as $\mathbf{P}_1$, the subscript 1 being used to distinguish between this and other forces acting on the structure. If we then wish to refer to just the magnitude of $\mathbf{P}_1$, it can be denoted in italic type by P_1. $\mathbf{P}_1$ can be resolved into its **vector components** $\mathbf{P}_{1x}$, $\mathbf{P}_{1y}$ and $\mathbf{P}_{1z}$ in the directions of the global coordinates where these are the sides of a parallelepiped containing $\mathbf{P}_1$ as a diagonal, as in Fig. 1.12(a). The magnitudes of these vector components, P_{1x}, P_{1y} and P_{1z}, are then called the **scalar components**. In structural analysis, it is usual to simplify this notation by treating the vector components as three *separate* forces $\mathbf{P}_2$, $\mathbf{P}_3$ and $\mathbf{P}_4$ say, acting on the structure at the point of application of $\mathbf{P}_1$, as in Fig. 1.12(b).

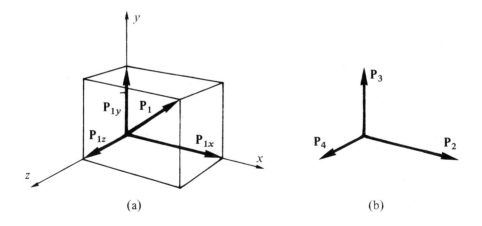

Fig. 1.12

A distributed force is a one-dimensional **vector field**. It can be distinguished from the concentrated force by denoting it by a lower case symbol such as p_1 say. The subscript 1 again distinguishes this force from other distributed forces on the framework. However, in the many cases in this text where only one such force is considered, the subscript will be dropped as an unnecessary complication.

A concentrated couple also behaves as a vector [1.8]. In structural analysis it is designated by the *same* symbol as the concentrated direct force, the distinction between the two types of vector always being clear from the context. Suppose therefore a couple is represented by a vector $\mathbf{P_5}$. The magnitude of $\mathbf{P_5}$ is then equal to the magnitude of the couple, and its direction coincides with the axis about which the couple forces act. When looking in the positive direction of the vector, the corresponding angular acceleration appears clockwise, as in Fig. 1.13(a). Couples can be resolved into their components in the three coordinate directions, each causing a clockwise acceleration about the coordinate axes. These components are again treated as *separate* couples $\mathbf{P_6}$, $\mathbf{P_7}$ and $\mathbf{P_8}$ say, acting at the point of application of $\mathbf{P_5}$. For clarity they will be depicted in figures by bold circular arrows about their axes, showing the directions of the corresponding angular accelerations as in Fig. 1.13(b).

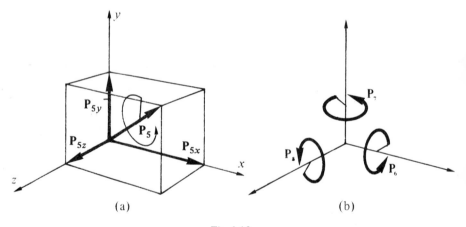

(a)   (b)

Fig. 1.13

In structural analysis it is usual to make a further modification in the classical vector notation. This is because the stiffness method in particular, involves operations on large numbers of variables that need to be described by matrices. For conciseness, matrices too are denoted by bold letters. Thus in order to avoid confusion, vectors are denoted by their symbolic *magnitudes*, that is, by subscripted letters in italic type such as P_1 say. The direction of the vector is always clear from its context. This convention will be employed in the present text, from Chapter 2 onwards.

Table 1.1
Reactions

1.5 REACTIONS

Certain joints in every framework have their movements restrained by **supports**. In restraining the joints, the supports apply forces to the structure called **reactions**, again taking the form of direct forces or couples. It might be helpful to consider the reactions as the passive forces on the framework, whereas the loading forces are the active forces. Reactions, whether direct forces or couples, will be designated by R_1, R_2 say, representing the components of the force vectors in the coordinate directions.

The reactions generated by a support depend on its type, the main types being a **roller support**, a **pinned support** and an **encastered support**. These are depicted in Table 1.1. Thus in a plane framework for example, a roller support restrains the movement in one direction, at right angles to the travel of the roller. A pinned support restrains the movement in two directions. Both supports allow the joint to rotate and the corresponding reactions take the form of direct forces in the restrained directions, as shown in Table 1.1. An encastered support restrains the movement of the joint in two directions and the rotation as well, generating the two direct force reactions and the couple reaction shown in the table.

1.6 INTERNAL FORCES

The internal forces generated by the loading on a framework, are measured by the **internal stresses**. To define internal stresses, it is helpful to imagine an infinitesimal parallelepiped element cut from a member as in Fig. 1.14. The faces of the element coincide with the local coordinate planes and **normal vectors** drawn outwards from the element at right angles to the faces are parallel to the local coordinates. The faces correspond to **coordinate surfaces**, which are defined as being positive or negative according to the direction of the normal vectors. Thus the **positive x_m coordinate surface** shown in Fig. 1.14(b) is a surface with the material situated on one side, such that the normal vector drawn outwards from the material is in the positive direction of the x_m coordinate.

Consider the intermolecular forces cut in the process of isolating the element from the member. These forces can be represented by **internal stress vectors** acting on the faces of the element giving the intensity of the forces (N/m^2), and their direction on each face. The stress vector on the positive x_m coordinate surface shown in Fig. 1.14(b) is then denoted by $\boldsymbol{\sigma}_x$. The scalar components of this vector, defined as in Section 1.4, are called the **internal stress components**. Strictly following the notation in Section 1.4, they are denoted by σ_{xx}, σ_{xy} and σ_{xz}, that is by symbols with *two* subscripts. σ_{xx} directed at right angles to the surface is called the **direct stress component**. σ_{xy} and σ_{xz} directed tangentially to the surface are called the **shear stress components**. In fact, the direct and shear stress components describe quite distinct physical behaviour in the member and, to emphasise this, in engineering texts the shear stresses are denoted by a different letter, τ. Thus the set of internal stress components σ_{xx}, τ_{xy}, τ_{xz}, acting on the positive x_m coordinate surface, are as shown in Fig. 1.14(c). In a similar way, the stress vectors $\boldsymbol{\sigma}_y$ and $\boldsymbol{\sigma}_z$ acting on the positive y_m and z_m co-ordinate surfaces respectively are represented by the direct stress components σ_{yy} and σ_{zz}, and the shear stress components τ_{yx}, τ_{yz}, and τ_{zy}, τ_{zx} as in the figure. These components then define the internal stress at a particular point in a member corresponding to the position of the infinitesimal element.

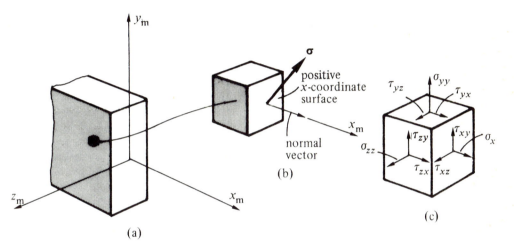

Fig. 1.14

An important result in the theory of elasticity is that the rotational equilibrium of the element requires the following pairs of **complementary shear stresses** to be equal; τ_{xy} and τ_{yx}, τ_{yz} and τ_{zy}, and τ_{zx} and τ_{xz} [1.9]. Also linear equilibrium implies that, to first order magnitude, the stress vectors on the negative coordinate surfaces are of equal magnitude but in the opposite direction to the stress vectors on the positive surfaces. We then note that if the direct stress component σ_{xx}, for example, is positive as in Fig. 1.15(a) the element is in **tension** in the x_m direction, while if it is negative, as in Fig. 1.15(b) the element is in **compression**.

In the long thin members of frameworks, it is usual to assume that the direct stresses at right angles to the member axes can be neglected. These are the stresses σ_{yy} and σ_{zz} shown in Fig. 1.14(c). The same assumption is also made for the stresses τ_{yz} and τ_{zy}. Therefore the significant stresses are those acting on the x_m coordinate surfaces, namely σ_{xx}, τ_{xy} and τ_{xz}. Analysing a framework for its internal forces therefore involves determining these stresses at every point in every member.

Fig. 1.15

1.7 STRESS RESULTANTS

It is possible to simplify the problem of determining the internal stress distribution in long thin members, by working in terms of the **stress resultants** acting on their cross-sections. Thus suppose we make a cut in a member at right angles to its axis, and consider the stresses on the *positive x_m coordinate surface* shown in Fig. 1.16(a). The whole surface is called the **cross-section** and it can be of a general shape as shown. The stress resultants are then defined as the direct forces and couples that are statically equivalent to the stresses σ_{xx}, τ_{xy} and τ_{xz} acting on the cross-section. Thus in Fig. 1.16(a), the direct forces are the **axial force** N in the x_m direction, and the two **shear forces** S_y and S_z in the y_m and z_m directions respectively. In Fig. 1.16(b), the couples are the **torque** T, clockwise about the x_m axis, and the **bending moments** M_y and M_z, clockwise about the y_m and z_m axes respectively.

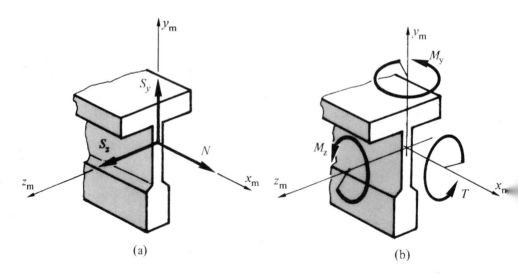

Fig. 1.16

The object of an analysis for the internal forces in frameworks is to determine the distribution of the stress resultants along the members. The stresses at any point in a cross-section can then be obtained from formulae derived by classical theories for the behaviour of one-dimensional elements. These theories form the basis of the extensive field of study traditionally called the 'Strength of Materials'. For conciseness in the present text, we shall simply summarise some important results.

The direct stresses σ_{xx} are related to the axial force and bending moments. They are derived by the **engineering theory of beams**, which assumes that, however a member deforms, plane cross-sections remain plane and at right angles to the axis. Thus if a member is made from a homogeneous material, the direct stress at point A distance y_m and z_m from the centroid C of the cross-section shown in Fig. 1.17, is given by the **engineering beam equation** as

$$\sigma_{xx} = \frac{N}{A} + M_y \frac{z_m}{I_y} - M_z \frac{y_m}{I_z} \qquad (1.2)$$

A is the area of the cross-section, and I_y and I_z are the second moments of area about the principal axes, y_m and z_m respectively [1.10].

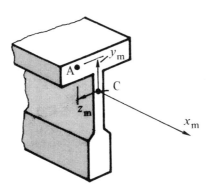

Fig. 1.17

The shear stresses τ_{xy} and τ_{xz} are related to the shear forces and the torque. Those due to the shear forces are derived by an extension of the engineering theory of beams due to St. Venant [1.11]. Thus the vertical shear stress τ_{xy} at point A in the web of the beam shown in Fig. 1.18(a) is given by St. Venant's formula as

$$\tau_{xy} = \frac{A_e c_e S_y}{b_e I_y} \cdot \qquad (1.3)$$

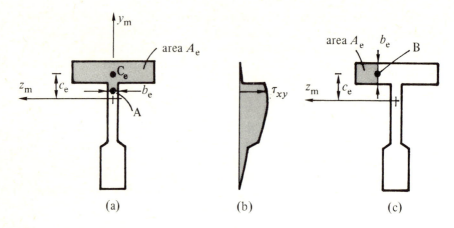

Fig. 1.18

The parameters in (1.3) distinguished by the subscript e relate to an element of the cross-section shown shaded in Fig. 1.18(a). Thus A_e is the area of the element, c_e is the distance of its centroid C_e from the centroid of the cross-section, and b_e is the breadth of the web at A. (1.3) can be shown to lead to the parabolic distribution of shear stress in the web, sketched in Fig. 1.18(b). The same formula, when applied to an element of the cross-section in the flange, as shown in Fig. 1.18(c), gives the horizontal shear stress τ_{xz} at point B. The shear stresses due to the torque, strongly depend on the shape of the cross-section of the member, and the nature of the restraints at its ends. Thus for other than circular cross-sections, plane sections distort out-of-plane as the member twists – a process called **warping**. If then warping is prevented at the ends, the shear stresses are altered throughout the member. All the analysis involving torsion in the subsequent text will be based on the assumption that the effects of this **end warping restraint** can be ignored. For members with circular cross-sections, such as the tube shown in Fig. 1.19, simple theory based on symmetry demonstrates that the shear stress τ at a point A in the section, is directed at right angles to the radius and is given by

$$\tau = \frac{Tr}{J} \tag{1.4}$$

J is the **torsional constant** of the cross-section, equal to the polar second moment of area about the x_m axis. Thus for a hollow tube say of external and internal radii r_1 and r_2 respectively,

$$J = \frac{\pi}{2}(r_1^4 - r_2^4) \tag{1.5}$$

The derivation of shear stresses for other shapes of cross-section is complicated and difficult, and cannot be concisely summarised. The interested reader is referred as a starting point to the account given by Timoshenko and Goodier [1.12].

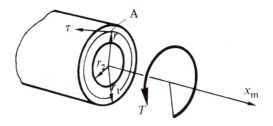

Fig. 1.19

For asymmetric cross-sections, there is interaction between the shear forces and the torque. In this case, a further member axis has to be considered passing through the **shear centres** of the cross-sections. If then a member is to be free of twist, the torque calculated about *this* axis has to be zero. The position of the shear centre in any section can be calculated from the torque produced by the St. Venant shear stresses [1.13]. Thus, for symmetric cross-sections, the shear centre coincides with the centroid; but for the channel shown in Fig. 1.20 for example, the shear centre is located on the z_m axis, at the distance from the web shown. Throughout the subsequent text it is assumed that the torque is calculated about the **shear centre axes** of members.

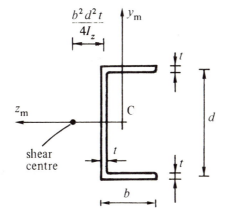

Fig. 1.20

The above summary of the determination of stresses from stress resultants, assumes that members are made of homogeneous material. Equations (1.2) and (1.3) are modified for other members such as composite or cracked reinforced concrete members [1.14]. Further changes occur if parts of the cross-section are thin. Then, plane sections no longer remain plane, and the direct stress distribution is altered by a phenomenon called **shear lag** [1.15]. This is particularly important, for example, in steel box girder bridges, and in aircraft wings.

The six stress resultants in Fig. 1.16 are all non-zero in rigid-jointed space frameworks under arbitrary loading. In other cases, many of the stress resultants are zero or can be neglected. Thus in pin-jointed frameworks, only the axial forces N are non-zero. In beams and unbraced plane frameworks composed of members with symmetrical sections about the $x-y$ plane, only the in-plane shear force S_y and the bending moment M_z need be considered. When analysing these latter structures, subscripts y and z can be omitted from the notation without causing ambiguity.

Finally in this section, we shall interpret the stress resultants physically. We first note that Newton's law of actions and reactions being equal and opposite [1.16], implies that the stress resultants acting on the negative x coordinate surface at a cut in a member are equal and opposite to those acting on the positive surface. Thus if N, S and M in a beam are positive, two sets of forces act on the cut surfaces as shown in Fig. 1.21. This concept then allows us to interpret the stress resultants physically. Consider a short section of the beam subject to stress resultants acting on its ends as in Fig. 1.22. It is then apparent that a positive axial force causes the section to be in tension. A positive shear force causes the section to rotate in a positive direction, that is clockwise about the z axis. A positive bending moment causes the section to form a curve which is convex downwards, this behaviour being called **sagging**. A negative bending moment causes a curve which is convex upwards, behaviour called **hogging**.

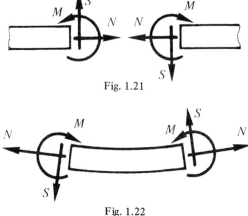

Fig. 1.21

Fig. 1.22

1.8 DEFLECTIONS

The deflections of structures are defined in terms of the **displacement vectors** d_1, d_2, etc. of particular points. In frameworks, composed as they are of one-dimensional elements, we need only consider the displacements of points on the axes of the members. Like the concentrated force vectors discussed in Section 1.4, the displacement vectors have two quite distinct meanings. Thus d_1 can be the direct linear displacement of a point, and it can be resolved into its components d_{1x}, d_{1y}, d_{1z} as in Fig. 1.23(a). Again for simplicity of notation in structural analysis, it is usual to treat these as three separate displacements d_2, d_3, d_4 say, of the point in the three coordinate directions, as in Fig. 1.23(b). (For clarity in the many figures in the subsequent text that contain both force and displacement vectors, displacement vectors will be depicted by fine arrows, as in Fig. 1.23.) Also, d_5 say, can refer to the rotational displacement at a point, and represents the clockwise rotation in radians of the member, about an axis parallel to d_5. This definition is exactly complementary to the definition of the

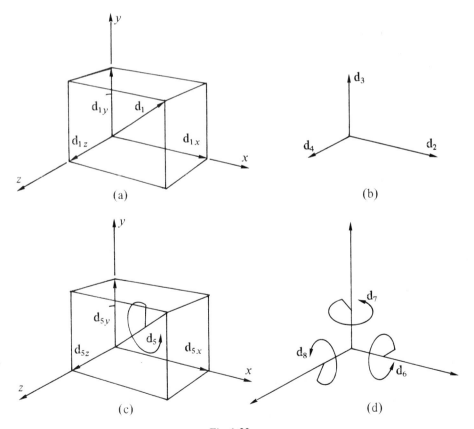

Fig. 1.23

concentrated couple $\mathbf{P}_5$ in Section 1.4. Again, $\mathbf{d}_5$ can be resolved into its components in the x, y and z directions, which, providing the rotations are *small*, are the corresponding rotations about the x, y and z axes as in Fig. 1.23(c) and (d). These are expressed as separate vectors $\mathbf{d}_6$, $\mathbf{d}_7$ and $\mathbf{d}_8$ say, and for clarity will be depicted by circular arrows, as shown in the figure. We can illustrate these definitions by considering a beam forming part of a plane framework as in Fig. 1.24. If point A moves from A_1 to A_2 in the figure, and the beam rotates, then defining the general linear and rotational displacements of A by the vectors $\mathbf{d}_2$, $\mathbf{d}_3$, $\mathbf{d}_4$, and $\mathbf{d}_6$, $\mathbf{d}_7$, $\mathbf{d}_8$ as above, the non-zero vectors for this particular movement would be $\mathbf{d}_2$, $\mathbf{d}_5$ and $\mathbf{d}_8$ as shown in the figure. We remind the reader that the displacement vectors will be denoted by their *magnitudes* from Chapter 2 of this text onwards.

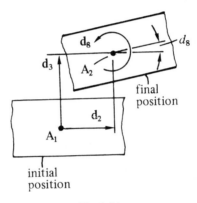

Fig. 1.24

1.9 INTERNAL DEFORMATION

The internal deformation in a framework corresponding to the internal stresses discussed in Section 1.6, is called **strain**. It is again defined by means of the infinitesimal parallelepiped element of Fig. 1.14. Thus the element deforms in distinct ways described by **strain components**. The **direct strain component** in the x direction ϵ_{xx}, for example, describes the **elongation** of the element in that direction, being the increase in its length divided by its original length. Thus an element dx long which is displaced and deformed as shown in Fig. 1.25(a), increases in length by $\epsilon_{xx}dx$. It is important to note that the rigid body displacements $\mathbf{d}_1$, $\mathbf{d}_2$ and $\mathbf{d}_3$ of the element are not considered. Similar elongations in the y and z directions are defined by the strain components ϵ_{yy} and ϵ_{zz}. The **engineering shear strain component** γ_{xy} describes the deformation of the element cross-section in the $x-y$ plane into the lozenge shape shown in Fig. 1.25(b). γ_{xy} is equal to the *reduction* in radians of the original right angle between the x and y coordinate surfaces as shown. Similar lozenge deformations of the $y-z$ and $z-x$ planes are defined by the shear strain components γ_{yz} and γ_{zx}.

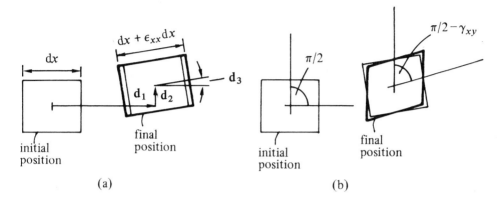

Fig. 1.25

1.10 STRESS–STRAIN RELATIONS

In members composed of **linear elastic material**, there is a unique linear relationship between the stress and strain components at each point. This relationship depends on the material and is derived experimentally by performing tensile tests and shear tests. If the material is **isotropic**, that is, if it shows no directional variation in its properties, the **stress–strain relations** take the form

$$\epsilon_{xx} = \frac{1}{E}(\sigma_{xx} - \nu(\sigma_{yy} + \sigma_{zz}))$$

$$\epsilon_{yy} = \frac{1}{E}(\sigma_{yy} - \nu(\sigma_{xx} + \sigma_{zz}))$$

$$\epsilon_{zz} = \frac{1}{E}(\sigma_{zz} - \nu(\sigma_{xx} + \sigma_{yy}))$$

$$\gamma_{xy} = \frac{1}{G}\tau_{xy}, \quad \gamma_{yz} = \frac{1}{G}\tau_{yz}, \quad \gamma_{zx} = \frac{1}{G}\tau_{zx}.$$

(1.6)

E, ν and G are **elastic constants** called **Young's modulus, Poisson's ratio** and the **shear modulus,** respectively. It can also be shown theoretically [1.17] that the elastic constants are related by a simple formula as follows

$$G = \frac{E}{2(1+\nu)}$$

(1.7)

Values for typical practical structural materials are given in Table 1.2.

In addition to the strains corresponding to the stresses given by (1.6), strains can also be generated by temperature changes. Thus if part of a framework undergoes a temperature rise of $\Delta\theta$ and expansion is unrestrained, the direct strains in the three coordinate directions each increase by $\alpha\Delta\theta$. α is called the **temperature coefficient of expansion** of the material, and typical values are included in Table 1.2.

Table 1.2
Material properties

Material	$E\,(\text{GN/m}^2)$	ν	$G\,(\text{GN/m}^2)$	$\alpha(^\circ\text{C}^{-1})$	$\rho\,(\text{kg/m}^3)$
steel	200.0	0.3	76.9	1.25×10^{-5}	7840.0
aluminium alloy	70.0	0.3	26.9	2.3×10^{-5}	2800.0
concrete	25.0–36.0	0.2	10.4–15.0	1.2×10^{-5}	2410.0
glass	60.0	0.26	23.8	0.7×10^{-5}	2580.0
timber (with grain)	7.0	–	–	0.6×10^{-5}	580.0

1.11 WORK

A fundamental theorem in structural analysis is related to the work done externally and internally, when the forces on a structure cause it to deflect. We shall conclude this chapter by deriving general expressions for these work terms, in preparation for discussing the theorem in Chapter 3.

1.11.1 External Work

The work done by the external forces on a structure is called the **external work**, W_E. Consider a framework loaded by N forces $\mathbf{P}_i$ ($i = 1, 2, \ldots, N,$) which may be either concentrated forces or couples as in Fig. 1.26 (with $N = 3$). Suppose the corresponding displacements are $\mathbf{d}_i$ ($i = 1, 2, \ldots, N$). By 'corresponding' is meant that if $\mathbf{P}_1$ say is a direct force at a point then $\mathbf{d}_1$ is the component in the direction of $\mathbf{P}_1$ of the total linear displacement of the point, as in Fig. 1.27(a). If $\mathbf{P}_3$ say is a couple, then $\mathbf{d}_3$ is the component about the same axis as $\mathbf{P}_3$ of the total rotation vector at the point as in Fig. 1.27(b). The external work done by the forces is then given simply as the sum of the products of the magnitudes of the forces and the magnitudes of the corresponding displacements. Thus

$$W_\text{E} = \sum_i (P_i d_i) \tag{1.8}$$

($\sum\limits_i$ indicates summing an expression over the range of values of i.)

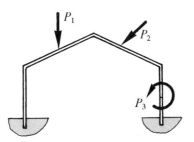

Fig. 1.26

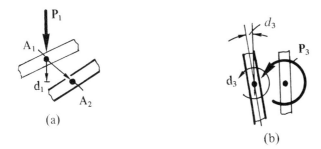

(a)

(b)

Fig. 1.27

1.11.2 Internal Work

The work done by the stresses within the members as they deform is called the **internal work**, W_I. It is found by considering the work done by the separate stress components on the infinitesimal parallelepiped element, in Fig. 1.14. Thus on the x_m coordinate surfaces, the forces due to σ_{xx} are dP_1 directed forwards, and dP_2 directed backwards as shown in Fig. 1.28(a). They are of the same magnitude, equal to σ_{xx} times the area $(dy\,dz)$ of the surfaces. These forces move through the displacements shown in Fig. 1.25(a), dP_1 doing positive work, and dP_2 doing negative work. The infinitesimal internal work, dW_I, done on the element is thus given by

$$dW_I = \sigma_{xx}\,dy\,dz\left(d_1 + \epsilon_{xx}\frac{dx}{2} - \left(d_1 - \epsilon_{xx}\frac{dx}{2}\right)\right)$$

$$= \sigma_{xx}\,\epsilon_{xx}\,dx\,dy\,dz = \sigma_{xx}\,\epsilon_{xx}\,dV$$

(1.9)

where dV is the volume of the element. Note that σ_{xx} does not do work as a result of the rigid body displacements. Similar expressions give the internal work done by the components σ_{yy} and σ_{zz}. The shear stress components τ_{xy} and τ_{yx} give rise to the shear forces dP_3, dP_4, dP_5 and dP_6 on the x and y coordinate surfaces as in Fig. 1.28(b). dP_3 and dP_5 are of equal magnitude $\tau_{xy}dydz$ and dP_4 and dP_6 are of equal magnitude $\tau_{yx}dxdz$. These forces move through the displacements shown in Fig. 1.25(b). As in the case of the direct stress components, work is not done as a result of the rigid body displacements. If then we superimpose the deformed cross-section onto the undeformed cross-section as in Fig. 1.29, the infinitesimal internal work done by the shear forces is seen to be given by

$$dW_I = \tau_{xy}dydz\left(2\frac{dx}{2}\phi_1\right) + \tau_{yx}dxdz\left(2\frac{dy}{2}\phi_2\right) \tag{1.10}$$

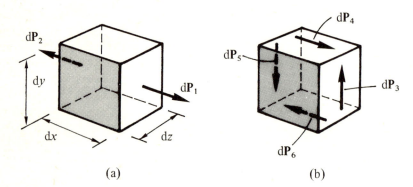

(a)   (b)

Fig. 1.28

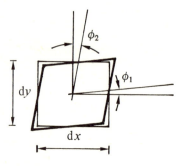

Fig. 1.29

However, since τ_{xy} and τ_{yx} are complementary shear stresses as discussed in Section 1.6, they are equal in magnitude. Thus

$$dW_I = \tau_{xy}(\phi_1 + \phi_2)dxdydz = \tau_{xy}\gamma_{xy}dV . \tag{1.11}$$

Similar expressions give the internal work done by the components τ_{yz} and τ_{zx}. Thus we finally obtain the work done by the six stress components acting simultaneously, by summing expressions of the type (1.9) and (1.11). This leads to the following general expression for the internal work done on an element of volume dV:

$$dW_I = (\sigma_{xx}\epsilon_{xx} + \sigma_{yy}\epsilon_{yy} + \sigma_{zz}\epsilon_{zz} + \tau_{xy}\gamma_{xy} + \tau_{yz}\gamma_{yz} + \tau_{zx}\gamma_{zx})dV \ .$$

(1.12)

The total internal work done in the complete framework has to be found by integrating dW_I through each member and summing. A method of evaluating this **volume integral** in terms of stress resultants will be discussed in Chapter 3.

1.12 PROBLEMS

The following introductory problems are exercises in the use of the engineering beam equation (1.2) and St. Venant's formula (1.3). Most of the problems include the calculation of the geometrical properties of sections. Definitions of the relevant properties and methods of calculating them are discussed in Appendix A1.1.

1.1   A member in a space framework has a cross-section in the form of the hollow rectangle shown in Fig. 1.30. Calculate the area A of the cross-section, and the second moments of area I_y and I_z about the y and z axes respectively. At a particular point, the member is subjected to an axial force N of 150.0 kN and a bending moment M_z of 100.0 kNm about the z axis. Calculate the maximum tensile and compressive stresses in the cross-section at the point.

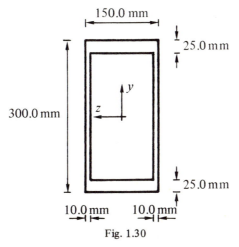

Fig. 1.30

1.2 The maximum permissible direct stresses in the member in Problem 1.1 are 100.0 MN/m² in compression and 150.0 MN/m² in tension. At a particular point, the member is again subjected to an axial force of 150.0 kN. What is the maximum permissible bending moment about the z axis that can be supported by the member at the point?

1.3 At a particular point, the member in Problem 1.1 is subjected to an axial force of 150.0 kN, and a bending moment about the z axis of 100.0 kNm. What is the maximum permissible bending moment about the y axis M_y, that can be supported by the member at the point? If this moment were applied at the point, where in the cross-section would the permissible stress be attained?

1.4 A steel flagpole is to be designed using a uniform-section thin-walled circular tube. The external diameter of the tube is to be 50.0 mm. For the purpose of limit state design, the collapse strength of the tube in bending is reached when the material at any point is subjected to a direct axial stress of 230.0 MN/m². Calculate the bending moment required to cause collapse in a tube of wall thickness t. Under the wind loading it is calculated that the flagpole is subjected to its maximum bending moment at ground level — a moment of 0.75 kNm. Determine the required thickness of the tube if the partial load factor is 1.5, and the partial strength factor is 1.2. (The second moment of area of a thin-walled tube of radius r and thickness t about a diameter passing through the centroid, is approximately equal to $\pi r^3 t$.)

1.5 A rolled steel joist with the cross-section shown in Fig. 1.31 has the following sectional properties: overall depth, 203.0 mm; breadth of flanges, 102.0 mm; $I_z = 19.57 \times 10^{-6}$ m⁴. Over a particular part of its length, the joist can be subjected to a maximum bending moment of 50.0 kNm. This is found to overstress the joist and it is proposed to

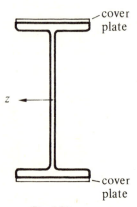

Fig. 1.31

reinforce the section with plates of breadths 102.0 mm, to be welded onto the top and bottom flanges as shown in the figure. What is the thickness of plating needed to limit the direct stresses to 100.0 MN/m² at any point in the new composite section?

1.6 The inverted U-beam with the cross-section shown in Fig. 1.32 is subjected to a sagging bending moment of 500.0 kNm at a particular point. Calculate (i) the distance of the centroid of the section from the top surface, (ii) the second moment of area about the z axis, and (iii) the maximum tensile and compressive stresses in the section at the point.

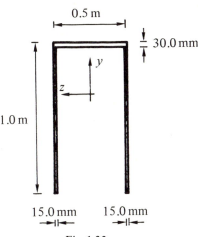

Fig. 1.32

1.7 The steel bridge section shown in Fig. 1.33 is to be analysed for its strength in bending. The following assumptions may be made:

(i) The contribution of the webs to the stress analysis may be ignored.
(ii) The stiffened top flange has the following effective properties:
  (a) area/metre = 50.0×10^3 mm²,
  (b) distance of the centroid of the flange from the top surface of the flange plate = 70.0 mm,
  (c) second moment of area/metre about a horizontal axis through the centroid = 100.0×10^6 mm⁴.
(iii) The position of the centroid of the entire bridge section is at 625.0 mm from the top surface of the top flange plate.
(iv) The collapse stress in compression of the top flange plate is 50.0 MN/m². The collapse stress in tension of the bottom flange plate is 200.0 MN/m². Assuming that the bridge can be analysed by simple engineering bending theory, calculate the positive bending moment that will cause the bridge to collapse.

(Southampton University)

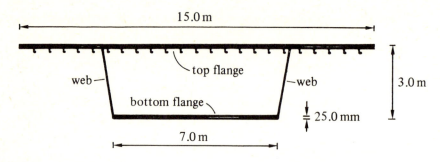

Fig. 1.33

1.8   A beam is of solid rectangular section, 25.0 mm in breadth and 150.0 mm in depth. At a particular point it is subjected to a vertical shear force of 50.0 kN. Obtain an expression for the shear stress τ_{xy} in the cross-section as it varies with y, the distance from the centroid. What is the maximum shear stress in the beam at the point?

1.9   Calculate and draw graphs of the distribution of shear stress in the flanges and web of the built-up section shown in Fig. 1.34(a), when it is subjected to a vertical shear force of 200.0 kN. Show that the maximum shear stress in the web occurs at the centroid of the section. Calculate the maximum shear stress. (The centroid of the section is at 257.5 mm from the top surface, and $I_z = 1.433 \times 10^9$ mm^4. The *direction* of the shear stresses in the flanges can be determined from the fact that St. Venant's theory predicts a continuous **shear flow** in sections, as indicated in Fig. 1.34(b).)

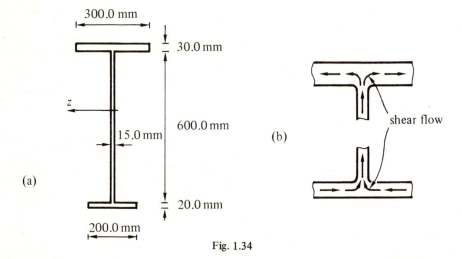

Fig. 1.34

1.10 A cold-formed channel fabricated from steel sheet of uniform thickness t, has the cross-section shown in Fig. 1.35. At a particular point it is subjected to a vertical shear force S. Show that the shear-stress distribution in the web is given by

$$\tau_{xy} = \frac{S}{2I_z}\left(bd + \frac{d^2}{4} - y^2\right)$$

and in the flanges, by

$$\tau_{xz} = \pm \frac{Sd}{2I_z}(b - z')$$

where z' is the distance from the web, shown in the figure.

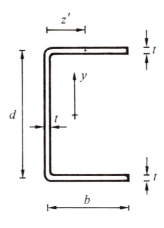

Fig. 1.35

1.11 The total shear force in a thin straight element forming part of a cross-section can be found by integrating the product (τt) along the length of the element, where τ is the shear stress in the direction of the element and t is the thickness. Use this fact to demonstrate that

(i) the shear force in the web of the channel in Problem 1.10 is equal to S,

(ii) the shear forces in the flanges are equal to $\pm (Sb^2dt/4I_z)$.

St. Venant's formula (1.3) applies only to members that are free of twist. Noting that the shear forces in the flanges together form a couple about the axis of the channel, confirm that the line of action of the external forces on the member producing the shear force, must pass through the shear centre shown in Fig. 1.20, if the member is to be free of twist.

1.12 A hollow circular tube has an external diameter of 50.0 mm and a wall
thickness of 5.0 mm. It is subjected at a particular point to a torque of
1.0 kNm. Calculate the maximum shear stress in the tube. The tube is
subjected in addition to a bending moment M at the point, generating the
usual direct bending stresses. If the maximum allowable combination of
direct stress σ_{xx} and shear stress τ in the tube material is given by

$$(\sigma_{xx}{}^2 + 3.0\,\tau^2)^{\frac{1}{2}} = 150.0\ \text{MN/m}^2$$

calculate the maximum allowable value of M. (The second moment of area
of a circular tube about a diameter passing through the centroid is equal
to $J/2$.)

APPENDIX A1.1 PROPERTIES OF SECTIONS

A1.1.1 Definitions
Consider a beam with a general cross-section containing a vertical y axis and a
horizontal z axis as shown in Fig. A1.1.1.

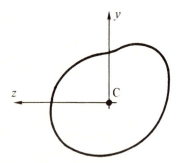

Fig. A1.1.1

The **first moment of area** G_z of the section about the z-axis is defined by the
equation

$$G_z = \int_A z\,\mathrm{d}A \ . \tag{A1.1.1}$$

For any particular case, G_z can be evaluated by considering horizontal strips of
thickness $\mathrm{d}y$, whose lengths b are some known function of y as in Fig. A1.1.2.
The first moment of area of the section is then given by

$$G_z = \int_{-d_2}^{d_1} b y\,\mathrm{d}y \ . \tag{A1.1.2}$$

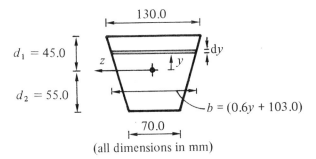

$d_1 = 45.0$

$d_2 = 55.0$

130.0

dy

$b = (0.6y + 103.0)$

70.0

(all dimensions in mm)

Fig. A1.1.2

The **centroid** of the section is the point C such that the first moment of area about any axis passing through C is zero. In particular, if C is the centroid of the section in Fig. A1.1.1, then

$$G_z = 0 . \tag{A1.1.3}$$

The vertical position of the centroid can be found by selecting an arbitrarily positioned horizontal z' axis as in Fig. A1.1.3. The distance c' of the centroid from this axis is obtained from the relation

$$Ac' = G_{z'} \tag{A1.1.4}$$

where A is the area of the section and $G_{z'}$ is the first moment of area about the z' axis. Whence

$$c' = G_{z'}/A . \tag{A1.1.5}$$

The horizontal position of the centroid can be found by selecting an arbitrarily positioned vertical axis and performing a similar calculation. If the section contains an axis of symmetry, the centroid lies on this axis.

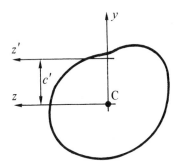

Fig. A1.1.3

The **second moment of area** I_z of the section about the z axis is defined by the equation

$$I_z = \int_A z^2 \, dA \ .$$

(A1.1.6)

Again for any particular case, I_z can be evaluated by considering horizontal strips as in Fig. A1.1.2. Then

$$I_z = \int_{-d_2}^{d_1} by^2 \, dy \ .$$

(A1.1.7)

For conciseness, if z is an axis passing through the centroid of the section as in Fig. A1.1.1, I_z is called the **self-inertia** of the section.

A1.1.2 Parallel Axis Theorems

The following two results known as the **parallel axis theorems**, are useful in calculating the properties of complicated cross-sections:

(i)   The first moment of area $G_{z'}$ of a section about an arbitrary z' axis is given by

$$G_{z'} = Ac'$$

(A1.1.8)

where A is the area of the section and c' is the distance of its centroid from the axis. (Note that this theorem has already been invoked in (A1.1.4).)

(ii) The second moment of area $I_{z'}$ of a section about an arbitrary z' axis parallel to the z axis passing through the centroid, is given by

$$I_{z'} = I_z + A(c')^2 \ .$$

(A1.1.9)

A1.1.3 Calculation of the Properties of Sections

For ease of calculation, a complicated cross-section is divided into elements whose section properties are known. For example, the section shown in Fig. A1.1.4(a) could be divided into the rectangular elements a, b, c shown in Fig. A1.1.4(b), the properties of the rectangular element e say in Fig. A1.1.5, being given by:

(i)   the centroid C_e positioned at the centre of symmetry,
(ii) the self-inertia I_{ze} about a horizontal axis z_e given by

$$I_{ze} = b_e d_e^3/12 \ .$$

(A1.1.10)

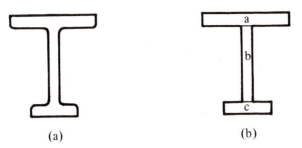

Fig. A1.1.4

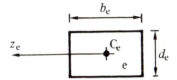

Fig. A.1.1.5

The properties of the complete cross-section are summed from the properties of the individual elements as follows:

1. *The position of the centroid* C
Consider the arbitrarily positioned z' axis in Fig. A1.1.6. Let c_e' be the distance of the centroid of an element e from this axis. Then from (A1.1.8)

$$G_{z'} = A_a c_a' + A_b c_b' + A_c c_c' = \sum_e (A_e c_e') \ ,$$

$$A = A_a + A_b + A_c = \sum_e A_e \ .$$

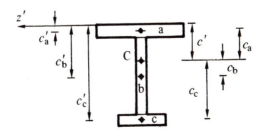

Fig. A1.1.6

Thus from (A1.1.5)

$$c' = G_{z'}/A = \sum_e (A_e c'_e)/\sum_e A_e \ . \tag{A1.1.11}$$

2. *The second moment of area about the centroidial axis* I_z, *(the self-inertia)*
Let c_e be the distance of the centroid of an element e from the z axis passing through the centroid of the complete cross-section. Then

$$c_e = c' - c'_e \tag{A1.1.12}$$

and from (A1.1.9) applied to each *element* in turn

$$I_z = I_{za} + A_a c_a^2 + I_{zb} + A_b c_b^2 + I_{zc} + A_c c_c^2$$

$$= \sum_e (I_{ze} + A_e c_e^2) \tag{A1.1.13}$$

As an example, consider the section shown in Fig. A1.1.7. The calculation of the section properties is carried out in tabular format as in Table A1.1.1. Thus the centroid is at 61.35 mm from the top surface of the section, and the second moment of area about the centroidial axis is 3.236×10^6 mm^4.

Table A1.1.1
Calculation of section properties

Element	a	b	c	$\sum_e$	Notes
A_e (mm^2)	450.0	1 050.0	1 000.0	2 500.0	
c'_e (mm)	7.5	50.0	97.5		
$A_e c'_e$ (mm^3)	3 375.0	52 500.0	97 500.0	153 375.0	$c'= 61.35$ mm (A1.1.11)
c_e (mm)	53.85	11.35	−36.15		$\equiv (c' - c'_e)$ (A1.1.12)
c_e^2 (mm^2)	2 899.82	128.82	1 306.82		
$A_e c_e^2$ (mm^4)	1 304 919.0	135 261.0	1 306 820.0	2 747 000.0	
I_{ze} (mm^4)	8 437.0	428 750.0	52 083.0	489 270.0	$b_e d_e^3/12$ (A1.1.10)

$$\sum_e (I_{ze} + A_e c_e^2) = 3\ 236\ 270.0$$

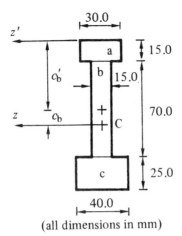

(all dimensions in mm)

Fig. A1.1.7

REFERENCES

[1.1]   B.S. 5400: Steel, Concrete and Composite Bridges, Part 1: General State-
        ment, British Standards Institution, London, 1978, pp. 1-2, (for example).
[1.2]   Blockley, D. (1980), *The Nature of Structural Design and Safety,* Ellis
        Horwood, Chichester.
[1.3]   Moy, S. S. J. (1981), *Plastic Methods for Steel and Concrete Structures,*
        Macmillan. Basingstoke.
[1.4]   Blockley, D., op. cit., p. 123.
[1.5]   Case, J., and Chilver, A. H. (1971), *Strength of Materials and Structures,*
        Edward Arnold, London.
[1.6]   Edmunds, H. G. (1980), *Mechanical Foundations of Engineering Science,*
        Ellis Horwood, Chichester.
[1.7]   Edmunds, H. G., op. cit., Sections 5.1 and 5.2.
[1.8]   Meriam, J. L. (1975), *Statics,* 2nd edn, Wiley International, New York,
        pp. 36–38.
[1.9]   Case, J., and Chilver, A. H., op. cit., p.52.
[1.10] Case, J., and Chilver, A. H., op. cit., Chapter 9.
[1.11] Case, J., and Chilver, A. H., op. cit., Chapter 10.
[1.12] Timoshenko, S., and Goodier, J. N. (1951), *Theory of Elasticity,* 2nd
        edn, McGraw-Hill, New York, Chapter 11.
[1.13] Case, J., and Chilver, A. H., op. cit., Section 10.8.
[1.14] Case, J., and Chilver, A. H., op. cit., Chapter 11.
[1.15] Williams, D. (1960), *An Introduction to the Theory of Aircraft Structures,*
        Edward Arnold, London, Chapter 8.
[1.16] Meriam, J. L., op. cit., p.6.
[1.17] Case, J., and Chilver, A. H., op. cit., Section 5.14.

CHAPTER 2

Analysis of Statically Determinate Frameworks I: Internal Forces

2.1 INTRODUCTION

In Chapter 1 we have shown that the response of structures to actions is measured by the internal forces and by the deflections. For certain relatively simple structures it is possible to find the internal forces simply by considering static equilibrium, these structures being called **statically determinate**. In this chapter we shall describe methods of finding the internal forces of statically determinate frameworks.

As a starting point, we shall consider in the next section, what is meant by static equilibrium. We shall review the basic concepts of this branch of classical mechanics in sufficient detail to give a complete basis for the structural analysis presented in the remainder of this book. However, the subject will be presented as concisely as possible. Classical mechanics is extensively described in a great number of texts, and a very thorough and clear treatment is given by Meriam [2.1].

2.2 STATIC EQUILIBRIUM

In order to understand the meaning of equilibrium, we recall the definition of forces introduced in Chapter 1, Section 1.4. We first consider forces acting on particles, which are defined as hypothetical bodies having finite masses, but being infinitesimally small in size. If then a single force P_1 acts on a particle of mass M, as in Fig. 2.1(a), the particle is given an acceleration of P_1/M in the direction of the force. If several forces, P_1, P_2 and P_3 say, act together on the particle, as in Fig. 2.1(b), the particle accelerates as though it were acted on by the resultant P_4 of these forces, P_4 being the vector sum of P_1, P_2 and P_3. P_4 can be found graphically by summing P_1 and P_2 using the parallelogram law of vector addition, and then summing the resulting vector with P_3 as shown in the figure. The concept of equilibrium is derived from the case when the resultant is zero, for then the particle does not accelerate. It thus remains either at rest, or moving in a straight line at any velocity it might have had before the application of the forces. In most structural problems we are only concerned with stationary states. The particle is then said to be in a state of **static equilibrium** under the forces P_1, P_2 and P_3.

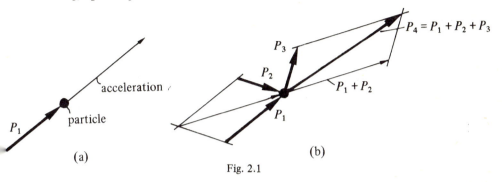

Fig. 2.1

Static equilibrium can be expressed as a simple relationship between the components of P_1, P_2 and P_3. Let us consider first for clarity the two-dimensional case, in which the forces lie in the x,y plane of the global coordinate system. The parallelogram law of vector addition, is equivalent to saying that the scalar components of the resultant P_4 in the x and y directions are equal to the sums of the scalar components of the three forces P_1, P_2 and P_3 in those directions. Thus

$$P_{4x} = P_{1x} + P_{2x} + P_{3x} = \sum_{i=1,2,3} P_{ix} \qquad (2.1a)$$

$$P_{4y} = P_{1y} + P_{2y} + P_{3y} = \sum_{i=1,2,3} P_{iy} \; . \qquad (2.1b)$$

Static equilibrium of the particle then requires that the resultant P_4 is zero, and is expressed by the condition that

$$\sum_{i=1,2,3} P_{ix} = 0, \qquad \sum_{i=1,2,3} P_{iy} = 0 \; . \qquad (2.2a,b)$$

Equations 2.2 are called the **equations of linear static equilibrium.**

The extension to three dimensions and any number of forces N simply involves adding an extra equation in the z components and summing from 1 to N. The corresponding linear static equilibrium equations can then be concisely expressed as

$$\sum_{i=1,\dots,N} \mathbf{P}_i = 0 \qquad (2.3)$$

where $\mathbf{P}_i$ is the column *matrix* of the scalar components P_{ix}, P_{iy} and P_{iz} of the force P_i.

Example 2.1

Suppose a particle is acted on by two forces P_1 and P_2 of magnitudes 10.0 kN and 15.0 kN respectively, and directed as shown in Fig. 2.2. We wish to determine the components of a third force P_3 which will maintain the particle in static equilibrium.

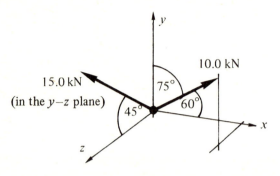

Fig. 2.2

The scalar components of P_1 and P_2 are calculated as follows:

$$P_{1x} = 10.0 \cos (60°) = 5.00 \text{ kN},$$

$$P_{1y} = 10.0 \cos (75°) = 2.59 \text{ kN},$$

$$P_{2x} = 0,$$

$$P_{2y} = 15.0 \sin (45°) = 10.61 \text{ kN},$$

$$P_{2z} = 15.0 \cos (45°) = 10.61 \text{ kN}.$$

Also since

$$\sqrt{P_{1x}^2 + P_{1y}^2 + P_{1z}^2} = P_1 = 10.0$$

then

$$P_{1z} = 8.26 \text{ kN} .$$

Thus, letting the unknown components of P_3 be P_{3x}, P_{3y} and P_{3z}, we obtain from the equilibrium equations (2.3)

$$\begin{bmatrix} 5.00 \\ 2.59 \\ 8.26 \end{bmatrix} + \begin{bmatrix} 0 \\ 10.61 \\ 10.61 \end{bmatrix} + \begin{bmatrix} P_{3x} \\ P_{3y} \\ P_{3z} \end{bmatrix} = 0 .$$

Therefore

$$\begin{bmatrix} P_{3x} \\ P_{3y} \\ P_{3z} \end{bmatrix} = \begin{bmatrix} -5.00 \\ -13.20 \\ -18.87 \end{bmatrix} \text{ kN}.$$

Up to now we have considered forces acting on a particle. If we go on to consider forces acting on a body of finite size, then, in addition to purely linear motion, it is possible for the body to rotate and undergo rotational acceleration. The corresponding equations of equilibrium then have to be extended to include this possibility.

Consider a body in space described by a global coordinate system x, y, z with an arbitrarily positioned origin O. Suppose the body is acted on by a single force P_1 in an x, y plane containing the centre of mass C, as shown in Fig. 2.3(a). The effect of this force on the linear and rotational motion of the body can be most

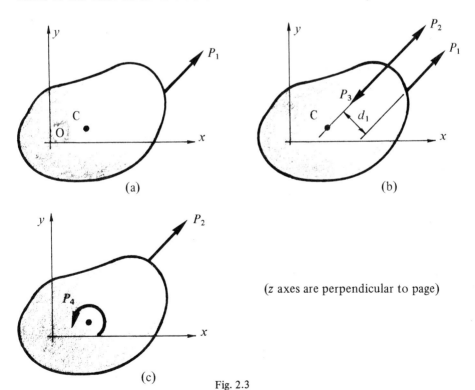

(a)

(b)

(z axes are perpendicular to page)

(c)

Fig. 2.3

clearly seen if we convert P_1 into an equivalent force acting through C together with a couple acting about a z axis through C. This is done by adding to the body two equal and opposite forces P_2 and P_3 of magnitudes equal to P_1, and with lines of action parallel to P_1 but passing through the centre of mass, as in Fig. 2.3(b). These two **self-equilibrating** forces have no effect on the motion of the body. However, by examining the combined system of forces we see that P_1 and P_3 comprise a couple P_4 say, of magnitude $P_1 d_1$ acting about the z axis, as in Fig. 2.3(c). If M is the mass of the body and I_z the rotational inertia of the body about the z axis passing through the centre of mass, then under the action of the force P_2 and couple P_4, the body undergoes a linear acceleration P_2/M in the direction of P_2 and a rotational acceleration P_4/I_z about the z axis. These then are the same as the accelerations imposed by P_1 alone.

The magnitude of the couple P_4 is given above in terms of d_1, which is the perpendicular distance between the line of action of the P_1 and the z axis. $P_1 d_1$ is also equal to the **moment** of P_1 about the z axis. This we call M_{1z}. However, M_{1z} can be more conveniently expressed in terms of the components of P_1 and the coordinates of its point of action. Thus, letting the coordinates of the point of action of P_1 be x_1, y_1 and the coordinates of C be x_c, y_c we can decompose the force P_1 into its two components in the x and y directions. M_{1z} is then simply the sum of the moments of these two components about the z axis through C. Thus, noting that a positive moment produces clockwise rotation in the positive direction of z about the z axis, (as viewed in the positive direction of z), we see from Fig. 2.4 that M_{1z} is given by

$$M_{1z} = -P_{1x}(y_1 - y_c) + P_{1y}(x_1 - x_c) \ . \tag{2.4}$$

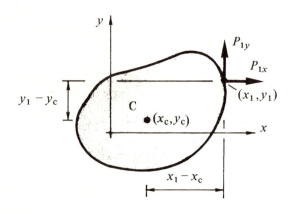

Fig. 2.4

If several forces P_1, P_2 and P_3, say, act together in the $x-y$ plane, as in Fig. 2.5, then each force P_i can be converted into an equivalent linear force passing through C, of the same magnitude and direction as P_i together with a couple of magnitude $P_i d_i = M_{iz}$. Thus the total equivalent linear force passing through C is the vector sum of P_1, P_2 and P_3, and the magnitude of the total equivalent couple about the z axis through C is the sum of M_{1z}, M_{2z} and M_{3z}. If, therefore, the body is in static equilibrium under the forces, and thus has zero linear or rotational acceleration, we must satisfy the same linear equilibrium equations as for the particle, namely (2.2) above, together with the **rotational equilibrium equation**, which can be derived by summing the moments M_{1z}, M_{2z} and M_{3z} and equating to zero. Thus from (2.4) we obtain

$$\sum_{i=1,2,3} (-P_{ix}(y_i - y_c) + P_{iy}(x_i - x_c)) = 0 . \tag{2.5}$$

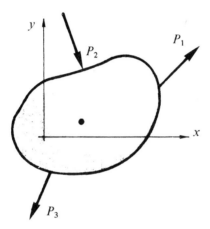

Fig. 2.5

Equation (2.5) can be expressed in a neater and more useful form by factorising out the constant coordinates x_c and y_c. Thus in (2.5)

$$\sum_{i=1,2,3} (P_{ix} y_c) = y_c \left(\sum_{i=1,2,3} P_{ix} \right), \quad \sum_{i=1,2,3} (P_{iy} x_c) = x_c \left(\sum_{i=1,2,3} P_{ix} \right) \tag{2.6a,b}$$

and since linear equilibrium is imposed, the summations of the scalar components on the right-hand sides of (2.6) are zero. The terms involving x_c and y_c in (2.5) can therefore be ignored and the rotational equilibrium equation takes its final form

$$\sum_{i=1,2,3} (-P_{ix} y_i + P_{iy} x_i) = 0 . \tag{2.7}$$

The importance of this simplification is that the left-hand side of (2.7) represents the sum of the moments of the forces about the z axis through the origin of the global coordinate system, the origin being in an entirely arbitrary position relative to the body. Any other position could have been used. This simplification therefore enables us to take moments about a z axis through *any* convenient point in order to check rotational equilibrium.

Example 2.2
A thin rectangular body shown in Fig. 2.6 is subjected to forces P_1 and P_2 of magnitude 10.0 kN and 15.0 kN directed as shown. P_3 is unknown in both magnitude and direction, P_4 is unknown in magnitude but acts parallel to the x axis. We wish to determine the unknown components of P_3 and P_4, to maintain the body in static equilibrium.

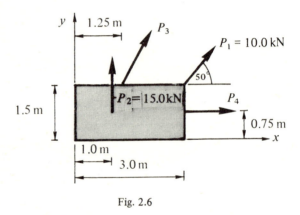

Fig. 2.6

The scalar components of P_1 and P_2 are calculated as follows:

$$P_{1x} = 10.0 \cos (50°) = 6.43 \text{ kN},$$

$$P_{1y} = 10.0 \sin (50°) = 7.66 \text{ kN},$$

$$P_{2x} = 0,$$

$$P_{2y} = 15.0 \text{ kN} .$$

Also

$$P_{4y} = 0 .$$

Thus linear equilibrium in the x and y directions is given by (2.2) as follows

$$\begin{bmatrix} 6.43 \\ 7.66 \end{bmatrix} + \begin{bmatrix} 0.0 \\ 15.0 \end{bmatrix} + \begin{bmatrix} P_{3x} \\ P_{3y} \end{bmatrix} + \begin{bmatrix} P_{4x} \\ 0.0 \end{bmatrix} = 0 .$$

Taking moments about the origin

$$M_{1z} = -6.43 \times 1.5 + 7.66 \times 3.0 = 13.34 \text{ kNm}$$

$$M_{2z} = 15.00 \times 1.0 = 15.0 \text{ kNm} .$$

Thus rotational equilibrium about the z axis is given by (2.7) as follows

$$13.34 + 15.0 + (-P_{3x} \times 1.5 + P_{3y} \times 1.25) + (-P_{4x} \times 0.75) = 0 .$$

Collecting the three equilibrium equations produces the following simultaneous equations for the unknowns P_{3x}, P_{3y} and P_{4x},

$$\begin{bmatrix} 1 & 0 & 1 \\ 0 & 1 & 0 \\ -1.5 & 1.25 & -0.75 \end{bmatrix} \begin{bmatrix} P_{3x} \\ P_{3y} \\ P_{4x} \end{bmatrix} = \begin{bmatrix} -6.43 \\ -22.66 \\ -28.34 \end{bmatrix}$$

giving the solution

$$\begin{bmatrix} P_{3x} \\ P_{3y} \\ P_{4x} \end{bmatrix} = \begin{bmatrix} 6.45 \\ -22.66 \\ -12.87 \end{bmatrix} \text{ kN} .$$

The equilibrium of a body subject to any number N, of applied forces in three dimensions can easily be deduced from the above discussion. Firstly, the linear static equilibrium of the body is represented by the same equations as for the particle (2.3). Rotational equilibrium is satisfied if the sums of the moments of the forces about all three axes are zero. These moments are again simple to formulate in three dimensions if we consider the components of the forces in the three coordinate directions. Thus for the force P_1 acting at the point (x_1, y_1, z_1) on the body relative to the global coordinate system shown in Fig. 2.7, the clockwise moments M_{1x}, M_{1y} and M_{1z} about the three coordinate axes through the origin are given by

$$M_{1x} = -P_{1y}z_1 + P_{1z}y_1 \qquad\qquad (2.8a)$$

$$M_{1y} = P_{1x}z_1 - P_{1z}x_1 \qquad\qquad (2.8b)$$

$$M_{1z} = -P_{1x}y_1 + P_{1y}x_1 \qquad\qquad (2.8c)$$

or in concise form

$$M_1 = \begin{bmatrix} 0 & -z_1 & y_1 \\ z_1 & 0 & -x_1 \\ -y_1 & x_1 & 0 \end{bmatrix} P_1 \qquad (2.9)$$

where M_1 is the column matrix of the three moments. The rotational equilibrium of the body is then expressed by summing the moments of all the forces about the three axes and equating to zero, giving

$$\sum_{i=1,\ldots,N} M_i = 0 . \qquad (2.10)$$

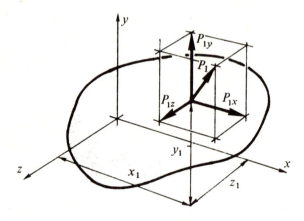

Fig. 2.7

Example 2.3
The body shown in Fig. 2.8 is subjected to forces P_1 and P_2 of magnitude 10.0 kN and 15.0 kN directed as shown, P_3 is unknown in both magnitude and direction, P_4 is unknown in magnitude but acts parallel to the x axis, P_5 is unknown in magnitude but acts in the y–z plane. We wish to determine the unknown components of P_3, P_4 and P_5 to maintain the body in static equilibrium.

The scalar components of P_1 and P_2 are as follows:

$$P_1 = \begin{bmatrix} 5.0 \\ 2.59 \\ 8.26 \end{bmatrix} kN , \quad P_2 = \begin{bmatrix} 0 \\ 10.61 \\ 10.61 \end{bmatrix} kN .$$

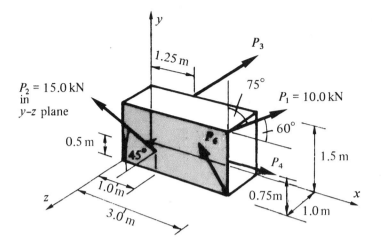

Fig. 2.8

The moments of P_1 and P_2 about the three coordinate axes are:

$$\mathbf{M_1} = \begin{bmatrix} 0 & -1.0 & 1.5 \\ 1.0 & 0 & -3.0 \\ -1.5 & 3.0 & 0 \end{bmatrix} \begin{bmatrix} 5.0 \\ 2.59 \\ 8.26 \end{bmatrix} = \begin{bmatrix} 9.81 \\ -19.79 \\ 0.26 \end{bmatrix} \text{ kNm}$$

$$\mathbf{M_2} = \begin{bmatrix} 0 & -1.0 & 0.5 \\ 1.0 & 0 & -1.0 \\ -0.5 & 1.0 & 0 \end{bmatrix} \begin{bmatrix} 0 \\ 10.61 \\ 10.61 \end{bmatrix} = \begin{bmatrix} -5.30 \\ -10.61 \\ 10.61 \end{bmatrix} \text{ kNm .}$$

The linear equilibrium equations (2.3) give

$$\begin{bmatrix} 5.0 \\ 2.59 \\ 8.26 \end{bmatrix} + \begin{bmatrix} 0 \\ 10.61 \\ 10.61 \end{bmatrix} + \begin{bmatrix} P_{3x} \\ P_{3y} \\ P_{3z} \end{bmatrix} + \begin{bmatrix} P_{4x} \\ 0 \\ 0 \end{bmatrix} + \begin{bmatrix} 0 \\ P_{5y} \\ P_{5z} \end{bmatrix} = 0 .$$

The rotational equilibrium equations (2.10) give

$$
\begin{bmatrix} 9.81 \\ -19.79 \\ 0.26 \end{bmatrix} + \begin{bmatrix} -5.30 \\ -10.61 \\ 10.61 \end{bmatrix} + \begin{bmatrix} 0 & 0 & 1.5 \\ 0 & 0 & -1.25 \\ -1.5 & 1.25 & 0 \end{bmatrix} \begin{bmatrix} P_{3x} \\ P_{3y} \\ P_{3z} \end{bmatrix} +
$$

$$
+ \begin{bmatrix} 0 & -1.0 & 0.75 \\ 1.0 & 0 & -3.0 \\ -0.75 & 3.0 & 0 \end{bmatrix} \begin{bmatrix} P_{4x} \\ 0 \\ 0 \end{bmatrix} + \begin{bmatrix} 0 & -1.0 & 0 \\ 1.0 & 0 & -3.0 \\ 0 & 3.0 & 0 \end{bmatrix} \begin{bmatrix} 0 \\ P_{5y} \\ P_{5z} \end{bmatrix} = 0 \; .
$$

Rearranging as six simultaneous equations for the six unknown components produces

$$
\begin{bmatrix} 1 & 0 & 0 & 1 & 0 & 0 \\ 0 & 1 & 0 & 0 & 1 & 0 \\ 0 & 0 & 1 & 0 & 0 & 1 \\ 0 & 0 & 1.5 & 0 & -1.0 & 0 \\ 0 & 0 & -1.25 & 1.0 & 0 & -3.0 \\ -1.5 & 1.25 & 0 & -0.75 & 3.0 & 0 \end{bmatrix} \begin{bmatrix} P_{3x} \\ P_{3y} \\ P_{3z} \\ P_{4x} \\ P_{5y} \\ P_{5z} \end{bmatrix} = \begin{bmatrix} -5.0 \\ -13.19 \\ -18.87 \\ -4.51 \\ 30.40 \\ -10.87 \end{bmatrix}
$$

giving the solution

$$
\begin{bmatrix} P_{3x} \\ P_{3y} \\ P_{3z} \\ P_{4x} \\ P_{5y} \\ P_{5z} \end{bmatrix} = \begin{bmatrix} 34.39 \\ -28.99 \\ 7.53 \\ -39.39 \\ 15.80 \\ -26.40 \end{bmatrix} \text{kN} \; .
$$

Up to now we have considered only concentrated direct forces acting on a body. However, as described in Chapter 1, Section 1.4, there are two other types of force which are of great importance in framework analysis, the couple and the distributed force. These also need to be included in the equilibrium equations.

Consider first a couple P_1 say that acts about a global z axis. P_1 is then composed of two equal and opposite parallel forces in the $x-y$ plane. Having no linear resultant, the couple does not effect the linear equilibrium equations in any way. However, P_1 does contribute directly to the moments causing rotational acceleration about the z axis, and it must therefore be added to the left-hand side of the rotational equilibrium equation (2.7). If next we consider a couple P_2 that acts about an arbitrarily oriented axis in space, then P_2 can be decomposed into the components P_{2x}, P_{2y} and P_{2z}, these components being equal to respective clockwise moments $\mathbf{M}_2$ about x, y, and z axes. The matrix $\mathbf{M}_2$ is again simply added to the left-hand side of (2.10).

Consider next a distributed force p/unit length that acts on a particular loaded length of the body. In order to see how this is included in the equilibrium equations, we can treat the force as a collection of infinitesimally small concentrated forces, each acting at the centre of an infinitesimal segment of the loaded length. Thus suppose for example that a uniformly distributed force acts downwards on a body on a loaded length parallel to the x axis as in Fig. 2.9(a). Then the force $\mathrm{d}P_1$ say, acting on the infinitesimal length $\mathrm{d}x$ in Fig. 2.9(b) has a single component $\mathrm{d}P_{1y}$ equal to $-p\,\mathrm{d}x$. Its moment $\mathrm{d}M_{1z}$ about the z axis is $-px\mathrm{d}x$. Integrating both quantities between the limits a and b leads to the following expressions for the total equivalent force P_{1y} and moment M_{1z}:

$$P_{1y} = \int_a^b -p\mathrm{d}x = -p(b-a) = -pl \qquad (2.11)$$

$$M_{1z} = \int_a^b -px\mathrm{d}x = -p\left(\frac{b^2}{2} - \frac{a^2}{2}\right) = -p(b-a)\frac{(b+a)}{2} = -plc\ . \qquad (2.12)$$

Thus, from the point of view of equilibrium, a *uniformly* distributed force is equivalent to a concentrated force of magnitude pl, acting in the same direction as p at the centre of the loaded length. This concentrated force is said to be **statically equivalent** to the distributed force.

If the distributed force in Fig. 2.9 were not uniform, then p would be some given varying function of x. The equivalent terms in the equilibrium equations, P_{1y} and M_{1z}, would then have to be worked out by performing the integrations in (2.11) and (2.12) from scratch.

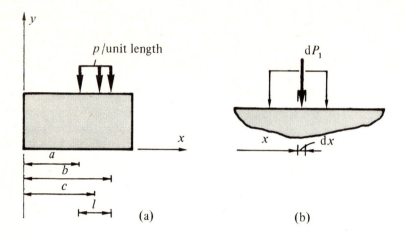

Fig. 2.9

Example 2.4

A simple extension of Example 2.2 illustrates the effects of a couple and a distributed force on the equilibrium equations. Thus suppose a thin rectangular body is loaded as in Fig. 2.6 but is subjected in addition to a couple of magnitude 7.5 kNm about the z axis and a uniformly distributed force of 3.0 kN/m in the x direction on its left-hand side as shown in Fig. 2.10. We again wish to find the unknown components P_{3x}, P_{3y} and P_{4x}.

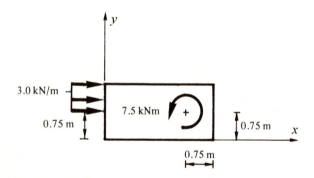

Fig. 2.10

The distributed force is equivalent to a concentrated force, P_5 say, in the x direction equal to $3.0 \times 0.75 = 2.25$ kN acting at 1.125 m from the origin. The linear equilibrium equations in Example 2.2 then change to

$$\begin{bmatrix} 6.43 \\ 7.66 \end{bmatrix} + \begin{bmatrix} 0.0 \\ 15.0 \end{bmatrix} + \begin{bmatrix} 2.25 \\ 0.0 \end{bmatrix} + \begin{bmatrix} P_{3x} \\ P_{3y} \end{bmatrix} + \begin{bmatrix} P_{4x} \\ 0.0 \end{bmatrix} = 0 \ .$$

The moment of P_5 about the z axis through the origin is given by

$$M_{5z} = -2.25 \times 1.125 = -2.53 \text{ kNm}$$

and the moment of the couple is 7.5 kNm. Thus the rotational equilibrium equation in Example 2.2 changes to

$$13.34 + 15.0 - 2.53 + 7.5 + (-P_{3x} \times 1.5 + P_{3y} \times 1.25) + (-P_{4x} \times 0.75) = 0 \ .$$

Solving the three simultaneous equations then gives

$$\begin{bmatrix} P_{3x} \\ P_{3y} \\ P_{4x} \end{bmatrix} = \begin{bmatrix} 15.31 \\ -22.66 \\ -23.99 \end{bmatrix} \text{ kN} \ .$$

Note that although the couple of 7.5 kNm is applied to the body about an axis passing through a particular point, perhaps by means of a crank mechanism welded to the body at the point, the position of the couple does not influence its effect on the equilibrium. The couple contributes the single moment of 7.5 kNm to the rotational equilibrium equation, wherever it acts on the body.

The production of the linear equilibrium equations is called **resolving** in particular directions. We can resolve in the x, y or z directions, or where the context makes the operation clear, horizontally or vertically. As stated in Chapter 1, Section 1.4, the majority of problems in framework analysis are analysed in terms of forces that are directed parallel to the global coordinates, with more than one force acting at a point if necessary. In this case, resolving in a particular direction involves summing all the forces in that direction. In the worked examples in the remainder of this book, it is convenient to symbolise this operation. Thus $\Sigma \underset{x}{\rightarrow}$ will indicate the equation produced by resolving in the x direction, and $\Sigma \rightarrow$, $\Sigma \uparrow$ will indicate resolving horizontally or vertically. The production of the rotational equilibrium equations is called **taking moments** about particular axes. Thus taking moments about a z axis passing through a particular point A say, involves summing the moments of all the forces and couples about this axis. This operation can be symbolised as $\Sigma \underset{z}{\text{Ⓐ}}$. The use of the above notation will be rapidly clarified in subsequent worked examples.

2.3 CALCULATION OF STRUCTURAL REACTIONS

The static equilibrium equations discussed in the previous section apply to the complete system of external forces acting on bodies. No reference was made to the complexity of the bodies in discussing their equilibrium, and the equations therefore apply to very complex structures. The equations are also called the **global equilibrium equations** to distinguish them from the equilibrium equations governing each separate part of the structure, which as we shall see in Section 2.4, are also considered in structural analysis. A feature of structures which differentiates them from the bodies discussed in the previous section is that they are fixed at various points by supports. In restraining the displacements of the structures, the supports impose unknown external forces R_i called **reactions**. In Chapter 1, Section 1.5, we have described the reactions, either direct forces or couples, which can be produced by particular types of support for frameworks. Note in particular, that for simplicity of notation, we also treat each component of each reaction as a separate reaction R_i.

Although unknown, the reactions contribute to the global equilibrium of frameworks in the same way as the applied external forces. A very important function of the global equilibrium equations is that they often enable us to calculate the reactions on statically determinate frameworks. In many cases this is a necessary first step towards calculating the internal forces. In this section, therefore, the problem of calculating the reactions on such frameworks is considered at some length. The procedure is described by a series of illustrative examples. In each case a clear conception of the reactions can be obtained by drawing the **free body diagram**, which depicts the structure as a free body in space acted on by the external loads and by the support reactions.

Example 2.5 Reactions on a simply supported beam
A single span beam is supported by a pinned support at the left-hand end and a roller support at the right-hand end. Such a beam is said to be **simply supported**. It is subjected to the forces shown in Fig. 2.11(a).

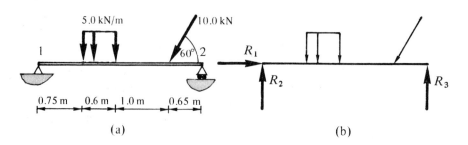

(a)   (b)

Fig. 2.11

The free body diagram shown in Fig. 2.11(b), contains the three unknown reactions R_1, R_2 and R_3 provided by the supports. In this example, in which all the forces are in the x–y plane, we are concerned only with linear equilibrium in the x and y directions, and rotational equilbrium about any convenient z axis. Thus resolving horizontally and vertically and taking moments about an axis through end 1 gives

$$\Sigma \rightarrow \quad R_1 - 10.0 \cos (60°) = 0$$

$$\Sigma \uparrow \quad R_2 - 5.0 \times 0.6 - 10.0 \sin (60°) + R_3 = 0$$

$$\Sigma \underset{z}{\circlearrowleft} \quad -5.0 \times 0.6 \times 1.05 - 10.0 \sin (60°) \times 2.35 + R_3 \times 3.0 = 0.$$

Thus

$$\begin{bmatrix} R_1 \\ R_2 \\ R_3 \end{bmatrix} = \begin{bmatrix} 5.0 \\ 3.83 \\ 7.83 \end{bmatrix} \text{kN} .$$

Example 2.6
A simply supported beam is subjected to a concentrated couple as shown in Fig. 2.12(a). The free body diagram is shown in Fig. 2.12(b). Whence

$$\Sigma \rightarrow \quad R_1 = 0$$

$$\Sigma \uparrow \quad R_2 + R_3 = 0$$

$$\Sigma \underset{z}{\circlearrowleft} \quad 7.0 + R_3 \times 3.0 = 0 .$$

Thus

$$\begin{bmatrix} R_1 \\ R_2 \\ R_3 \end{bmatrix} = \begin{bmatrix} 0 \\ 2.33 \\ -2.33 \end{bmatrix} \text{kN} .$$

Note that in this case the reactions R_2 and R_3 themselves form a couple as in Fig. 2.12(c), of the same magnitude but in the opposite direction to the applied external couple.

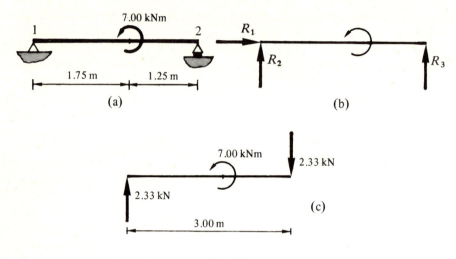

Fig. 2.12

Example 2.7 Reactions on a cantilever
A cantilever is loaded by the combination of load types shown in Fig. 2.13(a).
The encastered left-hand support provides the three reactions, two direct forces
R_1 and R_2 and a couple R_3 shown in the free body diagram in Fig. 2.13(b).

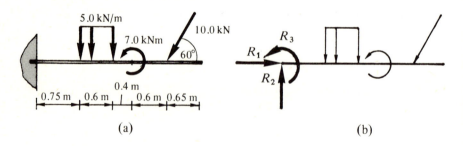

Fig. 2.13

Whence

$$\Sigma \rightarrow \quad R_1 - 10.0 \cos(60°) = 0$$

$$\Sigma \uparrow \quad R_2 - 5.0 \times 0.6 - 10.0 \sin(60°) = 0$$

$$\Sigma \underset{z}{①} \quad R_3 - 5.0 \times 0.6 \times 1.05 + 7.0 - 10.0 \sin(60°) \times 2.35 = 0.$$

Thus

$$\begin{bmatrix} R_1 \\ R_2 \\ R_3 \end{bmatrix} = \begin{bmatrix} 5.00 \\ 11.66 \\ 16.50 \end{bmatrix} \begin{matrix} kN \\ kN \\ kNm \end{matrix} .$$

The above examples demonstrate the two types of single-span beam that are statically determinate; the simply supported beam and the cantilever. Each beam has just three support reactions which are obtainable from the three equilibrium equations. The addition of other support reactions makes the beam **statically indeterminate** because it is not then possible to find the reactions by considering equilibrium alone. Examples of statically indeterminate beams are shown in Fig. 2.14. Thus the beams in Fig. 2.14(a), (c) and (d) each have four support reactions while the beam in Fig. 2.14(b) has six. Note that, in the absence of horizontal loads, it might appear possible to assume that both the horizontal reactions on the beam in Fig. 2.14(a) are zero, and hence that the beam is statically determinate. However, such an assumption is not generally valid because temperature changes or construction errors can induce compressive forces in a beam such as this, whose elongation is restrained by the supports.

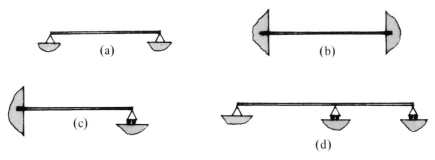

Fig. 2.14

Equilibrium can be used to obtain the reactions of any plane framework supported by just three support reactions. An important class of such frameworks comprises **pin-footed arches** with feet supported by a pinned and a roller support in the same way as the simply supported beam considered in Example 2.5. An example of such an arch is considered next, taking the form of a portal frame, of geometry typical of a large commercial building. The loading is idealised, and represents concentrated forces applied on the roof beams by purlins, and horizontal wind load applied as a distributed force to one of the columns by the wall cladding. The distributed forces due to self-weight are ignored.

Example 2.8 The pin-footed portal frame

A pin-footed portal frame is loaded by concentrated and distributed forces as in Fig. 2.15(a). Its free body diagram is shown in Fig. 2.15(b). Whence resolving and taking moments about an axis through joint 1 gives

$$\Sigma \rightarrow \quad R_1 + 2.0 \times 4.0 = 0$$

$$\Sigma \uparrow \quad R_2 - 5.0 - 10.0 - 10.0 - 8.0 - 4.0 + R_3 = 0$$

$$\Sigma_z \; ① \quad -2.0 \times 4.0 \times 2.0 - 10.0 \times 3.0 - 10.0 \times 6.0 -$$
$$- 8.0 \times 9.0 - 4.0 \times 12.0 + R_3 \times 12.0 = 0 \,.$$

Thus

$$\begin{bmatrix} R_1 \\ R_2 \\ R_3 \end{bmatrix} = \begin{bmatrix} -8.0 \\ 18.17 \\ 18.83 \end{bmatrix} \text{ kN} \,.$$

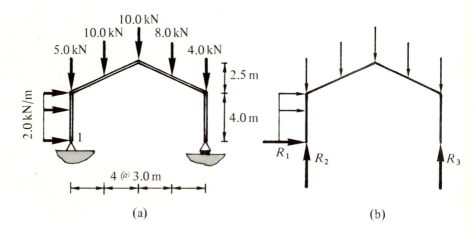

Fig. 2.15

The single-span beams and the portal frame considered above, are plane frameworks with forces in the $x-y$ plane and their behaviour is governed by the three equilibrium equations (2.2) and (2.7). If next we consider space frameworks, we then have six equilibrium equations (2.3) and (2.10) with which to calculate the reactions. Nevertheless, statically *determinate* space frameworks are limited in scope, because any complexity of structure usually introduces far more than just six reactions. Perhaps the most easily recognised space framework is the **space cantilever**, which is defined as any singly connected framework supported at one fully encastered support, and subjected to forces in any arbitrary direction. By **singly connected**, we mean a framework with a tree-like structure that does not contain any closed rings of members, a concept that will be discussed in more detail in Chapter 4. For the purpose of this section we only have to recognise that the fully encastered support imposes six reactions on the framework, three linear reactions in the coordinate directions and three couples about the coordinate axes. We conclude the section by considering two such cantilever problems, the second demonstrating the vector representation of couples.

Example 2.9 Cranked cantilever

The cantilever shown in Fig. 2.16(a) is fully encastered at point 1 where the support supplies the three linear reactions and three rotational reactions shown in Fig. 2.16(b). The six equilibrium equations are as follows:

$$\sum_x \rightarrow \quad R_1 - 15.0 = 0$$

$$\sum_y \rightarrow \quad R_2 - 10.0 - 6.0 \times 2.0 = 0$$

$$\sum_z \rightarrow \quad R_3 = 0$$

$$\sum_x \circlearrowleft \quad R_4 + 10.0 \times 4.0 + 6.0 \times 2.0 \times 3.0 = 0$$

$$\sum_y \circlearrowleft \quad R_5 - 15.0 \times 4.0 = 0$$

$$\sum_z \circlearrowleft \quad R_6 - 10.0 \times 0.75 - 6.0 \times 2.0 \times 1.5 - 15.0 \times 2.0 = 0 \ .$$

As for all cantilevers, the equilibrium equations are uncoupled, each leading to an explicit equation in a particular reaction. The reactions are therefore as shown below:

$$\begin{bmatrix} R_1 \\ R_2 \\ R_3 \\ R_4 \\ R_5 \\ R_6 \end{bmatrix} = \begin{bmatrix} 15.0 \\ 22.0 \\ 0 \\ -76.0 \\ 60.0 \\ 55.5 \end{bmatrix} \begin{matrix} \text{kN} \\ \text{kN} \\ \text{kN} \\ \text{kNm} \\ \text{kNm} \\ \text{kNm} \ . \end{matrix}$$

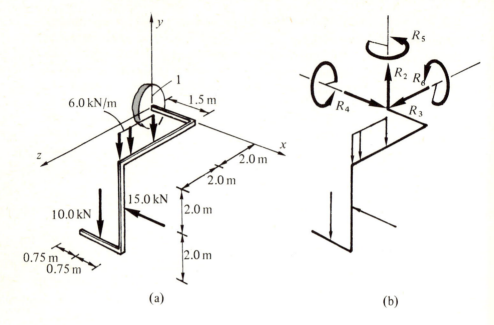

Fig. 2.16

Example 2.10

A cranked cantilever of the same geometry as that in Example 2.9, is subjected
to equal and opposite 10.0 kN forces, as in Fig. 2.17(a). These will be treated as
a single couple P_1, say. From the geometry of the figure, the perpendicular
distance between the forces is 2.236 m. Thus the magnitude of P_1 is 22.36 kNm,
and the axis about which it acts is horizontal and directed at right angles to the
line joining the points of action of the two forces. P_1 can therefore be represented
by the vector shown in the plan view in Fig. 2.17(b). From this figure, $\theta = \tan^{-1}$
$(1.0/2.0) = 26.57°$. Thus the components of P_1, equal to its clockwise moments
M_{1x}, M_{1y} and M_{1z} about the three coordinate axes, are

$$M_{1x} = 22.36 \cos(\theta) = 20.0 \text{ kNm}$$

$$M_{1y} = 0$$

$$M_{1z} = -22.36 \sin(\theta) = -10.0 \text{ kNm} .$$

Thus from the rotational equilibrium equations (2.10)

$$R_4 = -20.0 \text{ kNm}$$

$$R_6 = 10.0 \text{ kNm} .$$

It is easy to show that these reactions are the same as those that are obtained by calculating the effects of the two forces separately. This simple example therefore confirms that couples may be represented aş vectors.

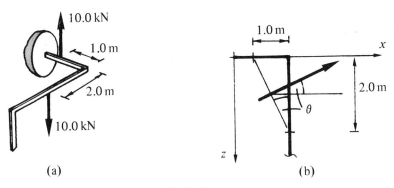

(a)   (b)

Fig. 2.17

2.4 INTERNAL EQUILIBRIUM

Up to now we have considered the global equilibrium equations that have to be satisfied if the structure as a whole remains stationary. However, an important concept that enables us to go on to calculate the internal forces, is that each separate sub-section of the structure, subject to external forces and to internal forces from other parts of the structure, must itself be in static equilibrium. The isolation of a sub-section of a structure is most easily conceived by supposing that it is cut from the structure, and then behaves as a free body under the forces acting on it. Where structural members are cut, we represent the internal forces by stress resultants acting on the cut surfaces. So consider again the simply supported beam of the same geometry and loading as in Fig. 2.11. We can isolate the sub-section of the beam between points A and B in Fig. 2.18(a) by making cuts at A and B. The stresses at the cuts are then represented by the stress resultants N_A, S_A, M_A and N_B, S_B and M_B shown in the free body diagram of Fig. 2.18(b), acting on the cut surfaces in their positive directions. The forces must then satisfy the equilibrium equations (2.2) and (2.7) if the sub-section of the structure between A and B remains stationary. Thus resolving and taking moments leads to

$$\Sigma \rightarrow \quad -N_A + N_B = 0$$

$$\Sigma \uparrow \quad -S_A - 5.0 \times 0.6 + S_B = 0$$

$$\Sigma \underset{z}{\circledA} \quad -M_A - 5.0 \times 0.6 \times 0.55 + S_B \times 1.00 + M_B = 0 \ .$$

Equations of this type relating to sub-sections of a structure, are called **internal equilibrium equations**.

The particular sub-section of the beam considered above is not convenient for calculating the stress resultants. If, however, we consider the sub-section of the beam to the left of A, the free body diagram shown in Fig. 2.18(c) includes the known reactions $R_1 = 5.00$ kN and $R_2 = 3.83$ kN calculated in Example 2.5. The internal equilibrium equations for this sub-section then are:

$$\Sigma \rightarrow \quad 5.00 + N_A = 0$$

$$\Sigma \uparrow \quad 3.83 + S_A = 0$$

$$\Sigma_z \text{ (A)} \quad -3.83 \times 0.5 + M_A = 0 \ .$$

Therefore

$$\begin{bmatrix} N_A \\ S_A \\ M_A \end{bmatrix} = \begin{bmatrix} -5.00 \\ -3.83 \\ 1.91 \end{bmatrix} \begin{matrix} \text{kN} \\ \text{kN} \\ \text{kNm} \ . \end{matrix}$$

and we have thus been able to calculate the stress resultants at A.

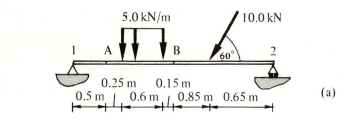

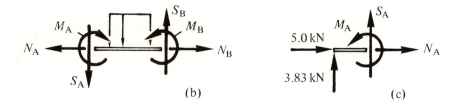

Fig. 2.18

By suitably choosing sub-sections of statically determinate frameworks we are always able to calculate the internal stress resultants. We thus accomplish one of the two objects of structural analysis, namely finding the internal forces. In the rest of this chapter we shall consider various examples of this.

2.5 STRESS RESULTANTS IN BEAMS

2.5.1 Stress Resultants at Particular Points

Let us consider further the simply supported beam in Fig. 2.18. We shall calculate the stress resultants at point B.

Example 2.11a
The free body diagram for the whole sub-section of the beam to the left of B is shown in Fig. 2.19(a). Whence the internal equilibrium equations for this sub-section are:

$$\Sigma \rightarrow \quad 5.00 + N_B = 0$$

$$\Sigma \uparrow \quad 3.83 - 5.0 \times 0.6 + S_B = 0$$

$$\Sigma \underset{z}{\textcircled{B}} \quad 3.83 \times 1.5 - 5.0 \times 0.6 \times 0.45 - M_B = 0 \, .$$

leading to the following stress resultants at B:

$$\begin{bmatrix} N_B \\ S_B \\ M_B \end{bmatrix} = \begin{bmatrix} -5.00 \\ -0.83 \\ 4.39 \end{bmatrix} \begin{array}{l} \text{kN} \\ \text{kN} \\ \text{kNm} \, . \end{array}$$

Clearly the forces on the sub-section of the beam to the right of point B in Fig. 2.18(a) also satisfy the equilibrium equations. It is therefore worth confirming that considering the equilibrium of the right-hand part will lead to the same results as those above for the stress resultants at B.

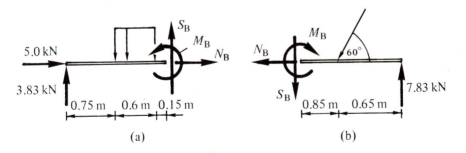

(a)   (b)

Fig. 2.19

Example 2.11b
Consider the part of the beam to the right of point B in Fig. 2.18, having the free body diagram shown in Fig. 2.19(b). This figure includes the known reaction

on the right-hand end $R_3 = 7.83$ kN calculated in Example 2.5. Note that the positive stress resultants acting on the negative coordinate surface at B, are reversed in direction. Whence the equilibrium equations take the form:

$$\Sigma \rightarrow \quad -N_B - 10.0 \cos(60°) = 0$$

$$\Sigma \uparrow \quad -S_B - 10.0 \sin(60°) + 7.83 = 0$$

$$\Sigma \underset{z}{\text{Ⓑ}} \quad -M_B - 10.0 \sin(60°) \times 0.85 + 7.83 \times 1.5 = 0 \ .$$

These again lead to

$$\begin{bmatrix} N_B \\ S_B \\ M_B \end{bmatrix} = \begin{bmatrix} -5.00 \\ -0.83 \\ 4.39 \end{bmatrix} \begin{matrix} \text{kN} \\ \text{kN} \\ \text{kNm} \end{matrix} \ .$$

It is apparent from Examples 2.11a and b that the stress resultants come out to have the same magnitude and sign, whichever side of the cut we consider in the analysis. Physically, both calculations predict a compressive axial force and a sagging bending moment at the point B. The fact that the results are independent of which of the two sub-sections of a structure we consider at any particular point, means that in practice we can freely choose whichever sub-section leads to the simpler equations.

2.5.2 Stress Resultant Diagrams

For design purposes, engineers need to know the values of the stress resultants at all points along the lengths of the members in a framework. They can then for example, position the reinforcement in a concrete framework, or select the size of sections to accommodate the maximum bending moments in a steel framework. The most convenient method for presenting this information is graphically in the form of **stress resultant diagrams**.

Stress resultant diagrams can obviously be obtained by evaluating the stress resultants at enough discrete points in each member, to enable the diagrams to be plotted. A practical way of doing this and interpolating between the calculated points will be introduced later in this section. However, we first demonstrate a method of producing stress resultant diagrams analytically. The method involves positioning the cut such as that at B in Fig. 2.19 at an arbitrary distance x from the left-hand end of the beam, so that the calculated value for the bending moment say, will come out to be a function of x. This then is the analytical expression for M from which a bending moment diagram can be drawn. The method is illustrated for various beams and load cases, in the next five worked examples.

Example 2.12
The simply supported beam in Fig. 2.20(a) is loaded by a single 10.0 kN concentrated force. Satisfying global equilibrium then leads to the two vertical
reactions at either end shown in the free body diagram in Fig. 2.20(b). Consider
a cut at point A, at a distance x from the left-hand end, where x is less than 2.0 m
as in Fig. 2.20 (c). Resolving and taking moments about an axis through A for
the section to the left of the cut leads to

$$S = -3.33 \text{ kN}$$

$$M = 3.33 \, x \text{ kNm} \ .$$

Consider next the cut positioned with x greater than 2.0 m. In this case, resolving and taking moments for the section to the right of the cut as in Fig. 2.20(d)
leads to

$$S = 6.67 \text{ kN}$$

$$M = 6.67 \, (3.0 - x) = (20.0 - 6.67 \, x) \text{ kNm} \ .$$

Thus we have two regions in the beam for which the stress resultants are known
functions of x, and this is all the information needed for plotting the shear force
and bending moment diagrams.

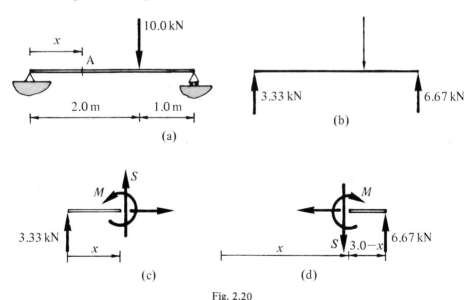

Fig. 2.20

Stress resultant diagrams are plotted, using a sign convention that is somewhat illogical from the mathematical viewpoint, but is of great engineering value.

Thus for the designers of concrete frameworks, it is helpful to have a clear visual impression of which side of each member is in tension. This is because the main steel reinforcement of a concrete member is situated on its tension side. In the case of a horizontal beam, a positive bending moment causes sagging, and produces tension on the bottom of the beam. Conversely a negative bending moment produces tension on the top. Thus if the bending moment diagram is drawn *positive downwards,* the diagram will always lie on the tension side of the beam. For the simply supported beam in Example 2.12, the bending moment diagram produced by this convention is shown in Fig. 2.21. It is then immediately apparent that if the beam were concrete, the reinforcement would need to be placed in the lower half of the section, and concentrated under the load position. We shall see that in more complicated frameworks, such as the pin-footed portal frame considered in Example 2.8, the equivalent convention is to draw the bending moment diagram for each member positive in the *opposite* direction to the member y_m coordinate direction. It should perhaps be noted that other writers on this subject have used the more natural mathematical convention of drawing stress resultant diagrams positive in the same direction as the y_m coordinate direction. The present writer, however, feels that this method of giving a clear impression of the structural action of concrete frameworks is too important to be disregarded.

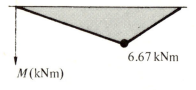

M (kNm)

6.67 kNm

Fig. 2.21

The sign of the shear force is not of the same importance to structural behaviour as the sign of the bending moment. However, in order to maintain consistency in the convention of drawing stress resultant diagrams, the shear force diagram is also drawn positive downwards. Thus the shear force diagram for the beam in Example 2.12 is shown in Fig. 2.22. The shear force has been obtained by satisfying linear equilibrium at right angles to the axis of the beam. This process does in fact lead to a simple 'rule of thumb' for drawing shear force diagrams which becomes apparent when we compare the shear force diagram in Fig. 2.22 with the corresponding free body diagram in Fig. 2.20(b). Thus we can construct the shear force diagram directly from the free body diagram, by starting from the left-hand end of the beam and progressing along the span, incrementing the shear force diagram upwards or downwards in the direction of the transverse forces as they are encountered. The arrows on the shear force diagram indicate the process, beginning with a step upwards in the direction of the left-hand reaction equal to 3.33 kN.

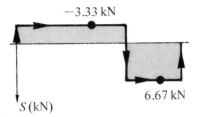

Fig. 2.22

Example 2.13
The cantilever in Fig. 2.23(a) is loaded by two concentrated forces. Satisfying global equilibrium then leads to the two reactions at the left-hand end shown in the free body diagram in Fig. 2.23(b). The rule of thumb then gives the shear force diagram shown in Fig. 2.23(c). The bending moment diagram is found by making a cut at a variable position A from the right-hand end of the cantilever, and taking equilibrium to the right of the cut. In fact, we note that the cantilever is one of the structural types for which the internal forces can be determined without calculating the reactions first. This is because it is always possible to consider the equilibrium of a sub-section of a cantilever which is free from support reactions; in this case, the sub-section to the right of the cut. For clarity in the arithmetic it is helpful to introduce the extra coordinate x', say, which defines the distance of the cut from the right-hand end. Considering rotational equilibrium about a z axis through A then leads to

$$M = 0, \qquad [0 \leqslant x' \leqslant 1.0]$$

$$M = -10.0\,(x' - 1.0) = (-10.0x' + 10.0)\,\text{kNm}, \qquad [1.0 \leqslant x' \leqslant 1.75]$$

$$M = -10.0\,(x' - 1.0) - 5.0\,(x' - 1.75) = (-15.0x' + 18.75)\,\text{kNm},$$

$$[1.75 \leqslant x' \leqslant 3.0].$$

These are linear functions in M, and as an aid to plotting the bending moment diagram, we can calculate M at particular points as follows

$$M(1.0) = 0, \quad M(1.75) = -7.5\,\text{kNm}, \quad M(3.0) = -26.25\,\text{kNm}.$$

Thence we obtain the final diagram shown in Fig. 2.23(d). Note again that the diagram gives a clear visual impression of the rapidly increasing hogging bending moment as A approaches the root of the cantilever, with tension on the top throughout.

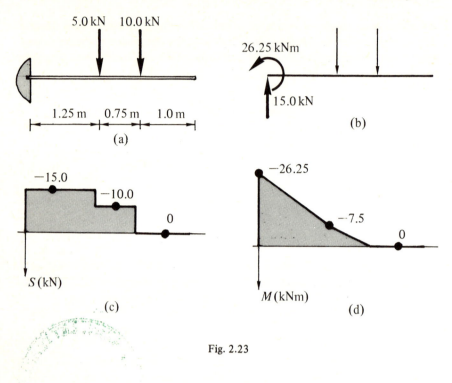

Fig. 2.23

In the next two examples, we consider the stress resultants corresponding to the two other types of loading discussed in Section 2.2: the couple and the distributed force.

Example 2.14
Consider again the concentrated couple on the simply supported beam shown in Fig. 2.24(a). Using the vertical reactions of ±2.33 kN calculated in Example 2.6, and shown in Fig. 2.24(b), the shear force diagram comes out to be the simple block diagram shown in Fig. 2.24(c). It should be carefully noted that since the couple has no linear resultant, there is no change in the shear force diagram at the loaded point. The bending moment diagram is more complicated. If we make a cut in the span to the left of the couple and take moments about the cut for the left-hand sub-section we find that

$$M = 2.33\,x \text{ kNm .}$$

If we make the cut to the right of the couple, then taking moments again for the left-hand sub-section leads to

$$M = (2.33\,x - 7.0)\,\text{kNm .}$$

These two expressions when plotted out, give a bending moment diagram in the form shown in Fig. 2.24(d). The discontinuity in the diagram at the loaded point, is equal to the moment being fed into the beam at this point by the couple.

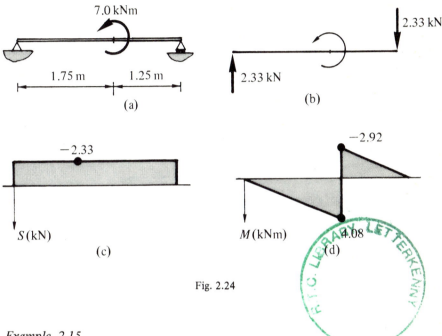

Fig. 2.24

Example 2.15
The simply supported beam in Fig. 2.25(a) is loaded by a uniformly distributed force p per unit length over its entire span l. By resolving and taking moments, the vertical reactions at each end come out to be $pl/2$. Making a cut at x from the left-hand end, we note that the sub-section of the beam to the left of the cut is loaded by the reaction and by the uniformly distributed force over the length x. This latter force has a total magnitude of px downwards, and a clockwise moment about the z axis through the cut of $px(x/2)$. Thus resolving and taking moments for the sub-section shown in Fig. 2.25(b), we get

$$S = px - pl/2$$
$$M = -px^2/2 + plx/2 = px(l-x)/2$$

giving the stress resultant diagrams shown in Figs. 2.25(c) and (d). The shear force diagram can be seen to follow the rule of thumb, if we consider the uniformly distributed force as a series of infinitesimal forces pdx acting at the centres of infinitesimal lengths of the beam. The bending moment diagram is a parabola convex downwards, and this is always the case under a vertical distributed force. The maximum value of the bending moment is at the centre and is $pl^2/8$. If we write the total vertical force on the beam (pl) as P, then the maximum bending moment M_{max} is given by

$$M_{max} = \frac{Pl}{8}.$$
(2.13)

This is an important result in structural engineering and is often used as an approximation in more complicated cases.

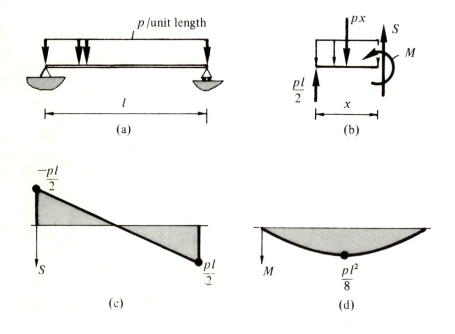

Fig. 2.25

As a final example in this section, we consider the case of a beam loaded by a concentrated force, and by couples at either end. This case which includes end couples, is important, because one of the main methods for the analysis of general frameworks, the stiffness method, yields a solution for the individual members in this form. It is thus necessary for the engineer to be able to interpret this solution by producing the corresponding stress resultant diagrams.

Example 2.16

A simply supported beam is loaded by a concentrated force and two end couples as shown in Fig. 2.26(a). This is the same problem as that dealt with in Example 2.12 but with the addition of the end couples. With the reactions R_1, R_2 and R_3 shown in the free body diagram in Fig. 2.26(b), the global equilibrium equations are

$$\Sigma \rightarrow \quad R_1 = 0$$

$$\Sigma \uparrow \quad R_2 - 10.0 + R_3 = 0$$

$$\Sigma_z \text{ Ⓐ} \quad 3.5 - 10.0 \times 2.0 - 4.0 + R_3 \times 3.0 = 0$$

giving

$$\begin{bmatrix} R_1 \\ R_2 \\ R_3 \end{bmatrix} = \begin{bmatrix} 0 \\ 3.17 \\ 6.83 \end{bmatrix} \text{ kN.}$$

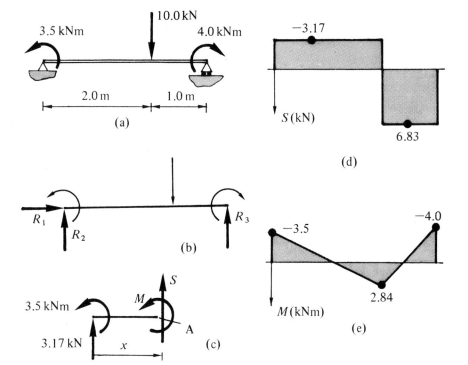

Fig. 2.26

We then make a cut in the beam at point A, distance x from the left-hand end, with x less than 2.25 m as in Fig. 2.26(c). Resolving and taking moments for the left-hand sub-section then leads to

$$S = -3.17 \text{ kN}$$

$$M = (-3.5 + 3.17 x) \text{ kNm.}$$

Again note that the couple does not appear in the equation for the shear force, but it does appear in the equation for the moment. If x is greater than 2.25 m, the 10.0 kN load can be included in the equilibrium equations to give

$$S = 6.83 \text{ kN}$$

$$M = -3.5 + 3.17 x - 10.0 (x - 2.0) = (16.5 - 6.83 x) \text{ kNm.}$$

These two functions, when plotted out, give the shear force and bending moment diagrams shown in Fig. 2.26(d) and (e).

If we compare the bending moment diagram in Fig. 2.26(e) with that for the concentrated force acting alone in Fig. 2.21, an important simplification in the method for constructing the bending moment diagram for a beam subject to end couples, becomes apparent. We first consider the couples and the concentrated force as two separate loading cases. The bending moment diagram for the couples is then simply a straight line between the bending moments at the end. The signs of the end bending moments can easily be found from physical reasoning. Thus a positive couple on the left-hand end produces tension on the top of the beam and therefore a negative bending moment. A positive couple on the right-hand end produces tension on the bottom, and therefore a positive bending moment. Examples of combinations of end-couples and their bending moment diagrams are given in Fig. 2.27. The bending moment diagram for the present example is shown in Fig. 2.28(a). We then take the bending moment diagram due to the concentrated force acting alone on the span shown in Fig. 2.28(b) and super-impose it graphically on the diagram due to the end couples to obtain the total bending moment diagram for the beam. The process is shown in the figure.

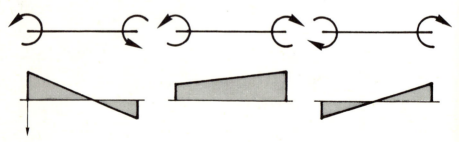

Fig. 2.27

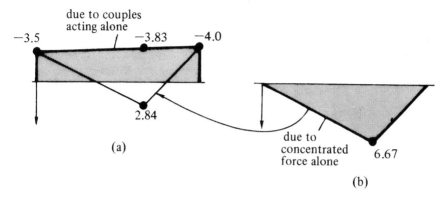

Fig. 2.28

2.5.3 Relationship between p, S and M

We have shown above that shear force and bending moment diagrams can be constructed from analytical expressions produced by making cuts at arbitrary distances x from the left-hand end of the beam. This method is unwieldy for complicated loading cases and we shall now describe how it can be simplified by considering relationships between the stress resultants themselves.

Consider the equilibrium of an infinitesimal element of a beam of length dx, at x from the left-hand end. The element is loaded by a vertical distributed force p per unit length. Treating the element as a free body cut from the beam, the free body diagram is shown in Fig. 2.29. Note that it has been assumed that the stress resultants vary with x and are thus infinitesimally different on the two cut surfaces. Whence resolving vertically gives

$$\Sigma \uparrow \qquad -S - p\,dx + S + dS \doteq 0$$

and therefore

$$\frac{dS}{dx} = p \,. \tag{2.14}$$

Taking moments about the z axis through A gives

$$\Sigma \,\textcircled{A}_z \quad S\frac{dx}{2} - M + (S + dS)\frac{dx}{2} + M + dM = 0$$

and therefore (ignoring second order infinitesimals)

$$\frac{dM}{dx} = -S \,. \tag{2.15}$$

These two important results can be summarised as follows:

(i) the slope of the shear force diagram is equal to the local intensity of the distributed force, and

(ii) the slope of the bending moment diagram is equal to minus the shear force.

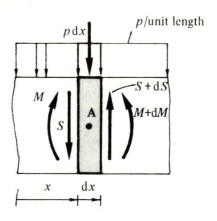

Fig. 2.29

Since the stress resultant diagrams are drawn positive downwards, we note that a positive slope is directed as shown in Fig. 2.30. Thus, referring back to Example 2.15 of a beam under a uniformly distributed force, we can see these relationships displayed graphically. The shear force diagram in Fig. 2.25(c) has a constant positive slope equal to the uniformly distributed force p. The bending moment diagram in Fig. 2.25(d) has its maximum positive slope at the left-hand end where the shear force has its maximum negative value, its slope is zero at the centre where the shear force is zero, and its maximum negative slope is at the right-hand end where the shear force has its maximum positive value.

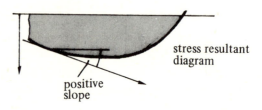

Fig. 2.30

A further important result is also evident in Example 2.15. This is that the bending moment has a local stationary value, either a maximum or a minimum, when the shear force is zero. We have seen that the shear force diagram is very easy to plot by the rule of thumb. It is therefore very easy to locate points of zero shear force in a beam. The corresponding maximum or minimum bending moments in the beam can thus be calculated immediately, by making cuts at these points and considering the rotational equilibrium of sub-sections in the usual way.

The relationship (2.15) between bending moment and shear force, forms the basis of a rapid method for plotting bending moment diagrams. It simplifies the interpolation between values of bending moment calculated at particular points in a beam. For brevity we shall call these values **spot values**, and they are calculated at the supports, at the positions of concentrated forces and couples, and at the ends of distributed forces. The method is illustrated in the following example.

Example 2.17
A cantilever is loaded by a uniformly distributed force, a concentrated couple and a concentrated force as in Fig. 2.31(a). The reactions at the root of the cantilever calculated from global equilibrium, are included in the free body diagram Fig. 2.31(b). Spot values of the bending moment are calculated at the ends 1 and 2 of the cantilever, and at the internal points A, B, C, D and E, where points C and D are just to the left and just to the right of the concentrated

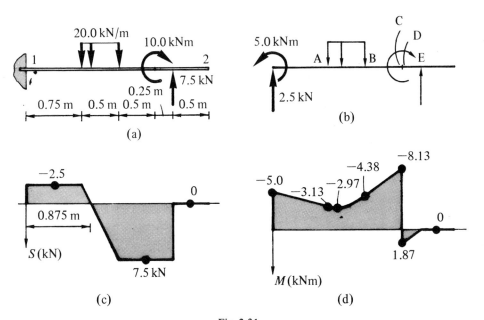

Fig. 2.31

couple respectively. The rotational equilibrium equations for the sub-sections of the cantilever to the right of each point are as follows, subscripts being used to indicate the particular spot values of bending moment,

$$\sum_z \text{①} \qquad -M_1 - 20.0 \times 0.5 \times 1.0 - 10.0 + 7.5 \times 2.0 = 0$$

$$\sum_z \text{Ⓐ} \qquad -M_A - 20.0 \times 0.5 \times 0.25 - 10.0 + 7.5 \times 1.25 = 0$$

$$\sum_z \text{Ⓑ} \qquad -M_B - 10.0 + 7.5 \times 0.75 = 0$$

$$\sum_z \text{Ⓒ} \qquad -M_C - 10.0 + 7.5 \times 0.25 = 0$$

$$\sum_z \text{Ⓓ} \qquad -M_D + 7.5 \times 0.25 = 0$$

$$\sum_z \text{Ⓔ}, \sum_z \text{②} \qquad M_E = M_2 = 0 .$$

Therefore

$$\begin{bmatrix} M_1 \\ M_A \\ M_B \\ M_C \\ M_D \\ M_E \\ M_2 \end{bmatrix} = \begin{bmatrix} -5.0 \\ -3.13 \\ -4.38 \\ -8.13 \\ +1.87 \\ 0 \\ 0 \end{bmatrix} \text{kNm} .$$

The shear force diagram constructed by the rule of thumb is shown in Fig. 2.31(c). The spot values of bending moments are plotted in Fig. 2.31(d). The interpolation between these values proceeds as follows. Between 1 and A, B and C, D and E, and E and 2, the shear force is constant indicating a bending moment diagram of constant slope and therefore a simple linear interpolation between the spot values. Note in particular, that on either side of the discontinuity corresponding to the couple, the shear force has the same positive value of 7.5 kN and that the slope of the bending moment diagram has the same negative value. Between points A and B the shear force changes continuously and is zero at a point which can be shown by simple proportion to be at $x = 0.875$ m. The bending moment diagram is a parabola convex downwards. The slope of the diagram is continuous at points A and B, and the parabola is therefore tangential to the slopes between 1 and A and between B and C. The local minimum value of the bending moment at $x = 0.875$ m is determined by making a cut at this point. Rotational equilibrium of the sub-section to the right of the cut then leads to:

$$M_{min} = -20.0 \times 0.375 \times (0.375/2) - 10.0 + 7.5 \times 1.125 = -2.97 \text{ kNm} .$$

All this information is sufficient to enable a sketch of the curved section of the bending moment diagram between A and B to be made, as shown in Fig. 2.31(d).

2.6 STRESS RESULTANTS IN FRAMEWORKS

2.6.1 Introduction
In the previous section, we have discussed methods for obtaining the stress resultants for single-span beams. Similar methods can be used for more complicated statically determinate frameworks. We first consider again the definition of stress resultants in relation to the member coordinate systems required for such frameworks.

2.6.2 Definition of Stress Resultants in Frameworks
As discussed in Chapter 1, Section 1.3, the notation for describing a framework is established by numbering the joints in some arbitrary order, and lettering the members. Thus the pin-footed portal frame in Example 2.8, might be described as in Fig. 2.32(a). Where helpful for clarity, parameters associated with a particular member m say, are distinguished by the subscript 'm'. The member coordinate systems are defined by taking the x_m coordinate in a particular member to run from the end with the lower joint number to the end with the higher. The y_m coordinate is parallel to the minor principal axis of the member, the z_m coordinate parallel to the major principal axis. Thus if the members of the portal frame are oriented for optimal structural performance, the member coordinate systems will be as shown in Fig. 2.32(b).

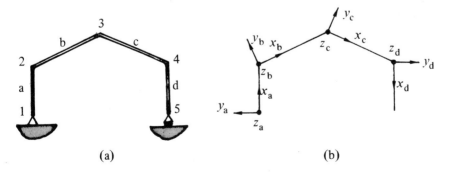

(a)   (b)

Fig. 2.32

Having established the member coordinate systems, the stress resultants can then be specified. Thus in member b of the portal frame, the stress resultants N_b, S_b and M_b on the positive x_b coordinate surface, act in the directions shown in Fig. 2.33(a). In drawing the stress resultant diagrams we take as their positive ordinates, directions perpendicular to the members, but opposite to the y_m coordinate directions. This again results in the bending moment diagram being on the tension side of the members. This is obvious in the case of the roof beam b. However, it is also worth demonstrating that the bending moment diagram lies on the tension side of the column d. Thus in Fig. 2.33(b) we see that a positive bending moment causes tension on the inside of the column, and this indeed is the side on which the bending moment would be plotted.

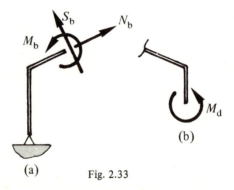

(a)   Fig. 2.33

(b)

Finally it should be noted that the sign given to the bending moment in a member depends on the orientation of the member coordinate system. Thus, if the joint numbering of joints 4 and 5 in Fig. 2.32(a) were reversed, the member coordinate system in d would be oriented as in Fig. 2.34. A positive bending moment M_d would then cause tension on the *outside* of the column. However, the bending moment, plotted in the opposite direction to y_d, would also be on the outside of the column, and thus the bending moment diagram would still be on the tension side of the member. The graphical significance of the bending moment diagram in indicating areas of tension is thus unaltered by the orientation of the member coordinates.

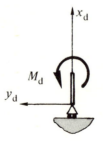

Fig. 2.34

2.6.3 Pin-footed Portal Frame

As an example of the analysis of statically determinate frameworks, we shall consider in detail the pin-footed portal frame discussed in Example 2.8.

Example 2.18

The pin-footed portal frame under the loading shown in Fig. 2.15 has been analysed for the three support reactions in Example 2.8. These are shown in the free body diagram in Fig. 2.35. The joint numbering and member coordinates are as shown in Figs 2.32(a) and (b). The stress resultants are found in a similar manner to the beams discussed in the previous section, by making cuts at points in the portal frame and satisfying equilibrium for the whole sub-section of the structure to one side of the cut.

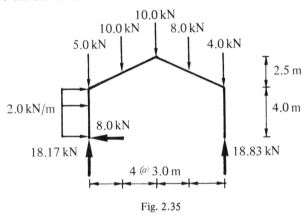

Fig. 2.35

Thus in column a, making a cut at point A, distance x_a from joint 1, the free body diagram for the lower sub-section of the column is shown in Fig. 2.36(a).

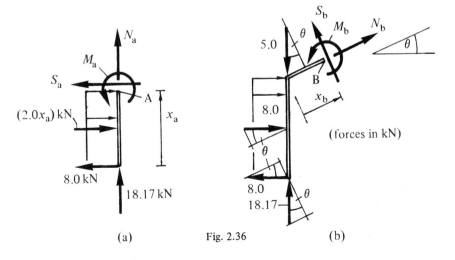

(a)   Fig. 2.36   (b)

Thence resolving in the x_a and y_a directions and taking moments about a z_a axis through A leads to

$$\underset{x_a}{\Sigma \rightarrow} \quad 18.17 + N_a = 0$$

$$\underset{y_a}{\Sigma \rightarrow} \quad 8.0 - 2.0x_a + S_a = 0$$

$$\underset{z_a}{\Sigma} \text{Ⓐ} \quad -8.0x_a + 2.0x_a (x_a/2) + M_a = 0 .$$

Therefore

$$\begin{bmatrix} N_a \\ S_a \\ M_a \end{bmatrix} = \begin{bmatrix} -18.17 \\ (-8.0 + 2.0x_a) \\ (8.0x_a - x_a{}^2) \end{bmatrix} \begin{matrix} \text{kN} \\ \text{kN} \\ \text{kNm} \end{matrix} .$$

In the roof beam b, making a cut at B to the left of the 10.0 kN load, the free body diagram for the sub-section to the left of the cut is as shown in Fig. 2.36(b). In order to obtain N_b and S_b we have to resolve all the forces on the sub-section in the x_b and y_b directions. Thus calling the slope of the beam θ and recalling that for the purposes of obtaining equilibrium, a uniformly distributed force p on the column can be replaced by the concentrated force pl at the centre of the loaded length, the equilibrium equations for the sub-section are as follows:

$$\underset{x_b}{\Sigma \rightarrow} \quad 18.17 \sin(\theta) - 8.0 \cos(\theta) + 8.0 \cos(\theta) - 5.0 \sin(\theta) + N_b = 0$$

$$\underset{y_b}{\Sigma \rightarrow} \quad 18.17 \cos(\theta) + 8.0 \sin(\theta) - 8.0 \sin(\theta) - 5.0 \cos(\theta) + S_b = 0$$

$$\underset{z_b}{\Sigma} \text{Ⓑ} \quad -18.17x_b \cos(\theta) - 8.0(4.0 + x_b \sin(\theta)) + 8.0(2.0 + x_b \sin(\theta)) + 5.0x_b \cos(\theta) + M_b = 0 .$$

We then note that $\theta = \tan^{-1}(2.5/6.0) = 22.62°$. Whence substituting into the above equations leads to

$$\begin{bmatrix} N_b \\ S_b \\ M_b \end{bmatrix} \begin{matrix} = \\ = \\ = \end{matrix} \begin{bmatrix} -5.07 \\ -12.16 \\ (16.0 + 12.16x_b) \end{bmatrix} \begin{matrix} \text{kN} \\ \text{kN} \\ \text{kNm} \end{matrix} .$$

We can continue this process, making a cut to the right of the 10.0 kN force in member b and cuts in c and d. The corresponding axial force and shear force diagrams for the complete framework, are shown in Figs. 2.37(a) and 2.37(b).

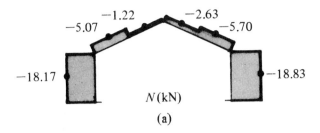

N (kN)

(a)

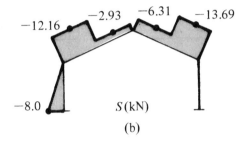

S (kN)

(b)

Fig. 2.37

The bending moment diagram, corresponding to the analytical functions of x_m given above, is again more easily plotted by calculating spot values for the bending moments at particular points in the framework and interpolating between these values using equation (2.15). Thus the spot values at the joints are calculated as

$M_1 = 0$

$M_2 = 8.0 \times 4.0 - 8.0 \times 2.0 = 16.0 \text{ kNm}$

$M_3 = 8.0 \times 6.5 - 8.0 \times 4.5 + (18.17 - 5.0) \times 6.0 - 10.0 \times 3.0 = 65.0 \text{ kNm}$

$M_4 = 0, \quad M_5 = 0$

and under the concentrated forces on the roof beams, at points C and D, as

$M_C = 8.0 \times 5.25 - 8.0 \times 3.25 + (18.17 - 5.0) \times 3.0 = 55.5 \text{ kNm}$

$M_D = (18.83 - 4.0) \times 3.0 = 44.5 \text{ kNm} .$

The spot values are plotted on the bending moment diagram in Fig. 2.38, and the interpolation between them, linear in members b, c, d and parabolic in a, is constructed by referring to the shear force diagram. In carrying out this interpolation it is helpful to view each member as a beam with the x_m coordinate running from left to right and the y_m coordinate upwards.

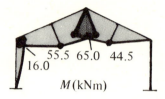

$M(kNm)$

Fig. 2.38

2.6.4 Frameworks with Curved Members

If a member of a framework is curved, the member is described by a curved coordinate system. Thus the x_m coordinate follows the axis of the member and the y_m and z_m coordinates are orthogonal to the x_m coordinate at any point, and are parallel to the principal axes of the cross-section. The coordinate system of a curved member in a plane framework would be as shown in Fig. 2.39(a). The stress resultants at any point act on the positive x_m coordinate surface which is orthogonal to the axis of the member, in the directions shown in Fig. 2.39(b). These stress resultants are calculated by the usual procedure of making cuts in the member at various points, and this is illustrated in the following example.

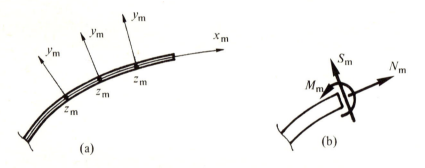

Fig. 2.39

Example 2.19

A pin-footed circular curved arch is shown in Fig. 2.40(a), and has a geometry that represents a large concrete arch bridge. For the purpose of the example, the loading will be taken as a live loading due to traffic of 30.0 kN/m, which is distributed to half the arch from the deck by hangers as shown. The loading will be considered to act vertically at all points on the curved member, and be uniformly distributed horizontally. Firstly we calculate the reactions R_1, R_2 and R_3 shown in the free body diagram in Fig. 2.40(b), by considering the global equilibrium of the arch. Thus

$$\Sigma \rightarrow \quad R_1 = 0$$

$$\Sigma \uparrow \quad R_2 + R_3 - 30.0 \times 50.0 = 0$$

$$\Sigma_z \textcircled{1} \quad -30.0 \times 50.0 \times 25.0 + R_3 \times 100.0 = 0$$

giving

$$\begin{bmatrix} R_1 \\ R_2 \\ R_3 \end{bmatrix} = \begin{bmatrix} 0 \\ 1125.0 \\ 375.0 \end{bmatrix} \text{kN} .$$

We then calculate the stress resultants by making cuts at points in the arch. In this particular case with the arch circular, the position of a point A say, is conveniently defined by the polar coordinates r and θ shown in Fig. 2.40(c).

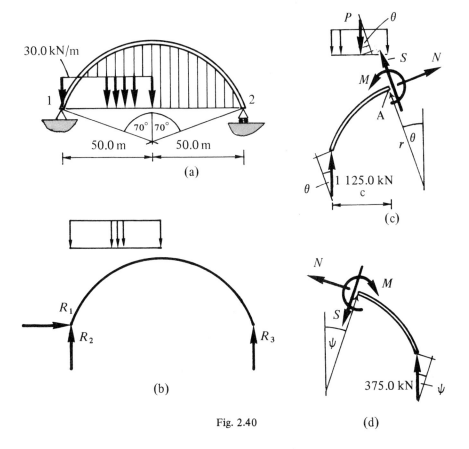

Fig. 2.40

The radius r is then calculated from the geometry of the arch as $50.0/\sin(70°) = 53.21$ m. Considering the sub-section to the left of A, we note that the horizontal loaded length c, say, is $r(\sin(70°) - \sin(\theta))$. Thus the total external force P on the sub-section is $30.0c$ kN acting at an effective horizontal distance of $c/2$ from A. Resolving in the coordinate directions at A and taking moments about a z axis through A, then leads to

$$\Sigma \underset{x}{\rightarrow} \qquad N_\theta - P\sin(\theta) + 1\ 125.0\sin(\theta) = 0$$

$$\Sigma \underset{y}{\rightarrow} \qquad S_\theta - P\cos(\theta) + 1\ 125.0\cos(\theta) = 0$$

$$\Sigma \underset{z}{A} \qquad -1\ 125.0 \times c + P(c/2) + M_\theta = 0\ .$$

Thus

$$N_\theta = (375.0\sin(\theta) - 1\ 596.3\sin^2(\theta))\ \text{kN}$$

$$S_\theta = (375.0\cos(\theta) - 1\ 596.3\sin(\theta)\cos(\theta))\ \text{kN}$$

$$M_\theta = (18\ 750.0 + 19\ 953.8\sin(\theta) - 42\ 469.6\sin^2(\theta))\ \text{kNm}\ .$$

When considering cuts to the right of the crown of the arch, it probably reduces the possibility of confusion if their positions are defined by ψ as shown in Fig. 2.40(d). Considering the equilibrium of sub-sections to the right of the cuts then leads to

$$N_\psi = (-375.0\sin(\psi))\ \text{kN}$$

$$S_\psi = (375.0\cos(\psi))\ \text{kN}$$

$$M_\psi = (18\ 750.0 - 19\ 953.8\sin(\psi))\ \text{kNm}.$$

In constructing the stress resultant diagrams corresponding to the above expressions for N, S and M, it is conventional to choose the ordinates of the diagrams so that they are at right angles to the axis of the curved member at any point, but in the opposite direction to the y_m coordinate direction. Thus the three stress resultant diagrams for the arch come out to be as shown in Figs. 2.41(a)–(c).

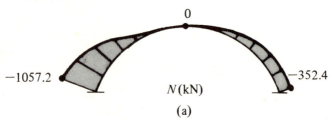

$$N(\text{kN})$$

(a)

Fig. 2.41 (*continued next page*)

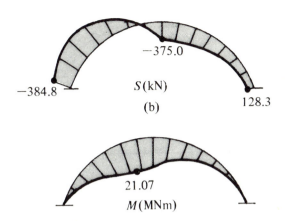

-375.0

-384.8 $S(\mathrm{kN})$

(b) 128.3

21.07

$M(\mathrm{MNm})$

Fig. 2.41 (*continued*)

2.6.5 Space Frameworks

Up to now we have considered plane frameworks in which the only non-zero stress resultants are the axial forces and shear forces in the plane of the framework, and the bending moments about axes at right angles to this plane. In the case of space frameworks under arbitrary loading, however, all six stress resultants discussed in Chapter 1, Section 1.7 can take values. For convenience, the stress resultants acting on a particular positive x_m coordinate surface are shown again in Fig. 2.42. Note that subscripts y and z are now needed to distinguish between the two shear forces and the two bending moments. The calculation of these stress resultants is usually a complicated problem and it is the purpose of this section to illustrate how they can be obtained for a particular case, namely the cranked cantilever shown in Fig. 2.16.

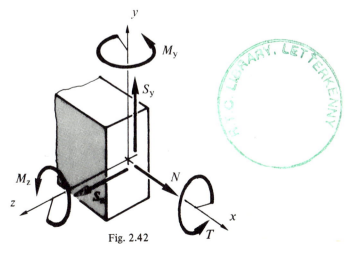

Fig. 2.42

Example 2.20

The notation for the cantilever and the corresponding member coordinate systems are established in the usual way. Thus with the x_m coordinate in any member again running from the end with the lower joint number to the end with the higher, and with the y_m and z_m coordinates parallel to the minor and major principal axes of the cross-section respectively, the cantilever joint and member notation and member coordinate systems are shown in Fig. 2.43(a).

Once again, the stress resultants are obtained by making cuts at various points in the members, and in this case, by considering the equilibrium of sub-sections of the cantilever remote from the root. The six stress resultants are found immediately by resolving in the three coordinate directions and by taking moments about the three coordinate axes. Thus making a cut at point A in member c say, the free body diagram for the sub-section remote from the root is shown in Fig. 2.43(b). Whence the equilibrium equations for the sub-section are as follows;

$$\underset{x}{\Sigma \rightarrow} \quad N = 0$$

$$\underset{y}{\Sigma \rightarrow} \quad S_y - 6.0 \times 2.0 - 10.0 = 0$$

$$\underset{z}{\Sigma \rightarrow} \quad S_z - 15.0 = 0$$

$$\underset{x}{\Sigma} \circledA \quad T - 10.0 \times 0.75 + 15.0 \times 2.0 = 0$$

$$\underset{y}{\Sigma} \circledA \quad M_y - 15.0 \times 2.0 = 0$$

$$\underset{z}{\Sigma} \circledA \quad M_z + 6.0 \times 2.0 \times 1.0 + 10.0 \times 2.0 = 0$$

giving the stress resultants

$$
\begin{bmatrix} N \\ S_y \\ S_z \\ T \\ M_y \\ M_z \end{bmatrix}
=
\begin{bmatrix} 0 \\ 22.0 \\ 15.0 \\ -22.5 \\ 30.0 \\ -32.0 \end{bmatrix}
\begin{matrix} \text{kN} \\ \text{kN} \\ \text{kN} \\ \text{kNm} \\ \text{kNm} \\ \text{kNm} \end{matrix} .
$$

The bending moment diagrams are again most conveniently constructed by the spot value method. Referring to Fig. 2.42, it is apparent that the positive

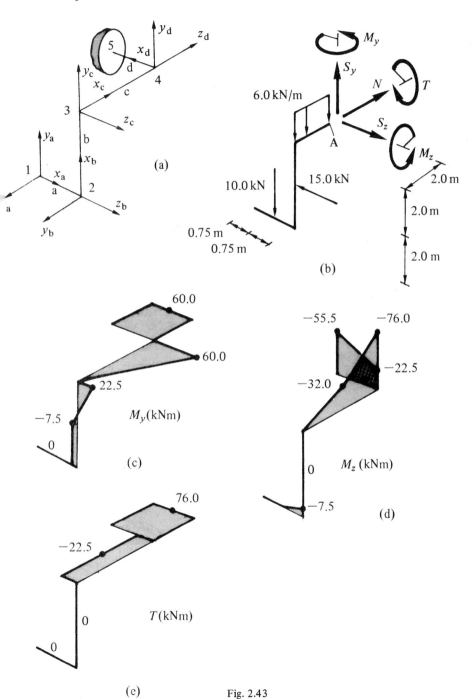

Fig. 2.43

bending moment M_y causes tension on the *positive* z_m side of the member. Thus, in order to retain the convention that the bending moment diagram lies on the tension side of the member, the ordinate of the M_y diagram must be chosen to be in the *same* direction as the z_m coordinate. The corresponding M_y diagram, and the M_z diagram are shown in Figs. 2.43(c) and (d). The torque diagram is particularly simple for this case, since the 10.0 kN and 15.0 kN concentrated forces produce a constant torque of $-$ 22.5 kNm in member c, and the 10.0 kN concentrated force and 6.0 kN/m uniformly distributed force produce a constant torque of 76.0 kNm in member d. The T diagram is then as shown in Fig. 2.43(e). The axial and shear force diagrams are equally simple to construct for this case, but for conciseness they are not illustrated.

2.7 ARTICULATION

2.7.1 Introduction

Structures are often designed with internal articulation by introducing hinges or pins at various points. Articulation can be important, for example in a bridge built in a mining area, since the hinges can allow the bridge to deform and accommodate support settlements. Strictly, for plane frameworks, we are only concerned with hinges, their axes being parallel to the z coordinate as in Fig. 2.44(a). A hinge at point A in a member cannot then resist a moment about a z axis through A. In space frameworks we are also concerned with pins, which can take the form of the ball and socket arrangement shown in Fig. 2.44(b). A pin at point A cannot then resist a moment about any axis through A.

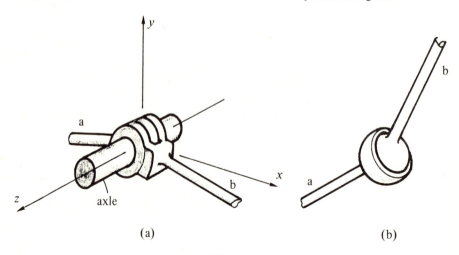

(a)   (b)

Fig. 2.44

In considering the analysis of frameworks containing articulation, it must first be emphasised that the global equilibrium equations derived in Section 2.2 make no reference to the internal articulation of bodies. They therefore remain unchanged. Articulation does, however, lead to extra internal equilibrium equations. This is simply because, in the case of a hinge for example, the moments of all the forces on the framework to one side or other of the hinge must be zero. These extra **hinge equations** enable extra unknown support reactions to be calculated. Therefore more complicated frameworks than have been considered hitherto, are statically determinate.

2.7.2 Compound Beams
Beams containing hinges are called **compound beams**. As an example of their analysis we shall consider the case of a compound propped cantilever.

Example 2.21
The compound cantilever in Fig. 2.45(a) is propped at end 2 by a vertical support. It contains a hinge at point A. The free body diagram is shown in Fig. 2.45(b), and in the absence of the hinge the cantilever would be statically indeterminate. The global equilibrium equations are as follows:

$$\Sigma \rightarrow \quad R_1 = 0$$

$$\Sigma \uparrow \quad R_2 - 5.0 - 10.0 + R_4 = 0$$

$$\Sigma \underset{z}{①} \quad R_3 - 5.0 \times 1.25 - 10.0 \times 2.0 + R_4 \times 3.0 = 0 \ .$$

The free body diagram of the sub-section of the cantilever to the right of the hinge is shown in Fig. 2.45(c), where we note that the hinge carries no bending moment but that it does transmit a shear force. Taking moments about the z axis through A for this sub-section leads to the extra equilibrium equation

$$\Sigma \underset{z}{Ⓐ}_r \quad -10.0 \times 0.25 + R_4 \times 1.25 = 0 \ .$$

(The subscript r has been added to the symbol for taking moments, to indicate that we are only concerned with the moments of the forces on a sub-section to the *right* of the axis through A.) The global equilibrium equations plus the hinge equation can then be solved for the reactions to give

$$\begin{bmatrix} R_1 \\ R_2 \\ R_3 \\ R_4 \end{bmatrix} = \begin{bmatrix} 0.0 \\ 13.0 \\ 20.25 \\ 2.0 \end{bmatrix} \begin{matrix} kN \\ kN \\ kNm \\ kN \end{matrix} \ .$$

The stress resultant diagrams are plotted using the methods discussed in the previous section. Thus using the rule of thumb we construct the shear force diagram shown in Fig. 2.45(d), from which we note that the shear force transmitted by the hinge is −8.0 kN. The bending moment diagram found from spot values and interpolation, is shown in Fig. 2.45(e). In this diagram we note that the bending moment at the hinge is zero.

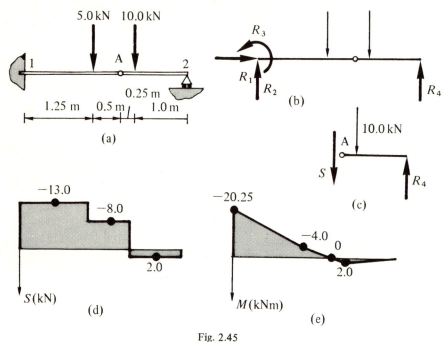

Fig. 2.45

We have seen in the above example, that the presence of a single hinge produces one internal equilibrium equation from which an extra support reaction can be calculated. Thus in a beam containing N hinges, N equations are available for calculating the reactions. Further simple examples of statically determinate compound beams containing one or more hinges, are shown in Fig. 2.46.

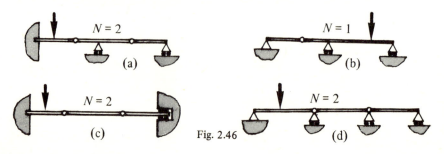

Fig. 2.46

2.7.3 Three-pinned Arches

An important group of articulated structures are the three-pinned arches. These are plane frameworks which consist of arches supported on pinned supports at the feet and containing a single internal hinge, usually at the crown. A portal frame example is shown in Fig. 2.47. Comparing this with the pin-footed portal frame discussed in Examples 2.8 and 2.18, an extra reaction is necessary to restrain the horizontal movement of the right-hand foot. The extra equilibrium equation is obtained by taking moments for a sub-section of the structure to one side or other of the hinge at the crown.

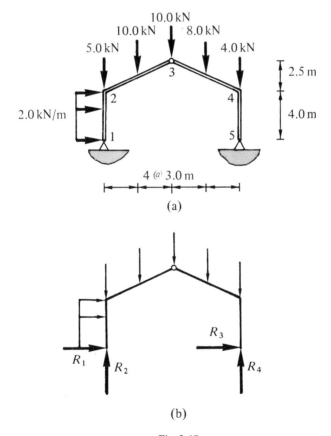

(a)

(b)

Fig. 2.47

Example 2.22 The three-pinned portal frame
Consider the three-pinned portal frame shown in Fig. 2.47(a), having the same geometry and loading as the portal frame in Examples 2.8 and 2.18. The free body diagram is shown in Fig. 2.47(b). Whence the global equilibrium equations

and the hinge-equation for the sub-section of the framework to the right of the hinge, take the form

$$\Sigma \rightarrow \quad R_1 + R_3 + 2.0 \times 4.0 = 0$$

$$\Sigma \uparrow \quad R_2 + R_4 - 37.0 = 0$$

$$\Sigma \underset{z}{①} \quad -2.0 \times 4.0 \times 2.0 - 10.0 \times 3.0 - 10.0 \times 6.0 - 8.0 \times 9.0 - \\ - 4.0 \times 12.0 + R_3 \times 12.0 = 0$$

$$\Sigma \underset{z}{③}_r \quad -8.0 \times 3.0 - 4.0 \times 6.0 + R_3 \times 6.5 + R_4 \times 6.0 = 0 \ .$$

These lead to

$$\begin{bmatrix} R_1 \\ R_2 \\ R_3 \\ R_4 \end{bmatrix} = \begin{bmatrix} 2.0 \\ 18.17 \\ -10.0 \\ 18.83 \end{bmatrix} \text{kN} \ .$$

The stress resultant diagrams are then plotted in the same way as for Example 2.18, and Fig. 2.48 is the bending moment diagram produced by the spot value and interpolation method. It is interesting that the change in support conditions and the internal hinge, completely change the bending moments within the framework, compared with those for the pin-footed portal frame in Fig. 2.38. In particular, we now have large hogging bending moments at the eaves, that are characteristic of frames dominated by vertical loading.

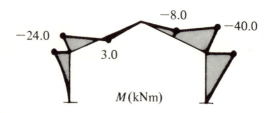

Fig. 2.48

2.8 PIN-JOINTED FRAMEWORKS

2.8.1 Introduction

We consider next a very important group of articulated structures called **pin-jointed frameworks**, which are statically determinate by virtue of the fact that they contain hinges or pins at the joints. We shall consider joint-loaded frameworks, which in the case of plane frameworks can be thought of as being loaded by forces acting on the axles of the hinged joints as in Fig. 2.49. It should be clear from this figure that the bending moments at the ends of the members framing into each joint are zero. A hypothetical ball and socket arrangement would give similar conditions at the ends of members framing into the joint of a three-dimensional space framework.

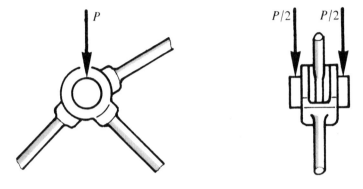

Fig. 2.49

Consider the free body diagram of a member of a pin-jointed framework, loaded by the end forces P_1, P_2, P_3 and P_4 shown in Fig. 2.50(a). Since the hinges or pins at either end cannot impose any couples on the member, rotational equilibrium about end 1 say, requires that the resultant P_6 of the forces acting on end 2 must pass through 1, that is, coincide with the axis of the member.

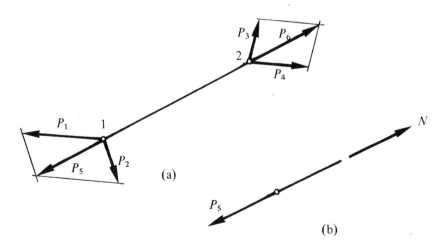

Fig. 2.50

Linear equilibrium in the axial direction then demonstrates that the resultant of the forces P_5 at end 1, is equal and opposite to P_6. If we next consider a cut in the member as in Fig. 2.50(b) and the equilibrium of a sub-section on one side of the cut, it follows that the only non-zero stress resultant is the axial force N and that this is constant at all points in the member. The analysis of pin-jointed frameworks, therefore involves determining these axial forces. They are again found by methods which involve isolating appropriate sub-sections of a framework by cutting its members, and considering the equilibrium of these sub-sections. We first describe the commonest method, which considers the equilibrium of sub-sections of the framework in the neighbourhood of individual joints.

2.8.2 Method of Joints

The method of joints involves considering in turn the equilibrium of the forces at the joints, usually in a particular sequence so that at each joint in a plane framework there are only two unknown member forces, and in a space framework three unknown forces. We illustrate the method with a simple example.

Example 2.23
Consider the joint loaded plane framework shown in Fig. 2.51. We begin the analysis by considering the equilibrium of joint 3. The free body diagram of a sub-section cut from the structure is shown in Fig. 2.52(a). Resolving horizontally and vertically for the sub-section then leads to

$$\Sigma \rightarrow \quad -N_b - N_c \cos(45°) = 0$$

$$\Sigma \uparrow \quad -N_c \sin(45°) - 10.0 = 0 \ .$$

Whence

$$\begin{bmatrix} N_b \\ N_c \end{bmatrix} = \begin{bmatrix} 10.0 \\ -14.14 \end{bmatrix} \text{kN} \ .$$

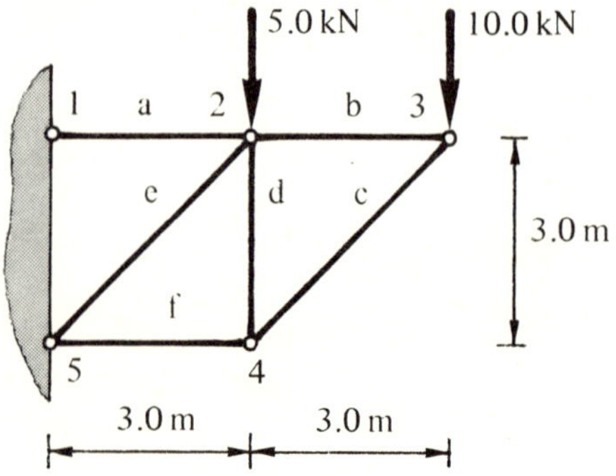

Fig. 2.51

We next consider joint 4, for which the free body diagram is shown in Fig. 2.52(b) and includes the known compressive force of 14.14 kN in member c. Linear equilibrium of the sub-section then leads to

$$\Sigma \rightarrow \qquad -N_f - 14.14 \cos (45°) = 0$$

$$\Sigma \uparrow \qquad N_d - 14.14 \sin (45°) = 0$$

giving

$$\begin{bmatrix} N_d \\ N_f \end{bmatrix} = \begin{bmatrix} 10.0 \\ -10.0 \end{bmatrix} \text{ kN }.$$

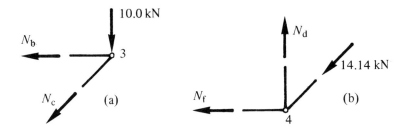

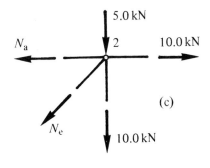

Fig. 2.52

Finally moving to joint 2, shown in Fig. 2.52(c) the linear equilibrium equations are

$$\Sigma \rightarrow \qquad -N_a - N_e \cos (45°) + 10.0 = 0$$

$$\Sigma \uparrow \qquad -N_e \sin (45°) - 10.0 - 5.0 = 0$$

giving

$$
\begin{bmatrix} N_a \\ N_e \end{bmatrix} = \begin{bmatrix} 25.0 \\ -21.21 \end{bmatrix} kN \; .
$$

We thus determine all the member forces in the framework. Note that in this case, it is not necessary to calculate the reactions at the supports. Indeed there are insufficient global equilibrium equations to enable the reactions to be calculated before solving internally. If the reactions *were* required, we should obtain them by considering the equilibrium of joints 1 and 5.

The sequence in which a framework is solved by the method of joints depends on the type of framework. It is also often necessary to calculate the reactions first. Thus for the truss bridge shown in Fig. 2.53, it would be convenient to start at the right-hand end and calculate the vertical reaction from the global rotational equilibrium equation. We should then proceed through the truss in the order of joints as they are numbered in the figure.

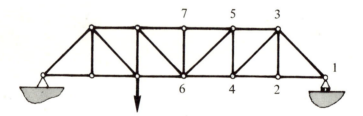

Fig. 2.53

2.8.3 Tension Coefficients

It is apparent from Example 2.23 that the method of joints requires the resolution of the axial forces in the members into components parallel to the global co-ordinates. For $45°$ plane frameworks, this is a simple problem. However, difficulties arise for irregular plane frameworks and for space frameworks. A systematic way of formulating and simplifying the problem is to use **tension coefficients**.

Consider the member shown in Fig. 2.54, joining joints i and j in a framework, the global coordinates of the joints being x_i, y_i and z_i and x_j, y_j and z_j. Making a cut in the member, the components of N acting on joint 1 are

$$
N_x = N \cos(\alpha_x), \quad N_y = N \cos(\alpha_y), \quad N_z = N \cos(\alpha_z) \; .
$$

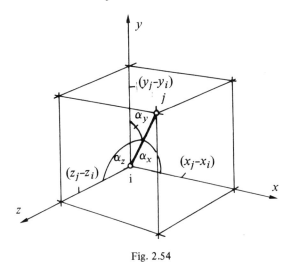

Fig. 2.54

From simple geometry, $\cos(\alpha_x)$, etc., are given in terms of the coordinates of joints i and j as follows

$$\cos(\alpha_x) = (x_j - x_i)/l$$
$$\cos(\alpha_y) = (y_j - y_i)/l$$
$$\cos(\alpha_z) = (z_j - z_i)/l \ .$$

If therefore we define a parameter t, called the *tension coefficient* of the member, as

$$t = N/l \tag{2.16}$$

the three components of the force N at joint i are given by

$$\mathbf{N} = \begin{bmatrix} (x_j - x_i) \\ (y_j - y_i) \\ (z_j - z_i) \end{bmatrix} t \ . \tag{2.17}$$

Structural frameworks are often specified by their joint coordinates relative to some global coordinate system. N can therefore be easily generated in terms of t to give the required resolved components of the internal forces at the joints.

Example 2.24

Consider the plane framework shown in Fig. 2.55. We write down and solve the equilibrium equations of joints 3, 4 and 2 in turn, in terms of the tension coefficients as follows:

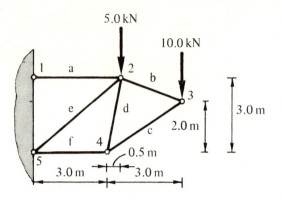

Fig. 2.55

At joint 3

$$\sum_x \rightarrow \quad (x_2 - x_3)\, t_b + (x_4 - x_3)\, t_c = 0$$

$$\sum_y \rightarrow \quad (y_2 - y_3)\, t_b + (y_4 - y_3)\, t_c - 10.0 = 0$$

$$-2.5\, t_b - 3.0\, t_c = 0$$

$$1.0\, t_b - 2.0\, t_c = 10.0 \ .$$

Therefore

$$\begin{bmatrix} t_b \\ t_c \end{bmatrix} = \begin{bmatrix} 3.75 \\ -3.13 \end{bmatrix} \text{kN/m} \ .$$

At joint 4

$$\sum_x \rightarrow \quad 3.0\, t_c - 3.0\, t_f + 0.5\, t_d = 0$$

$$\sum_y \rightarrow \quad 2.0\, t_c + 3.0\, t_d = 0 \ .$$

Therefore

$$\begin{bmatrix} t_d \\ t_f \end{bmatrix} = \begin{bmatrix} 2.08 \\ -2.78 \end{bmatrix} \text{kN/m} \ .$$

At joint 2

$$\sum_x \rightarrow \quad -3.5\, t_a - 3.5\, t_e - 0.5\, t_d + 2.5\, t_b = 0$$

$$\sum_y \rightarrow \quad -3.0\, t_e - 3.0\, t_d - 1.0\, t_b - 5.0 = 0 \ .$$

Therefore

$$\begin{bmatrix} t_a \\ t_e \end{bmatrix} = \begin{bmatrix} 7.38 \\ -5.0 \end{bmatrix} \text{ kN/m .}$$

We then compile Table 2.1 to evaluate the lengths of the members and thence the member forces.

Table 2.1
Calculations for Example 2.24

Member	$x_j - x_i$	$y_j - y_i$	l (m)	t (kN/m)	N (kN)
a	3.5	0.0	3.5	7.38	25.83
b	2.5	1.0	2.69	3.75	10.09
c	3.0	2.0	3.61	−3.13	−11.30
d	0.5	3.0	3.04	2.08	6.32
e	3.5	3.0	4.61	−5.00	−23.05
f	3.0	0.0	3.0	−2.78	−8.34

Example 2.25
We illustrate the use of tension coefficients in three dimensions by analysing the simple tripod under an inclined load in the $x-y$ plane shown in Fig. 2.56. The three equilibrium equations for joint 1 are then given in terms of the tension coefficients as

$$0.5\, t_b - 2.0\, t_c + 8.66 = 0$$

$$-1.5\, t_a - 1.5\, t_b - 1.5\, t_c + 5.00 = 0$$

$$-1.5\, t_a + 1.5\, t_b - 0.5\, t_c = 0$$

giving

$$\begin{bmatrix} t_a \\ t_b \\ t_c \end{bmatrix} = \begin{bmatrix} -1.25 \\ 0.21 \\ 4.38 \end{bmatrix} \text{ kN/m .}$$

Whence the member forces are evaluated in Table 2.2.

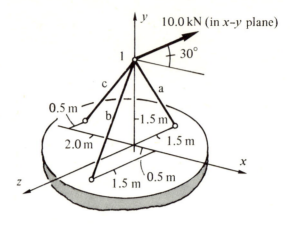

Fig. 2.56

Table 2.2
Calculations for Example 2.25

Member	$x_j - x_i$	$y_j - y_i$	$z_j - z_i$	l (m)	t (kN/m)	N (kN)
a	0	−1.5	−1.5	2.12	−1.25	−2.65
b	0.5	−1.5	1.5	2.18	0.21	0.46
c	−2.0	−1.5	−0.5	2.55	4.38	11.17

2.8.4 Method of Sections

The method of joints is one of the ways of isolating sub-sections of a pin-jointed framework to produce useful equilibrium equations. Another way is to take sections through the complete framework to yield information about the axial forces in selected members. Thus in general, if a sub-section of a plane framework containing more than one joint can be isolated by making cuts in just three members, the axial forces in these members can be determined by considering the linear and rotational equilibrium of the sub-section. Similarly, if a sub-section of a space framework can be isolated by making cuts in just six members, the axial forces in these members can be determined. The procedure is called the **method of sections.**

Example 2.26
Consider again the framework in Fig. 2.51, and suppose cuts are made through the members a, e and f. The free body diagram of the sub-section to the right of the cuts is shown in Fig. 2.57. Whence resolving horizontally and vertically and taking moments about a z axis through 2 yields the equations

$$\Sigma \rightarrow \quad -N_a - N_e \cos(45°) - N_f = 0$$

$$\Sigma \uparrow \quad -N_e \sin(45°) - 5.0 - 10.0 = 0$$

$$\Sigma_z \text{②} \quad -N_f \times 3.0 - 10.0 \times 3.0 = 0 \ .$$

Therefore

$$\begin{bmatrix} N_a \\ N_e \\ N_f \end{bmatrix} = \begin{bmatrix} 25.0 \\ -21.21 \\ -10.0 \end{bmatrix} \text{kN} \ .$$

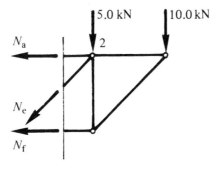

Fig. 2.57

The method of sections does not appear to be particularly advantageous when applied to the simple framework above. It can, however, be extremely useful if the forces in a few selected members in a long framework are required. Thus if the forces in members a, b and c in the truss bridge in Fig. 2.58 are required, they can be obtained immediately, by making cuts in the members, and considering the equilibrium of the right-hand sub-section of the structure.

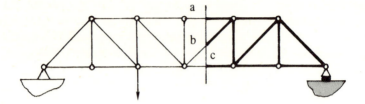

Fig. 2.58

2.8.5 Formulae for Statical Determinacy

The methods of joints and sections only work if a framework contains a particular number of members and joints. If, for example, the framework in Fig. 2.59 contains the diagonal member g, the method of joints breaks down at joints 2 or 4, where there are too many unknowns. Similarly there is no way of using the method of sections.

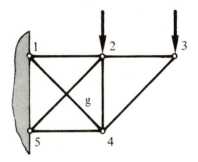

Fig. 2.59

The number of members needed to make a pin-jointed framework statically determinate, can be obtained simply by considering an attempted application of the method of joints. Thus for a plane framework containing J joints, there are two equilibrium equations at each joint, giving a total of $2J$ equations available for the solution of the structure. If the framework contains M members, and is supported by R reactions, the total number of unknown forces acting on the joints is $M + R$. Thus for a statically determinate pin-jointed framework, the following equation must be satisfied

$$M + R = 2J .$$ (2.18)

Some examples of statically determinate frameworks are given in Fig. 2.60. (Note that the method of joints applied to the framework in Fig. 2.60(d) will only produce *coupled* simultaneous equations for the internal forces.) The addition of two members and one joint to a statically determinate framework retains the equality in (2.18), so that frameworks are often built up of simple triangles of members in the so called **truss** form as shown in Fig. 2.61.

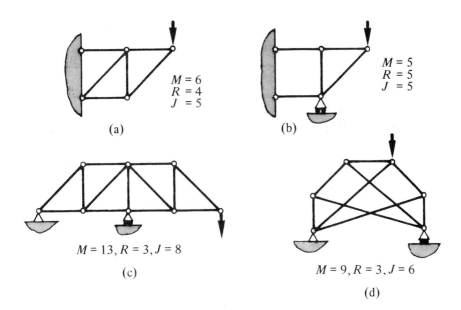

Fig. 2.60

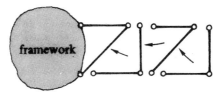

Fig. 2.61

If $M + R$ is greater than $2J$, there are insufficient equations available for the number of unknowns. Such frameworks are called **statically indeterminate**.

If $M + R$ is less than $2J$, there are more equations available than there are unknowns and non-unique solutions are possible for the forces and reactions. Physically this means that the structure is a **mechanism** and is incapable of maintaining equilibrium under arbitrarily directed loading. Some examples of mechanisms are shown in Fig. 2.62.

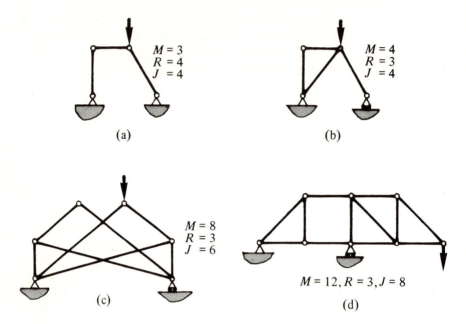

Fig. 2.62

Equation (2.18) has to be used with discrimination for although a particular framework might satisfy it, it is still possible for the framework to contain a mechanism in one part and be statically indeterminate in another. A mechanism also exists, if the framework as a whole, or any sub-section cut from it that contains more than one joint, is subjected to reactions or member forces that are all parallel or all intersect at a single point. This is because the reactions or member forces then cannot resist any external forces that either act at right angles to their parallel direction, or exert a moment about an axis through their point of intersection. Examples of such problems are shown in Fig. 2.63.

For pin-jointed space frameworks, three equations are available at each joint. Thus the corresponding equation for statical determinacy is

$$M + R = 3J .$$
(2.19)

Examples of statically determinate space frameworks are given in Fig. 2.64. Again, it should be noted that the addition of three members and one joint to a statically determinate framework retains the equality in (2.19), so space trusses are often composed of tetrahedra as in Fig. 2.65. Again it is possible for a space framework to satisfy (2.19) and still contain a mechanism. For a satisfactory framework, the whole structure or any part of it should be capable of resisting forces in any direction or moments about any axis.

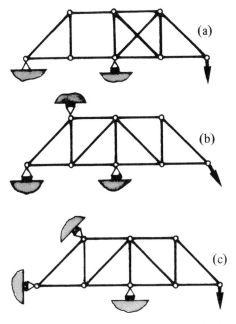

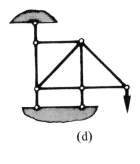

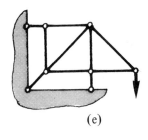

Fig. 2.63

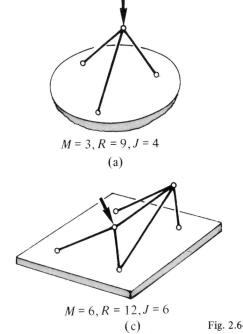

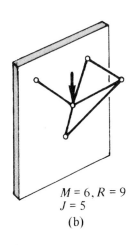

$M = 3, R = 9, J = 4$

(a)

$M = 6, R = 9$
$J = 5$
(b)

$M = 6, R = 12, J = 6$

(c)   Fig. 2.64

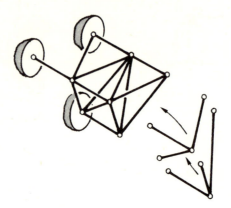

Fig. 2.65

2.9 PROBLEMS

2.1   Find the support reactions and draw the shear force and bending moment diagrams for the beams and cantilevers shown in Figs. 2.66(a) to (i).

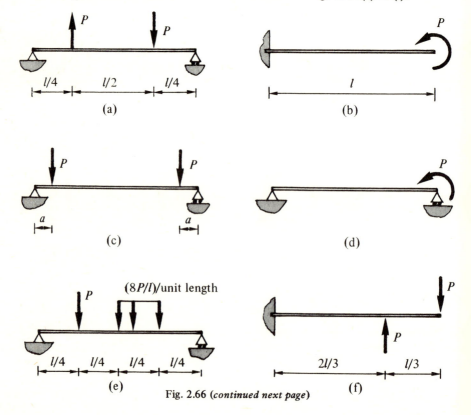

Fig. 2.66 (*continued next page*)

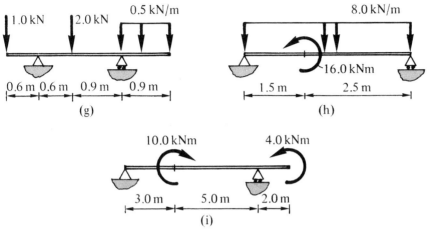

Fig. 2.66 (*continued*)

2.2   Find the support reactions and draw the shear force and bending moment
      diagrams for the compound beams shown in Figs 2.67(a) to (e).

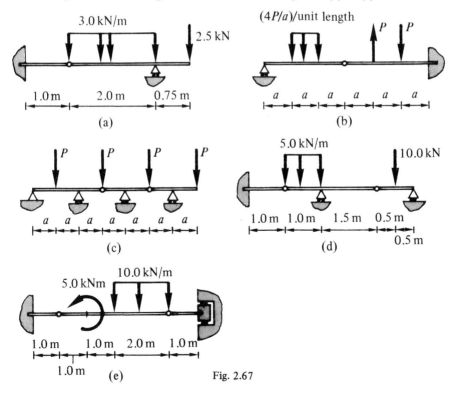

Fig. 2.67

2.3   The continuous beam shown in Fig. 2.68(a) has been analysed by the
       stiffness method. The solution is given in the table below in terms of the
       end couples acting on each span treated as a separate member, as in Fig.
       2.68(b).

Span	P_{m1} (kNm)	P_{m2} (kNm)
a	0.0	−3.44
b	3.44	−0.35
c	0.35	−3.00
d	3.00	0

Calculate the vertical support
reactions at the ends of each
span considered again as a separate member. Hence draw the
shear force and bending moment diagrms for the continuous beam.

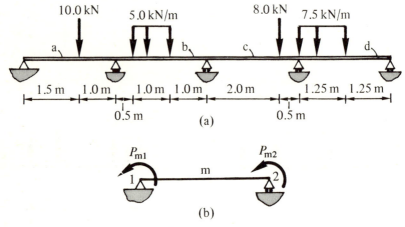

(a)

(b)

Fig. 2.68

2.4   Find the support reactions and draw the axial force, shear force and
       bending moment diagrams for the plane frameworks shown in Figs.
       2.69(a) to (h).

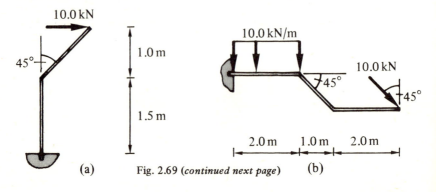

(a)   Fig. 2.69 (*continued next page*)   (b)

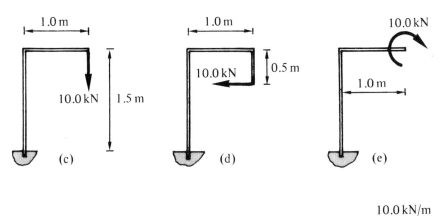

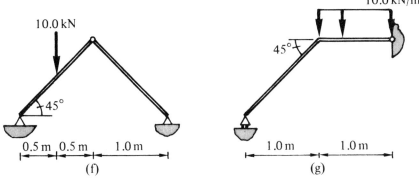

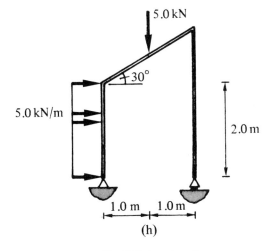

Fig. 2.69 (*continued*)

2.5   A street sign is supported by a cantilever, part of which forms a circular curve of 7.0 m radius as shown in Fig. 2.70. The weight of the sign is equivalent to a 4.0 kN force acting at its centre. Determine the bending moments in the curved part of the cantilever as a function of angle θ shown in the figure.

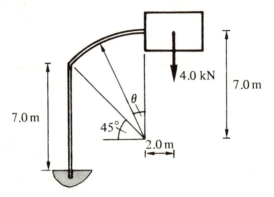

Fig. 2.70

2.6   A three-pinned arch bridge has the parabolic profile shown in Fig. 2.71. The curve of its centre-line is given by $y = 12.0 - 0.01333 (x-30.0)^2$. It is subjected to loading comprising a horizontally distributed force of 50.0 kN/m over half the bridge plus a concentrated force of 250.0 kN at the centre. Calculate (i) the reactions, (ii) the bending moments in the two halves of the bridge as functions of the horizontal coordinate x.

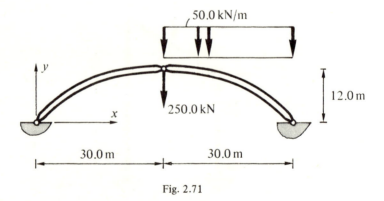

Fig. 2.71

2.7   A rigid-jointed space framework lies in a horizontal plane, and has its joints numbered from 1 to 5 as in Fig. 2.72. It is supported on a pinned support at 1, and on axially rigid pin-ended members at 3 and 5. These members are parallel to the coordinate directions and effectively supply direct force reactions in those directions. The framework is subjected to the uniformly distributed vertical and horizontal forces shown in the figure. Calculate the reactions. Draw the stress resultant diagrams corresponding to (i) the bending moments about horizontal axes and (ii) the torques.

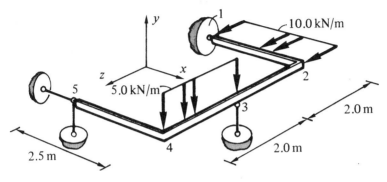

Fig. 2.72

2.8   A horizontal semicircular cantilever is subjected to a uniformly distributed vertical force of p/unit length over its entire length as shown in Fig. 2.73. Determine the bending moments and torques as functions of the angle θ shown in the figure.

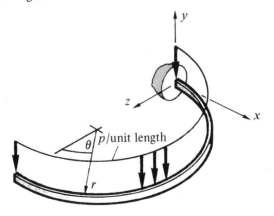

Fig. 2.73

2.9   Use the method of joints, and where necessary the method of tension coefficients, to obtain the forces in the members of the pin-jointed frameworks shown in Figs. 2.74(a) to (h). (In Fig. 2.74(b), the inclined members are at $60°$ to the vertical.)

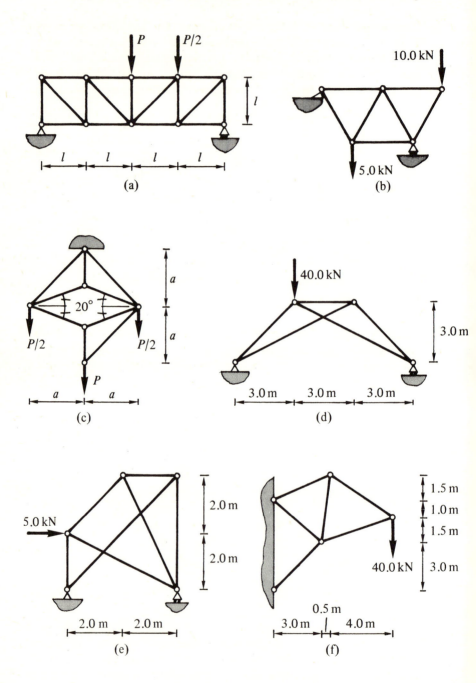

Fig. 2.74 (*continued next page*)

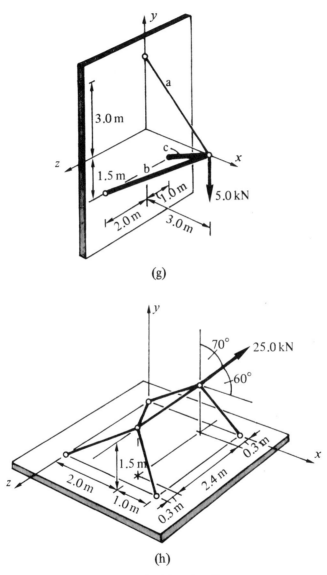

(g)

(h)

Fig. 2.74 (*continued*)

The following problems should be solved principally by the method of sections.

2.10 Figures 2.75(a) and (b) show respectively **Pratt** and **Howe** truss bridges. Noting that long thin members in compression are likely to buckle, which is the most efficient structure for supporting the central vertical force shown? The central vertical member in the Pratt truss carries no force. What is its purpose?

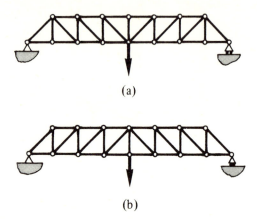

(a)

(b)

Fig. 2.75

2.11 A cantilever structure is shown in Fig. 2.76. Show that if the vertical force P acts at the tip as shown, then the forces in all the internal members a, b, c, etc. are zero. Show, however, that if P acts at joint 1, the internal members contain forces which gradually decrease in magnitude the further the members are from the tip. With P acting at joint 1, use a single equilibrium equation to determine the force in member c.

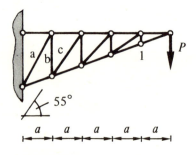

Fig. 2.76

2.12 A **Warren** truss is loaded by the forces shown in Fig. 2.77. Show that the forces in all but two of the vertical members are zero. What then is the purpose of these vertical members? Calculate the forces in the inclined members a to f and in the top chord members g to i.

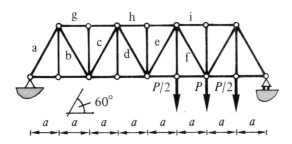

Fig. 2.77

2.13 Calculate the forces in the members a and b of the **k** truss shown in Fig. 2.78. (Note that the method of sections can be applied using *curved* sections through structures.)

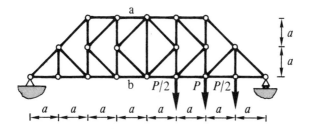

Fig. 2.78

2.14 Show that the forces in members a, b, c and d of the roof truss shown in Fig. 2.79 are zero. Calculate the force in member e.

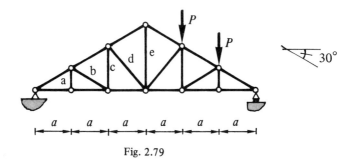

Fig. 2.79

2.15 A **Fink** roof truss is loaded as shown in Fig. 2.80. Inclined members are either at 30° or 60° to the horizontal. Calculate the forces in members a, b, c and d by considering in turn two appropriate sections through the structure.

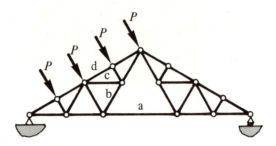

Fig. 2.80

2.16 The **compound** pin-jointed framework shown in Fig. 2.81 is composed of two triangulated sub-structures linked in such a way that the whole framework is statically determinate. By considering the overall equilibrium of the sub-structures, calculate the forces in members a, b and c.

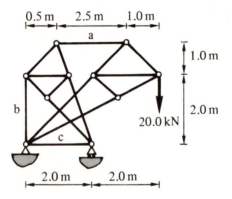

Fig. 2.81

2.17 A pylon is symmetrical in plan view tapering from an equilateral-triangular base with 5.0 m sides. It is composed of three space trusses 3.0 m high as shown in Fig. 2.82. It is subjected to a force of 5.0 kN at the top, and a force of 10.0 kN at point A in the x direction. Show that the force in member a is independent of the 5.0 kN force. Calculate the force in member a.

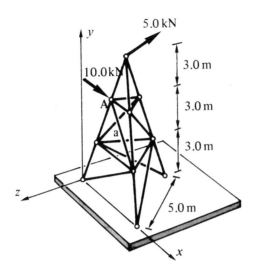

Fig. 2.82

2.18 Show that the formulae (2.18) and (2.19) for testing the statical deter-
     minancy of pin-jointed structures demonstrate that the structures in Figs.
     2.78, 2.81 and 2.82 are indeed statically determinate.

REFERENCE

[2.1] Meriam, J. L., (1975), *Statics,* 2nd edn., Wiley International, New York.

CHAPTER 3

Analysis of Statically Determinate Frameworks II: Deflections

---

3.1 INTRODUCTION

In Chapter 2 we discussed the analysis of statically determinate frameworks for their internal forces. In this chapter we shall consider the second objective of structural analysis, the calculation of the deflections. Calculating the deflections is important for two reasons: firstly because a knowledge of the deflections is obviously important in its own right, particularly when a structure is supporting deflection sensitive equipment; secondly, because we are able to determine structural flexibility. We shall see in Chapter 4 that the calculation of flexibility is a necessary step in using compatibility conditions to solve statically indeterminate frameworks.

In principle, it would be possible to calculate the deflections of frameworks from the internal forces and stresses obtained as in Chapter 2, by determining the local strains, and integrating up through the members to obtain the deflections. A graphical method, widely used before the advent of computers, was based on this idea. Thus the extensions of the members in a pin-jointed framework were calculated, and the final deflected shape of the structure found by means of a graphical construction called the Williott diagram [3.1]. A much more powerful and general method for calculating deflections, however, is based on the equation of virtual work, and this is considered in the present chapter.

3.2 EQUATION OF VIRTUAL WORK

The equation of virtual work relates the work done externally on a structure by forces moving through a set of displacements, to the internal work done by the stresses moving through the corresponding strains. It is of fundamental importance in structural analysis. Its derivation for the general case is an advanced theoretical problem. It is not, however, necessary to know the derivation in order to understand the application of the equation to structural analysis, and here we shall merely state the equation and refer the interested reader to Appendix A3.1.

In order to state the virtual work equation clearly in its structural context, we first need to define the independent systems of forces and displacements to which it refers.

3.2.1 Equilibrium Force System

Consider a set of N external forces P_i^* $(i = 1, \ldots, N)$ acting on a framework, as in Fig. 3.1 (with $N = 3$), where these are either direct forces or couples. The asterisk is used merely to distinguish between these forces and their effects, and the forces producing the compatible displacement system discussed below. P_i^* then give rise to sets of stresses $\sigma_{xx}^*, \sigma_{yy}^*, \sigma_{zz}^*, \tau_{xy}^*, \tau_{yz}^*, \tau_{zx}^*$ at all internal points in the members, such as point A in Fig. 3.1; stresses which are in equilibrium with P_i^*. These forces and stresses are called the **equilibrium force system**.

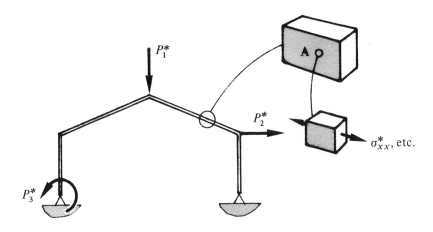

Fig. 3.1

3.2.2 Compatible Displacement System

Consider next the displacements produced by actions such as forces and temperature changes, that are *entirely independent* of the forces P_i^* of the equilibrium force system. The displacements are defined by d_i, which are either linear displacements of particular points of the framework or rotations about particular axes. They 'correspond' with the forces of the equilibrium force system in the way described in Section 1.11.1. Thus for the forces P_1^*, P_2^*, P_3^* shown in Fig. 3.1, the corresponding displacements are d_1, d_2 and d_3 shown in Fig. 3.2. At all internal points in the members, such as point A in Fig. 3.2, the deformation of the material is described by sets of strains $\epsilon_{xx}, \epsilon_{yy}, \epsilon_{zz}, \gamma_{xy}, \gamma_{yz}, \gamma_{zx}$. The displacements and strains are continuous since they obviously do not involve any tearing of the material of the framework, and they are called the **compatible displacement system**.

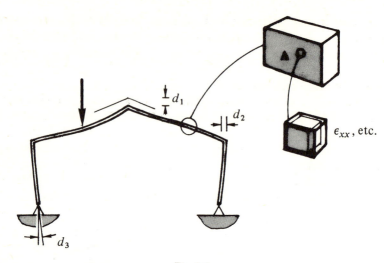

Fig. 3.2

3.2.3 Equation of Virtual Work

The equation of virtual work relates the parameters of the equilibrium force system and compatible displacement system as follows

$$\sum_i (P_i^* d_i) = \int_V (\sigma_{xx}^* \epsilon_{xx} + \sigma_{yy}^* \epsilon_{yy} + \sigma_{zz}^* \epsilon_{zz} + \tau_{xy}^* \gamma_{xy} + \tau_{yz}^* \gamma_{yz} + \tau_{zx}^* \gamma_{zx}) \, \mathrm{d}V.$$

(3.1)

This result is proved mathematically for a general body, by starting with the internal equilibrium equations governing the equilibrium force system and taking the product with the displacements of the compatible displacement system and integrating by parts through the body to its boundaries. An outline of the derivation is given in Appendix A3.1. Note also the clear account given by Bisplinghoff, Mar and Pian [3.2]. Strictly the equation of virtual work only applies if the displacements and strains are infinitesimally small. However, in structural analysis we assume that the actual displacements d_i and strains ϵ_{xx}, etc., are always small enough for the equation to be satisfied.

Referring again to Chapter 1, Section 1.11, where expressions for external and internal work done on a structure are derived, and comparing (3.1) with (1.8) and (1.12), we see that the term on the left-hand side of (3.1) is the work that would be done by the forces of the equilibrium force system *if* they were to move through the corresponding displacements of the compatible displacement system. It is not *actual* work done, because the systems are entirely independent. The term on the left-hand side of (3.1) is therefore called the **external virtual work** and is denoted by W_E^*, the asterisk in this case indicating that the work is virtual. In the same way, the term on the right-hand side is the work that would be done by the stresses of the equilibrium force system if they were to move through the strains of the compatible displacement system. This is called the **internal virtual work** W_I^*. Equation (3.1) is therefore justifiably called the **equation of virtual work.**

In the remainder of this chapter we shall show how the equation of virtual work when applied to frameworks can be cast in a simpler form, and then be used to obtain their deflections. We begin by considering the simplest case of the pin-jointed framework.

3.3 DEFLECTIONS OF PIN-JOINTED FRAMEWORKS

3.3.1 Equation of Virtual Work for a Pin-jointed Framework

We first simplify the general expression for the internal virtual work W_I^* on the right-hand side of (3.1), using the particular conditions within a pin-jointed framework. As we have seen in Chapter 2, a joint loaded pin-jointed framework only contains axial forces. Thus we shall be concerned here with the axial forces N^* of the equilibrium force system, and N of the compatible displacement system. At a particular point in a member in the framework, where the cross-sectional area is A, the corresponding direct stresses σ_{xx}^*, σ_{xx} are given by the engineering beam equation (1.2) as

$$\sigma_{xx}^* = \frac{N^*}{A}, \qquad \sigma_{xx} = \frac{N}{A} . \qquad (3.2 \text{ a, b})$$

We remind the reader that for one-dimensional members we ignore the stresses σ_{yy}, σ_{zz} and τ_{yz}. Further in the absence of shear forces and torques, τ_{xy} and τ_{xz} are also zero. Since therefore only the direct stresses given by (3.2) are non-zero, we can reduce the expression for W_I^* on the right-hand side of (3.1) to

$$W_I^* = \int_V (\sigma_{xx}^* \, \epsilon_{xx}) \, dV . \qquad (3.3)$$

ϵ_{xx} in the compatible displacement system immediately follows from the stress–strain relations including temperature changes, discussed in Section 1.10. Thus

$$\epsilon_{xx} = \frac{\sigma_{xx}}{E} + \alpha \Delta \theta = \frac{N}{EA} + \alpha \Delta \theta . \qquad (3.4)$$

The volume integral for the framework is then evaluated as follows.

Firstly consider the infinitesimal element of a member of length dx as in Fig. 3.3. Over the entire volume of this element, the stresses and strains given in (3.2) and (3.4) are constant. We thus obtain the internal virtual work done in the element, dW_I^* as the product $\sigma_{xx}^* \, \epsilon_{xx}$ times the infinitesimal volume $A \, dx$.

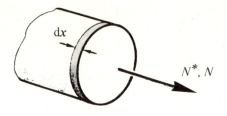

Fig. 3.3

Thus

$$dW_I^* = \sigma_{xx}^* \, \epsilon_{xx} \, A \, dx = N^* \left(\frac{N}{EA} + \alpha \Delta \theta \right) dx \quad . \tag{3.5}$$

The total internal virtual work can be found by integrating this expression along the length of each member and summing for all the members in the framework. Thus denoting the summation by $\sum\limits_m$ we have

$$W_I^* = \sum_m \int_0^{l_m} N_m^* \left(\frac{N_m}{E_m A_m} + \alpha_m \Delta \theta_m \right) dx_m \tag{3.6}$$

where subscripts have been introduced to indicate the values of the parameters associated with member m, all of which may be different for the different members.

In the subsequent text we shall frequently wish to refer to summations of groups of parameters which carry the subscript 'm' as in (3.6) above. It is helpful for clarity to introduce the following notation

$$f(N_m, M_m, E_m, \text{etc.}) = \{f(N, M, E, \text{etc.})\}_m \quad . \tag{3.7}$$

(3.6) then takes the form

$$W_I^* = \sum_m \int_0^{l_m} \left\{ N^* \left(\frac{N}{EA} + \alpha \Delta \theta \right) \right\}_m dx_m \quad . \tag{3.8}$$

Further, in Chapters 3 and 4, subscripts such as 'm' will be omitted in the general discussion, and they will only be used when referring to summations such as (3.8), and to properties of particular members in the worked examples.

For conciseness, we now introduce restrictions on the frameworks considered here. We shall assume that all the members are **prismatic**, meaning that their areas are constant along their lengths. We shall assume that they are of uniform material, so that Young's modulus and the temperature coefficient of expansion are constant. Any temperature changes that occur will be assumed to be uniform in each member. The framework is also assumed to be joint loaded, so that as shown in Chapter 2, Section 2.8.1, the axial force N in any member is constant along its length. The result of all these assumptions is that in the member integrals of (3.8), all the terms are constant. We are thus able to evaluate these integrals to produce

$$W_I^* = \sum_m \left\{ N^* \left(\frac{Nl}{EA} + \alpha l \Delta\theta \right) \right\}_m \quad . \tag{3.9}$$

Equation (3.9) can be presented more concisely. Thus

$$\left(\frac{Nl}{EA} + \alpha l \Delta\theta \right)$$

is the *total extension* of a member in the compatible displacement system. Suppose we call this Δ, say. Δ then contains two components defined by

$$\Delta_E = \frac{Nl}{EA} \tag{3.10}$$

$$\Delta_\Theta = \alpha l \Delta\theta \tag{3.11}$$

where Δ_E is the elastic extension of the member and Δ_Θ is the temperature expansion. The corresponding expressions for W_I^* are

$$W_I^* = \sum_m \{ N^*\Delta \}_m = \sum_m \{ N^*(\Delta_E + \Delta_\Theta) \}_m \quad . \tag{3.12}$$

Substituting (3.9) and (3.12) for the right-hand side of (3.1) then produces the simplified form of the equation of virtual work used in pin-jointed framework analysis:

$$\sum_i (P_i^* d_i) = \sum_m \left\{ N^* \left(\frac{Nl}{EA} + \alpha l \Delta\theta \right) \right\}_m = \sum_m \{ N^*\Delta \}_m \quad . \tag{3.13}$$

It should perhaps be noted that many writers have produced the same equation for pin-jointed frameworks by considering the virtual work done by the forces on the joints. The method is much more direct than that described here, but cannot be satisfactorily generalised to include frameworks containing bending moments, shear forces or torques.

3.3.2 Unit Force Method for the Deflection of Frameworks
We have seen that the equation of virtual work combines the external forces and displacements and internal stresses and strains of two entirely independent systems of actions on the body, one giving rise to an equilibrium force system,

the other to a compatible displacement system. Because these systems are independent, we are afforded a very powerful method for calculating structural displacements. We shall demonstrate the method by means of an example.

Example 3.1

Suppose we require the particular displacement d_1 of the pin-jointed framework analysed in Example 2.23 and shown in Fig. 3.4. All the members of the framework are of steel with Young's modulus 200.0 GN/m². Their cross-sectional areas in square millimetres are shown in the figure. In the virtual work equation of (3.13), we shall take as the unstarred displacements of the compatible displacement system, the actual structural displacements of the pin-jointed framework under the loading shown in Fig. 3.4, and the corresponding member extensions. Then in order to determine d_1, we choose for the equilibrium force system a single unit vertical force acting on the structure at joint 3 in the required direction d_1. The corresponding internal forces we shall call n_1. This lower case notation is used throughout the text to represent equilibrium force systems due to unit forces, and the asterisk can therefore be dropped as an unnecessary complication. The subscript 1 indicates which displacement is being sought. Thus in the virtual work equation (3.13)

$$P_1^* = 1.0, \quad P_2^* = P_3^* = \ldots = P_N^* = 0, \quad N^* = n_1 \qquad (3.14)$$

and the equation reduces to

$$1.0 \, d_1 = \sum_m \{n_1 \Delta\}_m \quad . \qquad (3.15)$$

This is an explicit expression for the required displacement d_1†. In calculating the right-hand side of (3.15), the member summation is carried out in tabular format.

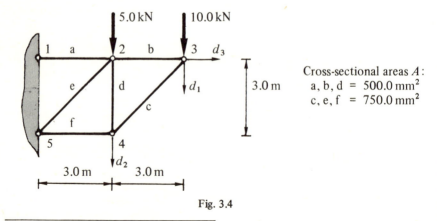

Cross-sectional areas A:
a, b, d = 500.0 mm²
c, e, f = 750.0 mm²

Fig. 3.4

---

† Note that the forces n_1 corresponding to a unit force, are considered to be dimensionless, (i.e. N/N).

This is shown in Table 3.1. (For conciseness in this and all subsequent tables, the appropriate basic units of the SI system, namely combinations of newtons and metres are used for all quantities, unless otherwise stated. Numbers are presented in engineering notation, and should be multiplied by the appropriate multiples of 10^3 shown in the left-hand column.) The extensions of the members Δ, are calculated from the axial forces which were determined in Example 2.23 by the method of joints. These are shown in Fig. 3.5(a), where the arrows in the members indicate the directions of the internal forces applied *on* the joints. The forces n_1 are also calculated by the method of joints and are shown in Fig. 3.5(b). (For clarity, the hypothetical unit forces of the equilibrium force system will be distinguished in figures from the actual forces on the framework, by drawing them with medium arrows.) Carrying out the summation horizontally across the table then gives $d_1 = 3.71$ mm.

The analysis can be repeated for other deflections, and the general form of (3.15) is

$$d_i = \sum_m \{n_i \Delta\}_m \tag{3.16}$$

where n_i are the axial forces due to $P_i^* = 1.0, P_j^* = 0, j \neq i$. Thus for d_2 in Fig. 3.4, we load the structure with a unit vertical force at joint 4 in the required vertical direction and reanalyse by the method of joints to give n_2 as in Fig. 3.5(c). The calculations are included in Table 3.1, giving $d_2 = 1.90$ mm.

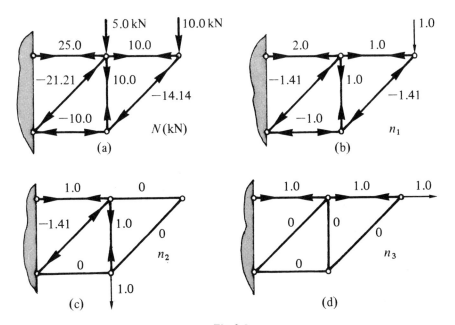

Fig. 3.5

Table 3.1

Calculations for Example 3.1

Member		a	b	c	d	e	f	Σ m
l		3.0	3.0	4.24	3.0	4.24	3.0	
A	10^{-6}	500.0	500.0	750.0	500.0	750.0	750.0	
N	10^3	25.0	10.0	−14.14	10.0	−21.21	−10.0	
Δ	10^{-3}	0.75	0.3	−0.4	0.3	−0.6	−0.2	
n_1		2.0	1.0	−1.41	1.0	−1.41	−1.0	
$n_1\Delta$	10^{-3}	1.5	0.3	0.57	0.3	0.85	0.2	3.71 $(=d_1)$
n_2		1.0	0	0	1.0	−1.41	0	
$n_2\Delta$	10^{-3}	0.75	0	0	0.3	0.85	0	1.90 $(= d_2)$
n_3		1.0	1.0	0	0	0	0	
$n_3\Delta$	10^{-3}	0.75	0.3	0	0	0	0	1.05 $(=d_3)$

It is important to note that as a result of the definition of external work, the unit force method isolates on the left-hand side of the equation of virtual work, the component of the total deflection of the point in the direction of the unit force. Thus it is not implied that the deflection of joint 3 of the framework is actually vertical. If we do need to calculate the total deflection at a point, it is necessary to determine the components in two orthogonal directions. Thus in the case of the total deflection of joint 3, we should also need to calculate the horizontal deflection d_3 shown in Fig. 3.4. This calculation is carried out in Table 3.1, using the axial forces n_3 corresponding to the unit horizontal load at joint 3 given in Fig. 3.5(d). Thus $d_3 = 1.05$ mm. The total deflection of joint 3 is then the vector sum of d_1 and d_3, being a displacement of magnitude 3.86 mm directed at 74.2° to the horizontal, as shown in Fig. 3.6.

Example 3.1 is a complete illustration of the unit force method for determining the deflections of pin-jointed frameworks and applies equally well to frameworks in two or three dimensions. The only difficulty with the latter is that the analysis for N and n_i will probably have to be carried out using tension coefficients as in Chapter 2, Section 2.8.3.

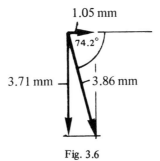

Fig. 3.6

3.3.3 Temperature Deflections
If all or part of a framework is subjected to temperature changes, then we merely add the temperature extensions Δ_Θ (3.11) to the elastic extensions of the relevant members.

Example 3.2
Suppose members a and b in Fig. 3.4 undergo a temperature rise of 20° C. Then

$$\Delta_{\Theta a} = \Delta_{\Theta b} = \{\alpha l \Delta\theta\}_{a,b} = 1.25 \times 10^{-5} \times 3.0 \times 20.0 = 0.75 \times 10^{-3} \text{ m} \ .$$

The calculations for finding the change in the displacements d_1, d_2 are given in Table 3.2, showing that both these points deflect downwards by respectively 2.25 mm and 0.75 mm.

Table 3.2
Calculations for Examples 3.2 and 3.3

Member		a	b	c	d	e	f	Σ m
Δ_Θ	10^{-3}	0.75	0.75	0	0	0	0	
n_1		2.0	1.0	−1.41	1.0	−1.41	−1.0	
$n_1 \Delta_\Theta$	10^{-3}	1.5	0.75	0	0	0	0	2.25 ($=d_1$)
n_2		1.0	0	0	1.0	−1.41	0	
$n_2 \Delta_\Theta$	10^{-3}	0.75	0	0	0	0	0	0.75 ($=d_2$)
Δ_C	10^{-3}	0	0	0	0	0	0.5	
$n_1 \Delta_C$	10^{-3}	0	0	0	0	0	−0.5	−0.5 ($=d_1$)
$n_2 \Delta_C$	10^{-3}	0	0	0	0	0	0	0 ($= d_2$)

3.3.4 Construction Errors

The pin-jointed framework in Fig. 3.4 has the dimensions shown. This is how it appears on the drawing board. The final shape of the framework, however, depends on how accurately it is constructed. Here we pose the question: what is the effect on the final form, of construction errors which involve particular members of the framework being cut to the wrong length? The question can in fact be answered by giving members extensions that are equivalent to the construction errors. The movement of any joint from its nominal position can then be evaluated by the unit force method as a deflection due to these extensions.

Example 3.3

Suppose member f in Fig. 3.4 is 5.0 mm longer than it should be. Treat this as an extra component in the extension of f, Δ_{Cf} say. The corresponding deflection d_1 is then determined by forming the sum

$$d_1 = \sum_m \{n_1 \Delta_C\}_m \ . \tag{3.17}$$

The calculation is included in Table 3.2, where it will be seen that due to the error, joint 3 is displaced upwards by 0.5 mm.

Deliberate changes from the nominal lengths of members can be introduced into structures to obtain a particular required camber in the final form. An example of this is included in the unworked problems at the end of this chapter (Problem 3.4).

3.4 DEFLECTIONS OF BEAMS

3.4.1 Equation of Virtual Work for a Beam

A simplified expression for the internal virtual work for a beam can be derived in a similar way to the expression for the pin-jointed member in the previous section. Let us assume we are dealing with a beam where the only non-zero stress resultants are the bending moments about the z axis; M^* in the equilibrium force system and M in the compatible displacement system†. The stresses are given by the engineering beam equation (1.2) as

$$\sigma_{xx}^* = -\frac{My^*}{I}, \ \sigma_{xx} = -\frac{My}{I} \ . \tag{3.18a, b}$$

Again in the absence of other stress resultants, these are the only stresses present so

$$W_I^* = \int_V (\sigma_{xx}^* \epsilon_{xx}) \, dV \ . \tag{3.19}$$

---

† For conciseness, the subscript z will be dropped in this section, and in Sections 3.5 to 3.10.

ϵ_{xx} in the compatible displacement system is given by

$$\epsilon_{xx} = \frac{\sigma_{xx}}{E} + \alpha\Delta\theta = -\frac{My}{EI} + \alpha\Delta\theta \ . \tag{3.20}$$

In order to evaluate the volume integral in (3.19) we again consider an infinitesimal element of a beam of length dx as in Fig. 3.7(a). However, in this case, the stresses and strains, being functions of y, are *not* constant over the volume of the element, so the internal virtual work dW_I^* cannot be found immediately. We therefore consider next a sub-element of the element of cross-sectional area dA at y from the z axis as shown in Fig. 3.7(b). Over this infinitesimal volume, $dA\,dx$, the stresses and strains *are* constant. Thus the internal virtual work done in the sub-element, $d(dW_I^*)$ say, is given by

$$d(dW_I^*) = \sigma_{xx}^* \epsilon_{xx}\,dA\,dx = \left(\frac{M^* M y^2}{EI^2} - \frac{M^*}{I} y\alpha\Delta\theta\right) dA\,dx \ . \tag{3.21}$$

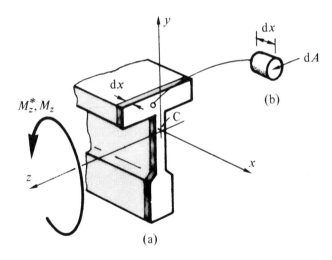

Fig. 3.7

If we then wish to find dW_I^*, we have to integrate $d(dW_I^*)$ over the area of the cross-section. In carrying out the area integration, the following terms are constants, that is, they do not change with position in the area; M^*, M, I, dx, E, α, $\Delta\theta$. Note that E and α *could* vary in a composite beam composed of different materials, and $\Delta\theta$ could vary if there were a temperature gradient across the beam. For conciseness we exclude these possibilities. Then

$$dW_I^* = \int_A d(dW_I^*) = \frac{M^*M}{EI^2} \, dx \int_A y^2 \, dA - \frac{M^*}{I} \, \alpha \Delta\theta \, dx \int_A y \, dA \ . \qquad (3.22)$$

We next recall that the engineering beam equation refers to bending about the z axis passing through the centroid C of the cross-section. $\int_A y^2 \, dA$ is therefore the *second moment of area* of the cross-section about an axis through the centroid, that is, it is equal to I. $\int_A y \, dA$ is the *first moment of area* about an axis through the centroid, and by definition is zero. Thus (3.22) reduces to

$$dW_I^* = \frac{M^*M}{EI} \, dx \ . \qquad (3.23)$$

We have therefore obtained the virtual work associated with an infinitesimal element of a beam. However, unlike the stress resultants in pin-jointed frameworks, we have seen from the bending moment diagrams derived in Chapter 2, that the bending moments in beams normally vary with x. Thus in the absence of information about the bending moment distributions, further integration cannot be carried out. The general expression for the total internal virtual work for the beam is therefore as follows

$$W_I^* = \int_0^l \left(\frac{M^*M}{EI} \right) dx \ . \qquad (3.24)$$

The equation of virtual work for a beam then takes the form

$$\sum_i (P_i^* d_i) = \int_0^l \left(\frac{M^*M}{EI} \right) dx \ . \qquad (3.25)$$

We remind the reader that (3.25) is derived by considering only bending moments and ignoring the shear forces that will also be present. The consequences of ignoring shear forces will be discussed in Section 3.5.

3.4.2 Deflections of a Cantilever
The virtual work equation (3.25) enables the unit force method to be used for finding the deflections of beams in bending. We illustrate this with the simple case of a cantilever loaded by a vertical end force P.

Example 3.4
The cantilever shown in Fig. 3.8(a) is loaded by a vertical force P at its free end. Suppose we require the vertical deflection d_1 at the centre, as shown in the figure. Again the actual displacements of the structure are taken as the compatible displacement system. Then in order to determine d_1 we choose for the

equilibrium force system, a single unit vertical force acting on the cantilever at the centre in the direction of d_1. Calling the corresponding moments m_1, the virtual work equation (3.25) becomes

$$1.0d_1 = \int_0^l \left(\frac{m_1 M}{EI}\right) dx \quad . \tag{3.26}$$

The bending moment diagrams corresponding to M and m_1 are as shown in Figs. 3.8(b) and (c). Thus

$$d_1 = \int_0^{l/2} \left(\frac{l}{2} - x\right) \frac{P(l-x)}{EI} dx + \int_{l/2}^l 0.\frac{P(l-x)}{EI} dx$$

$$= \int_0^{l/2} \frac{P}{EI} \left(\frac{l^2}{2} - \frac{3xl}{2} + x^2\right) dx = \frac{5}{48} \frac{Pl^3}{EI} \quad .$$

Note that the integration has to be split into two parts to accommodate the discontinuous function m_1.

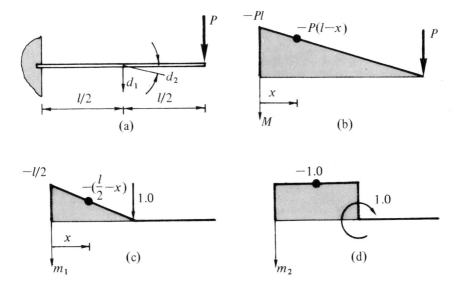

Fig. 3.8

Again, the analysis can be repeated for other deflections and the general form of (3.26) is

$$d_i = \int_0^l \left(\frac{m_i M}{EI}\right) dx \quad . \tag{3.27}$$

In particular note that d_i can either be a linear displacement *or* a rotation. If we thus require d_2, the slope at the centre of the cantilever as shown in Fig. 3.8(a), the corresponding unit force for the equilibrium force system is a unit concentrated couple applied at the centre in the direction of the rotation d_2. The resulting bending moment diagram is shown in Fig. 3.8(d) and

$$d_2 = \int_0^{l/2} 1.0 \frac{P(l-x)}{EI} \, dx + \int_{l/2}^{l} 0.\frac{P(l-x)}{EI} \, dx = \frac{3}{8} \frac{Pl^2}{EI} \,.$$

It is perhaps worth noting that the integration in (3.27) can be performed in any direction. In this case, a considerable simplification in the arithmetic in calculating d_1 can be achieved by working in terms of a coordinate x' measured backwards from the right-hand end.

3.4.3 Deflected Shapes and Influence Lines

Hitherto we have illustrated the use of the unit force method to give the displacements of particular chosen points of frameworks. However, the method can be used to generate complete deflected shapes if we take the position of the chosen point to be defined by a variable coordinate. Thus if we consider the vertical deflection d_1 of the cantilever in Fig. 3.9(a) at a point x_1 say from the left-hand end, then d_1 can be found by taking the unit vertical force of the equilibrium force system to act at this point. The corresponding bending moment diagram is shown in Fig. 3.9(b). Thus

$$d_1 = \int_0^{x_1} (x_1 - x) \frac{P(l-x)}{EI} \, dx = \int_0^{x_1} \frac{P}{EI} (lx_1 - (l + x_1) x + x^2) \, dx$$

$$= \frac{P x_1^2}{2EI} \left(l - \frac{x_1}{3} \right) \,.$$

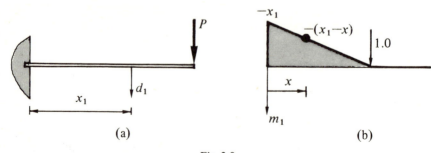

(a)   (b)

Fig. 3.9

The above is an expression for the vertical deflection of the beam as a function of x_1. Plotted out it gives the deflected shape of the beam as shown in Fig. 3.10.

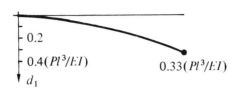

Fig. 3.10

In a similar way we can produce a deflection **influence line**, which depicts graphically the vertical deflection of a particular point of a beam as a unit load travels across the span, the deflection being plotted at the load position. Suppose we wish to obtain the influence line for the vertical deflection at the centre of the cantilever. The position of the unit force for the compatible displacement system is variable, while the position of the unit force for the equilibrium force system is fixed at $x = l/2$. Thus let the moving unit force be at x_1 from the left-hand end. M and m_1 are then as shown in Figs. 3.11(a) and (b). Thus

$$M = -(x_1 - x) \quad [x \leqslant x_1], \qquad m_1 = -\left(\frac{l}{2} - x\right) \quad [x \leqslant l/2]$$
$$= 0 \qquad\qquad [x \geqslant x_1] \qquad\qquad = 0 \qquad\qquad [x \geqslant l/2]$$

Thus if

$$x_1 \leqslant l/2, \quad d_1 = \int_0^{x_1}\left(\frac{l}{2} - x\right)\frac{(x_1 - x)}{EI}\, dx = \frac{x_1^2}{2EI}\left(\frac{l}{2} - \frac{x_1}{3}\right)$$

$$x_1 \geqslant l/2, \quad d_1 = \int_0^{l/2}\left(\frac{l}{2} - x\right)\frac{(x_1 - x)}{EI}\, dx = \frac{l^2}{8EI}\left(x_1 - \frac{l}{6}\right).$$

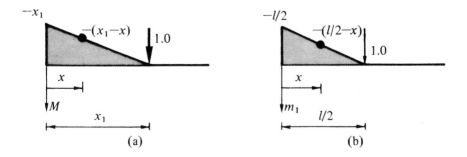

(a)   (b)

Fig. 3.11

These two functions when plotted out as in Fig. 3.12, give the vertical deflection of the centre of the cantilever as the unit load crosses the span.

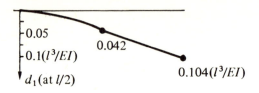

Fig. 3.12

3.5 INFLUENCE OF SHEAR FORCES

The deflections of the cantilever in Section 3.4 have been worked out considering only bending moments. However, we have seen in Chapter 2 that in general, shear forces will be present as well. We shall examine their effect in this section.

Again we have to produce a simplified expression for the internal virtual work resulting from the shear forces; S^* in the equilibrium force system and S in the compatible displacement system. As a start, we make the assumption that the corresponding shear stresses τ_{xy}^*, τ_{xy} are uniformly distributed across the cross-section, so that

$$\tau_{xy}^* = \frac{S^*}{A}, \quad \tau_{xy} = \frac{S}{A}.$$
(3.28a, b)

These are the only stresses considered so the right-hand side of (3.1) reduces to

$$W_I^* = \int_V (\tau_{xy}^* \, \gamma_{xy}) \, dV$$
(3.29)

γ_{xy} in the compatible displacement system is given by the shear stress–strain relations in (1.6) as

$$\gamma_{xy} = \tau_{xy}/G .$$
(3.30)

The internal virtual work dW_I^* associated with an element of the beam of length dx is then

$$dW_I^* = \tau_{xy}^* \, \gamma_{xy} \, A \, dx = \frac{S^*S}{GA} \, dx$$
(3.31)

and the total internal virtual work for the beam is

$$W_I^* = \int_0^l \left(\frac{S^*S}{GA} \right) dx .$$
(3.32)

The above derivation is based on the assumption that the shear stresses are uniformly distributed across the cross-section. In fact, they are not uniformly distributed, and we have seen that St. Venant's formula (1.3) in Chapter 1, Section 1.7, gives parabolic distributions of τ_{xy} in the webs of beams. Equation (3.32) is therefore corrected by introducing a correction factor k, which has different values depending on the type of cross-section considered. These are listed for a few common cases in Table 3.3. The internal virtual work for shear forces is then given by

$$W_I^* = \int_0^l \left(\frac{k\, S^*S}{GA} \right)\, dx \qquad (3.33)$$

In a beam in which bending and shear coexist, the virtual work equation is found by adding (3.33) to the right-hand side of (3.25) to give

$$\sum_i (P_i^*\, d_i) = \int_0^l \left(\frac{M^*M}{EI} + \frac{k\, S^*S}{GA} \right)\, dx \; . \qquad (3.34)$$

We shall now examine the influence of the shear terms for the simple case of the end-loaded cantilever.

Table 3.3
Correction factors for internal virtual work for shear

Section	k_y	k_z
rectangle	1.2	
circle	1.11	
hollow circle	2.0	
I-section or hollow rectangle (approx).	A/A_{webs}	A/A_{flanges}

Example 3.5
We recalculate the deflection d_1 of the centre of the cantilever in Fig. 3.8(a), this time taking shear into account. The general relation (3.27) produced by the unit force method, is modified to

$$d_i = \int_0^l \left(\frac{m_i M}{EI} + \frac{k\, s_i S}{GA} \right)\, dx \qquad (3.35)$$

where s_i refers to the shear force produced by the unit force corresponding to d_i and S is the shear force due to the actual loading. The shear force diagrams for calculating d_1 are shown in Fig. 3.13. Thus using the bending results derived in Example 3.4, we have

$$d_1 = \frac{5}{48} \frac{Pl^3}{EI} + \int_0^{l/2} \frac{k(-1.0)(-P)}{GA} \, dx = \frac{5}{48} \frac{Pl^3}{EI} + \frac{k}{2} \frac{Pl}{GA} .$$

We can then demonstrate the relative importance of the two terms, by considering the practical case of a typical rolled steel section supporting a 10.0 kN load. The properties of the section are listed in Table 3.4. Whence d_1 is given by

$$d_1 = (4.69 + 0.06) \text{ mm}$$

the first term being the bending and the second the shear displacement. The latter is about 1.0% of the former and this result is typical of the influence of shear in most practical cases. It only becomes important in particularly shear-flexible structures, such as cellular bridge decks. For this reason, shear will be ignored in the remainder of this text.

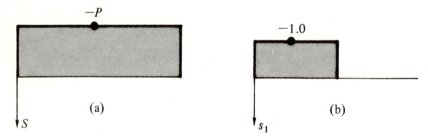

Fig. 3.13

Table 3.4

Properties of the cantilever in Example 3.5

Property	Value
length (l)	3.0 m
area (A)	4.0×10^{-3} m^2
second moment of area (I)	30.0×10^{-6} m^4
Young's modulus (E)	200.0×10^9 N/m^2
shear modulus (G)	77.0×10^9 N/m^2
k	1.2

3.6 PRODUCT INTEGRALS

We have seen that the evaluation of the displacements of a beam in bending involves integrating the product of two functions of x, namely m_i and M. Here we discuss a method for simplifying this problem.

Example 3.6
The simply supported beam in Fig. 3.14(a) carries a force P at the one-third span position. The vertical deflection d_1 of P is required. The bending moment diagrams M and m_1 are shown in Fig. 3.14(b) and (c). Firstly, it is worth noting that if the integration is being carried out analytically, it is often helpful to use different coordinates for different parts of the beam. We thus use x to describe the left-hand side of the beam and x' to describe the right-hand side, and form the appropriate analytical expressions for m_1 and M. Therefore

$$d_1 = \int_0^{l/3} \frac{1}{EI} \frac{2x}{3} \frac{2Px}{3} dx + \int_0^{2l/3} \frac{1}{EI} \frac{x'}{3} \frac{Px'}{3} dx' = \frac{4}{243} \frac{Pl^3}{EI} .$$

However, this direct analytical evaluation of product integrals rapidly becomes a very tedious problem. Thus suppose we wish to find the central deflection d_2 of the beam. We then need to integrate the product of m_2 in Fig. 3.14(d) with M in Fig. 3.14(b). The integration has to be split up into three parts and

$$d_2 = \int_0^{l/3} \frac{1}{EI} \frac{x}{2} \frac{2Px}{3} dx + \int_{l/3}^{l/2} \frac{1}{EI} \frac{x}{2} \frac{P(l-x)}{3} dx +$$

$$+ \int_0^{l/2} \frac{1}{EI} \frac{x'}{2} \frac{Px'}{3} dx' = 0.0177 \frac{Pl^3}{EI} .$$

Here for conciseness, we have omitted the considerable amount of algebraic manipulation needed to obtain the solution.

Engineers can avoid the above problem by using tables of **product integrals**. These are widely available in structural handbooks and list typical diagrams of functions of x representing the stress resultant diagrams, defined in terms of algebraic parameters. The product integrals are given at the intersections of rows and columns. An abbreviated version of such a table is given in Table 3.5. Its use is illustrated by evaluating the product integrals of Example 3.6. Thus for d_1 we have the two triangular diagrams in Figs, 3.14(b) and (c), with apexes at $x = l/3$, and heights $2Pl/9$ and $2l/9$. Thus in Table 3.5, $\alpha = \gamma = 1/3$, $\beta = \delta = 2/3$, $h_i = 2Pl/9$, $h_j = 2l/9$ and

$$\int_0^l f_i f_j \, dx = \left(\frac{2Pl}{9}\right) \left(\frac{2l}{9}\right) l \left(1 - \frac{1}{9} - \frac{4}{9}\right) \Big/ \left(6 \times \frac{2}{9}\right) = \frac{4}{243} Pl^3$$

giving the same value for d_1 as above. For d_2 we again have two triangular diagrams as in Figs. 3.14(b) and (d), with apexes at $x = l/3$ and $x = l/2$ respectively, and heights $2Pl/9$ and $l/4$. Thus in the table, $\alpha = 1/3$, $\beta = 2/3$, $\gamma = \delta = 1/2$, $h_i = 2Pl/9$, $h_j = l/4$ and

$$\int_0^l f_i f_j \, dx = \left(\frac{2Pl}{9}\right)\left(\frac{l}{4}\right) l \left(1 - \frac{1}{9} - \frac{1}{4}\right) \Big/ \left(6 \times \frac{2}{3} \times \frac{1}{2}\right) = 0.0177 Pl^3$$

and again d_2 is the same as above.

The scope of Table 3.5 is considerably widened by the fact that the integration of products of functions is distributive. Thus if f_i and f_j are each the sum of two functions f_{ai}, f_{bi} and f_{aj}, f_{bj} say then

$$\int_0^l f_i f_j \, dx = \int_0^l (f_{ai} + f_{bi})(f_{aj} + f_{bj}) \, dx = \int_0^l f_{ai} f_{aj} \, dx + \int_0^l f_{ai} f_{bj} \, dx +$$

$$+ \int_0^l f_{bi} f_{aj} \, dx + \int_0^l f_{bi} f_{bj} \, dx \ . \tag{3.36}$$

Complicated stress resultant diagrams can therefore be expressed as sums of the simpler diagrams presented in Table 3.5.

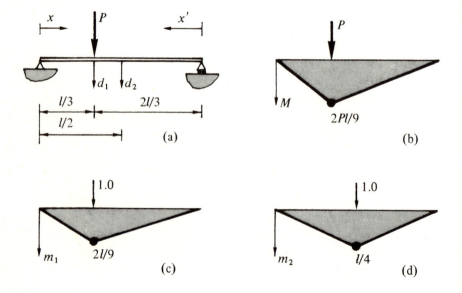

Fig. 3.14

Table 3.5

Product Integrals: $\displaystyle\int_0^l f_i(x) f_j(x)\, dx$

$f_i(x)$ → $f_j(x)$ ↓	Triangle (αl, βl, h_i)	Parabola (l, h_i)	Trapezoid (l, a_i, b_i)
Triangle (δl, δl, h_j)	$\alpha < \gamma:\ \dfrac{h_i h_j l}{6\beta\gamma}(1-\alpha^2-\delta^2)$ $\alpha > \gamma:\ \dfrac{h_i h_j l}{6\alpha\delta}(1-\beta^2-\gamma^2)$	$\dfrac{h_i h_j l}{3}(1+\alpha\beta)$	$\dfrac{h_i l}{6}[a_j(1+\beta)+b_j(1+\alpha)]$
Parabola (l, h_j)	$\dfrac{h_i h_j l}{3}(1+\gamma\delta)$	$\dfrac{8 h_i h_j l}{15}$	$\dfrac{h_i l}{3}(a_j+b_j)$
Trapezoid (l, a_j, b_j)	$\dfrac{h_j l}{6}[a_i(1+\delta)+b_i(1+\gamma)]$	$\dfrac{h_j l}{3}(a_i+b_i)$	$\dfrac{l}{6}[2(a_i a_j + b_i b_j)+b_i a_j + a_i b_j]$

3.7 DEFLECTIONS OF RIGID-JOINTED PLANE FRAMEWORKS

In calculating the deflections of beams, we have seen that shear forces can be ignored and only bending moments need be considered. However, when we go on to analyse more complicated frameworks, axial forces are generated as well. This is apparent for example, in the pin-footed portal frame in Example 2.18. There it is also evident that the axial forces in a rigid-jointed framework generally cannot be considered constant in each member. The internal virtual work for a member in a plane framework is therefore found by summing the expressions for the work due to varying axial forces (3.8) (for a single member) and due to bending (3.24). The total internal virtual work for the framework then follows by summing for the members; and the corresponding virtual work equation becomes

$$\sum_i (P_i^* d_i) \;=\; \sum_m \int_0^{l_m} \left\{ N^* \left(\frac{N}{EA} + \alpha \Delta \theta \right) + \frac{M^* M}{EI} \right\}_m dx_m \cdot (3.37)$$

In the subsequent text we shall frequently wish to indicate as in (3.37), the process of integrating an expression along each member in a framework, and then summing for all the members. For conciseness it is helpful to introduce the following notation

$$\sum_m \int_0^{l_m} f(N_m, M_m, \text{etc.}) \, dx_m \equiv \oint f(N, M, \text{etc.}) \, dx \qquad (3.38)$$

where the contour integral sign $\oint$ merely indicates integration along all the members of the framework, using local parameters in each particular member. (3.37) then takes the form

$$\sum_i (P_i^* d_i) \;=\; \oint \left(N^* \left(\frac{N}{EA} + \alpha \Delta \theta \right) + \frac{M^* M}{EI} \right) dx \qquad . \qquad (3.39)$$

The deflections of rigid-jointed plane frameworks, are derived from (3.39) using the unit force method and are given by the general equation

$$d_i \;=\; \oint \left(n_i \left(\frac{N}{EA} + \alpha \Delta \theta \right) + \frac{m_i M}{EI} \right) dx \qquad . \qquad (3.40)$$

We consider now a simple example of the application of (3.40).

Example 3.7
A cranked cantilever is shown in Fig. 3.15. It is made of steel for which $E = 200.0$ GN/m^2, and it is of uniform cross-section with $A = 4.0 \times 10^3$ mm^2, $I = 30.0 \times 10^6$ mm^4. Suppose we wish to calculate the vertical deflection d_1 at the tip, due to the 10.0 kN load. The stress resultant diagrams are shown in Fig. 3.16, and carrying out the product integrals for the two members a and b gives

$$d_1 = \frac{1}{200.0 \times 10^9} \left[\frac{1.0 \times 10.0 \times 10^3 \times 3.0}{4.0 \times 10^{-3}} + \right.$$

$$\left. + \frac{(-2.0) \times (-20.0) \times 10^3 \times 2.0}{3.0 \times 30.0 \times 10^{-6}} + \frac{2.0 \times 20.0 \times 10^3 \times 3.0}{30.0 \times 10^6} \right]$$

(axial)   (bending)
$$= (0.0375 + 24.4) \text{ mm}.$$

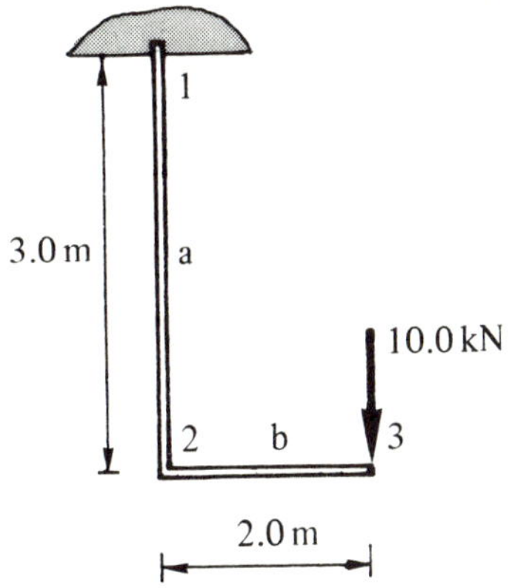

3.0 m

1

a

10.0 kN

2   b   3

2.0 m

Fig. 3.15

−20.0

−20.0

M (kNm)

+ 10.0

N (kN)

−2.0

−2.0   1.0

m_1

1.0

1.0

n_1

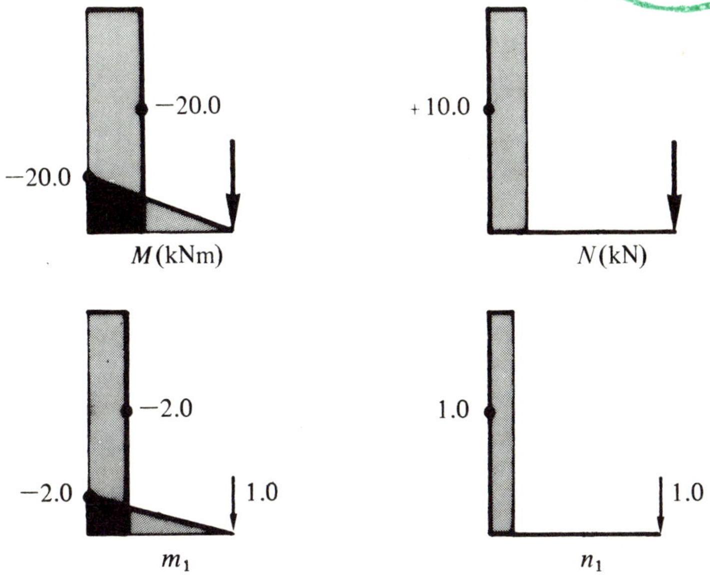

Fig. 3.16

Example 3.7 illustrates the important point that if a framework derives its stiffness from the bending of members, then the bending deflections dominate. Thus in analysing such frameworks, we can ignore the effects of the axial forces. The cranked cantilever is a very simple example, but an important group of practical frameworks also fall into this category. These frameworks, called **building frames**, are frameworks in the form of rectangular grids of members. In particular they are free of diagonal bracing members that are difficult to reconcile with architectural details. Analytically, building frames can be recognised by a very simple technique. If all the rigid joints are replaced by pinned joints, then they form mechanisms. Examples of such frameworks are shown in Fig. 3.17(a) and (b). An example of a framework where axial forces may *not* be ignored, is the rigid-jointed **braced frame** shown in Fig. 3.17(c), because replacing the joints

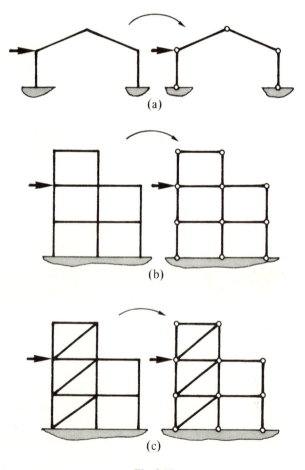

Fig. 3.17

of this framework by pins produces a stable structure. The braced frame derives its strength from the tension and compression of members and is generally much stiffer than an equivalent building frame. It is often introduced into structures to provide stability against wind loading.

In the remainder of this chapter we shall consider more complicated examples of building frames and ignore the effects of the axial forces.

3.8 THREE-PINNED PORTAL FRAME

Example 3.8
As an illustration of a more complicated plane framework, we shall calculate the deflections of the three-pinned portal frame analysed in Example 2.22. For convenience, the structure is reproduced in Fig. 3.18. Suppose all the members are of steel and of the same cross-section as the cantilever in Example 3.7, so that $A = 4.0 \times 10^3$ mm^2, $I = 30.0 \times 10^6$ mm^4. These are typical properties of a rolled-steel section suitable for this structure. We shall determine the vertical deflection d_1 of the crown, the horizontal deflection d_2 of the right-hand eave, and the rotation d_3 of the foot of the left-hand column. Firstly, we note that ignoring axial forces, modifies the general deflection equation (3.40) to

$$d_i = \oint \left(\frac{m_i M}{EI} \right) dx \ . \tag{3.41}$$

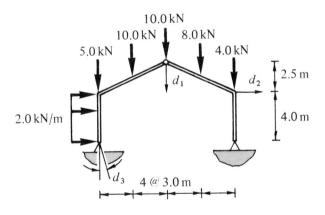

Fig. 3.18

The bending moments M, due to the applied loading have already been calculated in Example 2.22. The bending moments m_i ($i = 1, 2, 3$) due to the unit forces corresponding to the three displacements are calculated in the same way, first by calculating the reactions, then by making cuts in the members. The resulting bending moment diagrams are shown in Fig. 3.19. The evaluation of the displacements can be carried out in the tabular format of Table 3.6, the product integrals being obtained from the bending moment diagrams, using Table 3.5. Note that in order to be able to use Table 3.5, the bending moment diagrams M have been split up into appropriate simple components. Also, since the section properties E and I are the same for all the members, it is not necessary for this case to include $(1/EI)$ in the table. The values of the displacements are thus found to be

$$\begin{bmatrix} d_1 \\ d_2 \\ d_3 \end{bmatrix} = \begin{bmatrix} 322.7 \\ 266.9 \\ 11.1 \end{bmatrix} \times \frac{10^3}{EI} = \begin{bmatrix} 0.054 \\ 0.044 \\ 0.0019 \end{bmatrix} \begin{matrix} m \\ m \\ rad \end{matrix} \ .$$

Table 3.6
Calculations for Example 3.8

member	a	b	c	d	Σ m
l	4.0	6.5	6.5	4.0	
$M \ 10^3$	−24.0 / + / 4.0	−24.0 / + / 15.0	−40.0 / + / 12.0	−40.0	
m_1	−1.85	−1.85	−1.85	−1.85	
$\int m_1 M dx \ 10^3$	49.2	51.0	124.0	98.5	322.7
m_2	1.23	1.23	−2.77	−2.77	
$\int m_2 M dx \ 10^3$	−32.8	−34.0	186.0	147.7	266.9
m_3	−1.0 / −0.69	−0.69	0.31	0.31	
$\int m_3 M dx \ 10^3$	29.1	19.1	−20.7	−16.4	11.1

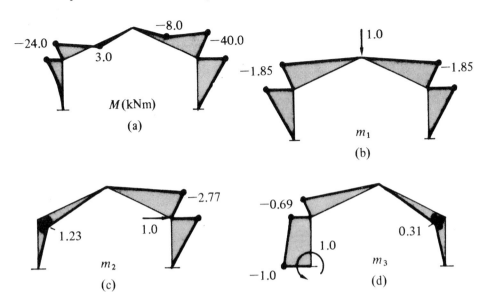

Fig. 3.19

3.9 TAPERED MEMBERS

In the previous example of a pin-footed portal frame, the members were assumed to be prismatic. However, modern design often incorporates tapered members to optimise structural efficiency, as in Fig. 3.20(a). We shall briefly describe how these can be dealt with. The deflections of a framework with tapered members are still given by equation (3.41), but in this case since I is also a variable in the integration, the product integral will usually need to be evaluated numerically. Thus if a tapered member is divided into N intervals of equal length Δx, where N is an even number, as shown in Fig. 3.20(b), the value for I at points $0, 1, 2, \ldots N$, can be found from the design drawing. Similarly m_i and M can be evaluated at these points by the spot value method. If the function

$$\left(\frac{m_i M}{EI}\right)$$

at point j is then called ϕ_j, the product integral for the member can be evaluated by Simpson's Rule [3.3] as

$$\int_0^l \left(\frac{m_i M}{EI}\right) dx = \frac{\Delta x}{3} \left[\phi_0 + 4\phi_1 + 2\phi_2 + 4\phi_3 + \ldots + 2\phi_{N-2} + 4\phi_{N-1} + \phi_N\right].$$

$$(3.42)$$

Satisfactory accuracy can usually be obtained by taking N equal to 10.

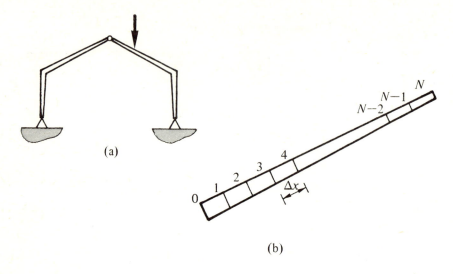

(a)

(b)

Fig. 3.20

3.10 CURVED MEMBERS

The derivation of the equation of virtual work for frameworks, does not require the members to be straight. Thus to obtain the deflections of frameworks containing curved members, the product integrals in those members simply have to be calculated round their arcs. For most curved members analytical functions for the stress resultants become too complicated for direct integration, and a numerical procedure of the type discussed in Section 3.9 has to be used. The curved member is then divided into equal length segments, Δx_m, and the bending moments evaluated at the points of division, enabling us to calculate ϕ_j in (3.42). However, in the case of circular curved members, analytical solutions can be obtained if the problem is described by polar coordinates.

Example 3.9
Consider again the pin-footed curved arch in Example 2.19, which for convenience is reproduced in Fig. 3.21(a). We shall assume that the arch is made from a uniform concrete box section of the dimensions shown in Fig. 3.21(b). Then $E = 30.0$ GN/m² and $I = 1.34$ m⁴. Suppose we require the vertical deflection d_1 of the crown, and the horizontal deflection d_2 of the right-hand foot. The governing equation ignoring axial and shear forces is again (3.41). The integration is carried out for the single member round the complete arc. The bending moments M, caused by the loading have already been calculated in Example 2.19 as separate functions of θ and ψ, the two coordinates chosen for clarity to define the parameters in the two sides of the arch, as in Fig. 3.21(c).

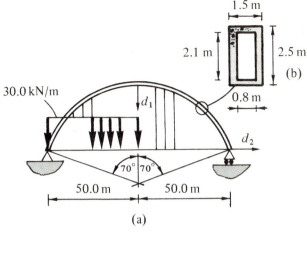

(a)

(b)

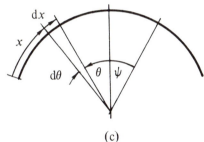

(c)

Fig. 3.21

Thus

$$M(\theta) = (18.75 + 19.95 \sin(\theta) - 42.47 \sin^2(\theta)) \text{ MNm}$$

$$M(\psi) = (18.75 - 19.95 \sin(\psi)) \text{ MNm} .$$

The arch is next analysed in the same way as in Example 2.19 for the bending moments m_1 due to a unit vertical load at the crown, and m_2 due to a unit horizontal load at the right-hand foot. The results are

$$m_1 = 25.0 - 26.60 \sin(\theta), \qquad m_1 = 25.0 - 26.60 \sin(\psi)$$

$$m_2 = 53.21 \cos(\theta) - 18.20, \qquad m_2 = 53.21 \cos(\psi) - 18.20 .$$

In calculating the integral in (3.41) we note that dx is an infinitesimal arc-length, and this is equal to $rd\theta$ in the left-hand side of the arch and $rd\psi$ in the right-hand side. We can therefore divide the integration into two parts and change the variables to θ and ψ in the two respective sides. The integrals are then evaluated between the limits 0 and 1.222 radians (70°). Thus with $r = 53.21$ m, the calculation for d_1 goes as follows:

$$d_1 = \frac{1}{EI} \int_0^{1.222} (25.0 - 26.60 \sin(\theta))(18.75 + 19.95 \sin(\theta) - 42.47 \sin^2(\theta)) \times$$

$$\times 10^6 \times 53.21 \, d\theta$$

$$+ \frac{1}{EI} \int_0^{1.222} (25.0 - 26.60 \sin(\psi))(18.75 - 19.95 \sin(\psi)) \times 10^6 \times 53.21 \, d\psi$$

The integrals can be evaluated analytically or numerically to give

$$d_1 = \frac{1.264 \times 10^{10} + 0.826 \times 10^{10}}{30.0 \times 10^9 \times 1.34} = 0.52 \text{ m} \; .$$

The similar calculation for d_2 is

$$d_2 = \frac{1}{EI} \int_0^{1.222} (53.21 \cos(\theta) - 18.20)(18.75 + 19.95 \sin(\theta) -$$

$$- 42.47 \sin^2(\theta)) \times 10^6 \times 53.21 \, d\theta$$

$$+ \frac{1}{EI} \int_0^{1.222} (53.21 \cos(\psi) - 18.20)(18.75 - 19.95 \sin(\psi)) \times$$

$$\times 10^6 \times 53.21 \, d\psi \; .$$

Thus

$$d_2 = \frac{2.519 \times 10^{10} + 1.547 \times 10^{10}}{30.0 \times 10^9 \times 1.34} = 1.01 \text{ m} \; .$$

3.11 DEFLECTIONS OF RIGID-JOINTED SPACE FRAMEWORKS

Up to now we have discussed the problem of finding the deflections of rigid-jointed plane frameworks, and shown that in many cases, we need only consider bending about the z axis. When we go on to deal with space frameworks, considerable complication ensues, because in the general case, all six stress resultants, N, S_y, S_z, M_y, M_z and T are non-zero. We first need to derive the corresponding equation of virtual work.

The internal virtual work in a member containing N, and both M_y and M_z, is a simple extension of the right-hand side of (3.37) for a single member. Thus

$$W_I^* = \int_0^l \left(N^* \left(\frac{N}{EA} + \alpha \Delta\theta \right) + \frac{M_y^* M_y}{EI_y} + \frac{M_z^* M_z}{EI_z} \right) dx \qquad (3.43)$$

where the subscripts y and z are now necessary to distinguish between bending about the y and z axes. Similarly when both shear forces S_y and S_z coexist, we extend (3.33) to give

$$W_I^* = \int_0^l \left(\frac{k_y S_y^* S_y}{GA} + \frac{k_z S_z^* S_z}{GA} \right) dx \quad . \qquad (3.44)$$

In (3.44) two correction factors, k_y and k_z, are necessary because in some sections, such as for example rolled-steel sections, the effects of shear are quite different in the y and z directions (see Table 3.3, p. 153).

We then need to determine the internal virtual work due to the torques in a member. Let us consider the simplest case of the circular hollow tube section. The shear stresses τ^*, τ due to the torque T^* in the equilibrium force system and T in the compatible displacement system respectively, are given by (1.4) as

$$\tau^* = \frac{T^* r}{J}, \qquad \tau = \frac{T r}{J} \quad . \qquad (3.45)$$

They are directed at right angles to the radial direction as shown in Fig. 3.22.

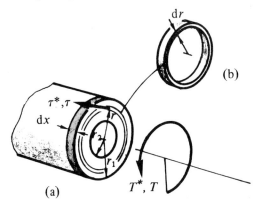

Fig. 3.22

The corresponding shear strain γ in the compatible displacement system is given by the shear stress-strain relations (1.6) as

$$\gamma = \tau/G \quad . \qquad (3.46)$$

Again we consider an infinitesimal element of the member of length dx as shown in Fig. 3.22(a). The internal virtual work $d(dW_I^*)$ done in the infinitesimal annulus of the element, thickness dr in Fig. 3.22(b), is then

$$d(dW_I^*) = \tau^* \gamma \, dA \, dx = \left(\frac{T^* T}{GJ^2} r^2\right) dA \, dx \tag{3.47}$$

where $dA = 2\pi r \, dr$. $\tag{3.48}$

Therefore

$$d(dW_I^*) = \frac{T^* T}{GJ^2} 2\pi \, r^3 \, dr \, dx \ . \tag{3.49}$$

The internal virtual work dW_I^*, done in the complete element is found by integrating $d(dW_I^*)$ between r_2 and r_1. Thus

$$dW_I^* = \frac{T^* T}{GJ^2} \frac{\pi}{2} (r_1^4 - r_2^4) \, dx = \frac{T^* T}{GJ} dx \tag{3.50}$$

(using the expression for the torsional constant J of the circular hollow tube section given in (1.5)). The total virtual work for the member is then given by

$$W_I^* = \int_0^l \left(\frac{T^* T}{GJ}\right) dx \ . \tag{3.51}$$

The calculation of torsional effects in other types of cross-section, is an advanced theoretical problem. Here we shall simply note that identical expressions to (3.51) are obtained provided appropriate values are taken for the torsional constants J. Values for typical cross-sections are given in Table 3.7.

The complete expression for the internal virtual work done in the space framework is found by collecting the three results (3.43), (3.44) and (3.51) and summing for all the members in the framework. The corresponding equation of virtual work is then given by

$$\sum_i (P_i^* d_i) = \oint \left(N^* \left(\frac{N}{EA} + \alpha \Delta\theta\right) + \frac{M_y^* M_y}{EI_y} + \frac{M_z^* M_z}{EI_z} + \right.$$
$$\left. + \frac{k_y S_y^* S_y}{GA} + \frac{k_z S_z^* S_z}{GA} + \frac{T^* T}{GJ} \right) dx \ . \tag{3.52}$$

Similarly the deflections of a space framework are given by

$$d_i = \oint \left(n_i \left(\frac{N}{EA} + \alpha \Delta\theta\right) + \frac{m_{yi} M_y}{EI_y} + \frac{m_{zi} M_z}{EI_z} + \right.$$
$$\left. + \frac{k_y s_{yi} S_y}{GA} + \frac{k_z s_{zi} S_z}{GA} + \frac{t_i T}{GJ} \right) dx \tag{3.53}$$

where the stress resultants in lower case letters, are those in the equilibrium force system due to the unit force corresponding to the required displacement d_i.

We consider now a simple example of the application of (3.53) for the cranked cantilever already analysed in Example 2.20.

Table 3.7

Torsional constants, J

section		J
	d/b	
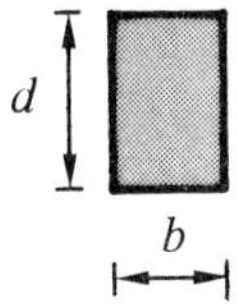	1.0	$0.1406\,(b^3d)$
	1.2	$0.166\;(b^3d)$
	1.5	$0.196\;(b^3d)$
	2.0	$0.229\;(b^3d)$
	2.5	$0.249\;(b^3d)$
	3.0	$0.263\;(b^3d)$
	4.0	$0.287\;(b^3d)$
	5.0	$0.291\;(b^3d)$
	10.0	$0.312\;(b^3d)$
	∞	$0.333\;(b^3d)$
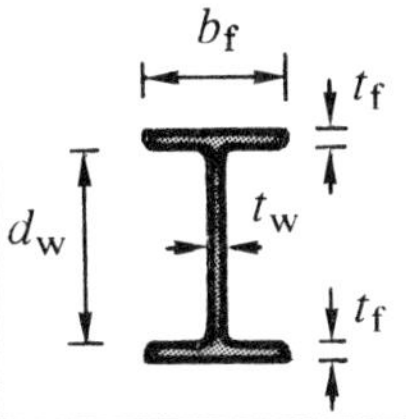 I-section (b_f, t_f, d_w, t_w)		$\dfrac{2b_f t_f^3 + d_w t_w^3}{3}$
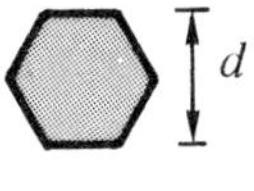 hexagon (d)		$0.133\,Ad^2$
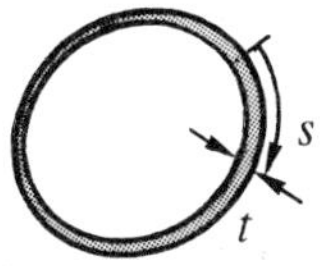 closed thin-walled section (A = gross area)		$\dfrac{4A^2}{\oint \dfrac{ds}{t}}$

Example 3.10

For convenience we reproduce the cranked cantilever in Fig. 3.23(a). We shall assume that the members are all of the same steel section, and are arranged as in Fig. 3.23(b) with $I_z = 30.0 \times 10^6$ mm^4 in all the members and $I_y = 15.0 \times 10^6$ mm^4. J is constant throughout and equal to 20.0×10^6 mm^4. Suppose we require the vertical deflection d_1 of the free end, and its rotation d_2 about the global z axis. Again the cantilever can be regarded as a building frame and the effects of N, S_y and S_z ignored. Thus (3.53) reduces to

$$d_i = \oint \left(\frac{m_{yi} M_y}{EI_y} + \frac{m_{zi} M_z}{EI_z} + \frac{t_i T}{GJ} \right) dx .$$ (3.54)

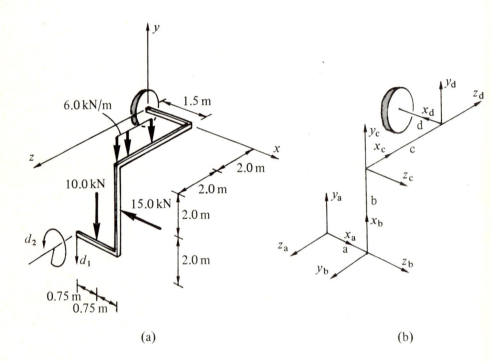

(a)   (b)

Fig. 3.23

The stress resultants M_y, M_z and T have already been calcuiated in Example 2.20 and are shown in Figs. 2.43(c), (d) and (e). The stress resultants of the equilibrium force systems for the two displacements are shown in Fig. 3.24. The evaluation of the product integrals in (3.54) is then carried out in Table 3.8, and the values of the displacements are given by

$$d_1 = \frac{343.2 \times 10^3}{200.0 \times 10^9 \times 30.0 \times 10^{-6}} + \frac{321.0 \times 10^3}{76.9 \times 10^9 \times 20.0 \times 10^{-6}} = 0.266 \text{ m}$$

$$d_2 = \frac{-55.7 \times 10^3}{200.0 \times 10^9 \times 30.0 \times 10^{-6}} + \frac{-90.0 \times 10^3}{76.9 \times 10^9 \times 20.0 \times 10^{-6}} = -0.068 \text{ rads} \; .$$

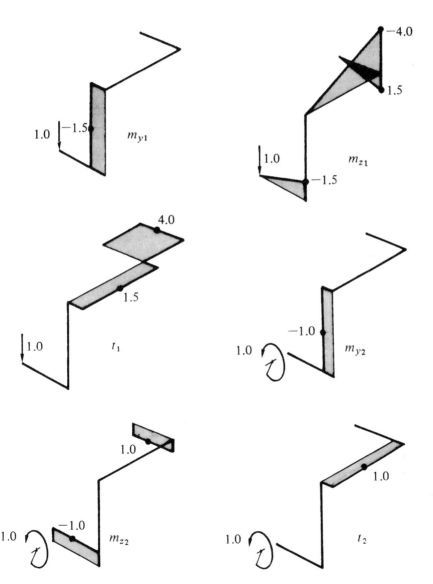

Fig. 3.24

Table 3.8
Calculations for Example 3.10

Member	a	b	c	d	Σ m
l	1.5	4.0	4.0	1.5	
M_z 10^3	-7.5 3.75	0	-76.0 -32.0 3.0	-55.5 -22.5	
M_y 10^3	0	-7.5 22.5	60.0	60.0	
T 10^3	0	0	-22.5	76.0	
m_{z1}	-1.5	0	-4.0	1.5	
m_{y1}	0	-1.5	0	0	
t_1	0	0	1.5	4.0	
$\int m_{z1} M_z \, dx \; 10^3$	3.5	0	377.3	-37.7	343.2
$\int m_{y1} M_y \, dx \, 10^3$	0	0	0	0	0
$\int t_1 T \, dx \quad 10^3$	0	0	-135.0	456.0	321.0
m_{z2}	-1.0	0	0	1.0	
m_{y2}	0	-1.0	0	0	
t_2	0	0	1.0	0	
$\int m_{z2} M_z \, dx \; 10^3$	2.8	0	0	-58.5	-55.7
$\int m_{y2} M_y \, dx \; 10^3$	0	0	0	0	0
$\int t_2 T \, dx \quad 10^3$	0	0	-90.0	0	-90.0

3.12 SUPPORT SETTLEMENTS

One of the 'actions' on a structure mentioned in Chapter 1, Section 1.2, was the settlement of its supports. Consider for example, a continuous bridge deck. If a central pier settles the deflected shape of the deck changes and also the distribution of bending moments. For statically determinate structures, support settlements do not alter the internal forces, this being clear from the fact that in solving such structures as in Chapter 2, we considered only their equilibrium. However, the displacements of the structures are changed, and it is the purpose of this section to show how these changes can be calculated.

Suppose a framework is subject to S known displacements at its supports, d_{sj} $(j = 1, \ldots, S)$. At each of these supports there is a 'corresponding' reaction R_j, being the component of the direct force reaction in the direction of d_{sj} if d_{sj} is a linear displacement, and the component of the couple reaction about the same axis as d_{sj} if d_{sj} is a rotational displacement. These reactions can be regarded as external forces on the framework and because of the support displacements, they now do external work. They are therefore included in the external virtual work W_E^* and the virtual work equation becomes

$$\sum_i (P_i^* d_i) + \sum_j (R_j^* d_{sj}) = \oint \left(N^* \left(\frac{N}{EA} + \alpha \Delta \theta \right) + \ldots + \frac{T^*T}{GJ} \right) dx \ . \quad (3.55)$$

If we then wish to determine a particular displacement d_i of the framework, we again use the equilibrium force system corresponding to the unit force $P_i^* = 1.0$. The external virtual work then becomes

$$W_E^* = 1.0 d_i + \sum_j (r_{ji} d_{sj}) \quad (3.56)$$

where r_{ji} is the reaction on the structure corresponding to support settlement d_{sj} due to a unit load on the structure corresponding to the required displacement d_i. Thus the equation for the displacement d_i given in (3.53) is modified to

$$d_i = \oint \left(n_i \left(\frac{N}{EA} + \alpha \Delta \theta \right) + \ldots + \frac{t_i^* T}{GJ} \right) dx - \sum_j (r_{ji} d_{sj}) \ . \quad (3.57)$$

We shall illustrate the application of (3.57) with a simple example of a pin-jointed framework.

Example 3.11
Consider the pin-jointed framework of Examples 2.23 and 3.1. Suppose supports at 1 and 5 undergo the displacements $d_{s1} = 5.0$ mm and $d_{s2} = 7.5$ mm as shown in Fig. 3.25(a). We shall recalculate the vertical deflection d_1 of joint 3. For the pin-jointed framework (3.57) reduces to

$$d_1 = \sum_m \{n_1 \Delta\}_m - \sum_{j=1,2} (r_{ji} d_{sj}) \ . \tag{3.58}$$

The forces n_1 due to the equilibrium force system $P_1^* = 1.0$, are shown in Fig. 3.25(b). We also have to calculate the reactions r_{11} and r_{21} at the supports.

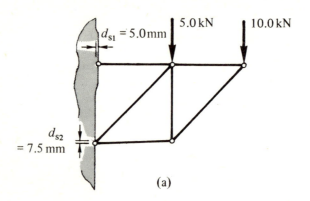

(a)

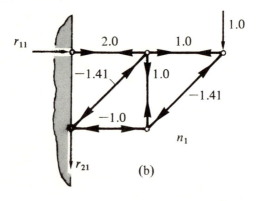

(b)

Fig. 3.25

These are found by resolving horizontally and vertically respectively. Thus at support 1

$$\Sigma \rightarrow \quad r_{11} + 2.0 = 0$$

and at support 2

$$\Sigma \uparrow \quad -r_{21} - 1.41 \cos{(45°)} = 0 \ .$$

Therefore

$$\begin{bmatrix} r_{11} \\ r_{21} \end{bmatrix} = \begin{bmatrix} -2.0 \\ -1.0 \end{bmatrix} \ .$$

Note that the signs of these reactions are negative indicating that their actual directions are opposite to the directions of the support displacements. The summation $\sum\limits_{m} \{n_1 \Delta\}_m$ has already been carried out in Table 3.1 and is equal to 3.71 mm. Thus from (3.58)

$$d_1 = 3.71 - ((-2.0) \times 5.0 + (-1.0) \times 7.5) = 3.71 + 17.5 = 21.21 \text{ mm} \ .$$

We then note that the vertical deflection of joint 3 is considerably increased by the support displacements.

3.13 FLEXIBILITY OF FRAMEWORKS

In calculating the displacements of a framework under loading we are determining its **flexibility**. This process is the basis of the method of analysing statically indeterminate frameworks to be discussed in the next chapter. In this section we shall consider a formal definition of the concept of flexibility.

We first note that the flexibility f of the spring shown in Fig. 3.26, is defined as the extension of the spring caused by a *unit* force applied as shown. If then a force P is applied to the spring, the corresponding extension d is given by

$$d = fP \ . \tag{3.59}$$

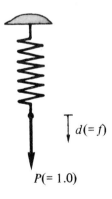

Fig. 3.26

Considering next a framework, we define N external forces P_i $(i = 1, \ldots, N)$ acting on the framework, being either direct forces or couples as in Fig. 3.27.

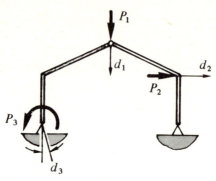

Fig. 3.27

The 'corresponding' external displacements are d_i. We then define **flexibility coefficients** f_{ii} and f_{ij} in a similar way to the single flexibility f of a spring. Thus f_{ii} is the displacement d_i caused by a unit force $P_i = 1.0$ with all the other forces zero. This definition can be written concisely as

$$f_{ii} = d_i \, [P_i = 1.0, \; P_k = 0, \; k \neq i] \; . \tag{3.60}$$

f_{ii} is called the **direct flexibility coefficient**. However the application of a single unit force to a framework also causes displacements to occur at other points, and this gives rise to the concept of **cross flexibility coefficients**. Thus in general, f_{ij} is the displacement d_i caused by a unit force $P_j = 1.0$ with all the other forces zero; that is

$$f_{ij} = d_i \, [P_j = 1.0, \; P_k = 0, \; k \neq j] \; . \tag{3.61}$$

The flexibility coefficients, being displacements caused by unit forces, can be calculated by the virtual work equation. For simplicity, we shall suppose that we are dealing with a plane building frame in which only the bending terms are important. We then recall that equation (3.41) for the deflection d_i is

$$d_i = \oint \left(\frac{m_i M}{EI} \right) \, dx \tag{3.41*\dagger}$$

where M are the bending moments due to any general loading. In order to find f_{ij}, we need to apply a unit load $P_j = 1.0$. In this case, strictly following the notation that lower case subscripted letters denote stress resultants corresponding to a single unit load on the framework, M is equal to m_j. Thus

$$f_{ij} = \oint \left(\frac{m_i m_j}{EI} \right) \, dx \; . \tag{3.62}$$

---

† For clarity, when original equation numbers are requoted, they will be denoted by an asterisk.

From the symmetry of this expression it is immediately obvious that the flexibility coefficients are symmetrical in i and j, so that

$$f_{ij} = f_{ji} . \tag{3.63}$$

This result is called the **Reciprocal Theorem**, which expresses the rather surprising fact that the displacement d_i due to a unit load P_j, is equal to the displacement d_j due to a unit load P_i.

For the N loaded points, we generate N^2 flexibility coefficients. The matrix of these coefficients $\mathbf{f}$, given by

$$\mathbf{f} = \begin{bmatrix} f_{11} & f_{12} & \cdots & f_{1N} \\ f_{21} & f_{22} & \cdots & f_{2N} \\ \vdots & \vdots & & \\ f_{N1} & f_{N2} & \cdots & f_{NN} \end{bmatrix} \tag{3.64}$$

is called the **flexibility matrix** of the structure.

Example 3.12
We shall calculate the flexibility matrix for the three-pinned portal frame of Example 3.8, in which three loads P_i ($i = 1, 2, 3$) are defined as in Fig. 3.27. The bending moment diagrams due to the unit forces have already been calculated in Example 3.8, and are shown in Figs. 3.19(b), (c) and (d). The corresponding summations of the product integrals in (3.62) are carried out in Table 3.9. The flexibility coefficients are then determined by dividing by EI, ($= 6.0 \times 10^6$) yielding the following flexibility matrix:

$$\mathbf{f} = \begin{bmatrix} 3.99 & 1.66 & 0.62 \\ 1.66 & 5.36 & -1.13 \\ 0.62 & -1.13 & 0.71 \end{bmatrix} \times 10^{-6} \quad \begin{array}{l} \text{m/N, rad/N,} \\ \text{m/Nm, and rad/Nm .} \end{array}$$

It should be noted that the dimensions of the flexibility coefficients are mixed and depend on the dimensions of the displacements and the forces. Thus referring to definition (3.61), if d_i and P_j are respectively a linear displacement and a direct force, then f_{ij} has the dimensions m/N, while if they are a rotation and a couple, f_{ij} has the dimensions rad/Nm, etc.

Suppose next that all the loads P_i ($i = 1, \ldots, N$) act simultaneously on the framework. d_1 say, is then given by the sum of the flexibility coefficients times the forces as

$$d_1 = f_{11} P_1 + f_{12} P_2 + \ldots + f_{1N} P_N . \tag{3.65}$$

Similar expressions can be written for d_2 to d_N. If $\mathbf{P}$ is the column matrix of the N forces P_i, and $\mathbf{d}$ the column matrix of the N displacemnts d_i, these expressions can be collected into the matrix equation

$$\mathbf{d} = \mathbf{f} \mathbf{P} \tag{3.66}$$

which can be regarded as the extended version of the one-dimensional equation (3.59) for the spring.

Finally we note that if the flexibility coefficients for a general space frame are required, the expression corresponding to (3.62) will be

$$f_{ij} = \oint \left(\frac{n_i n_j}{EA} + \frac{m_{yi} m_{yj}}{EI_y} + \frac{m_{zi} m_{zj}}{EI_z} + \frac{k_y \, s_{yi} s_{yj}}{GA} + \frac{k_z \, s_{zi} s_{zj}}{GA} + \frac{t_i t_j}{GJ} \right) dx . \tag{3.67}$$

Table 3.9

Calculation of the flexibility matrix of a pin-footed portal frame

Member	a	b	c	d	Σ m
l	4.0	6.5	6.5	4.0	
$\int m_1 m_1 \, dx$	4.56	7.41	7.41	4.56	23.95 ($= f_{11} \times EI$)
$\int m_1 m_2 \, dx$	−3.03	−4.93	11.10	6.83	9.97 ($= f_{12} \times EI$)
$\int m_1 m_3 \, dx$	2.94	2.77	−1.24	−0.76	3.70 ($= f_{13} \times EI$)
$\int m_2 m_2 \, dx$	2.02	3.28	16.62	10.23	32.15 ($= f_{22} \times EI$)
$\int m_2 m_3 \, dx$	−1.95	−1.84	−1.86	−1.14	−6.80 ($= f_{23} \times EI$)
$\int m_3 m_3 \, dx$	2.89	1.03	0.21	0.13	4.26 ($= f_{33} \times EI$)

3.14 MATRIX FORMULATION

The equation of virtual work (3.55) and the corresponding equation for the displacements (3.57) of a general space framework with support settlements are somewhat unwieldy. In this final section we recast them in a simpler form using matrices. For this purpose, in addition to the column matrices $\mathbf{P}$ and $\mathbf{d}$ of the external loads and displacements defined in Section 3.13, we denote the column matrix of the S support reactions R_j and corresponding support settlements d_{sj} by $\mathbf{R}$ and $\mathbf{d}_s$ respectively. The column matrix of the support reactions r_{ji} due to a unit force $P_i = 1.0$ is denoted by $\mathbf{r}_i$. The stress resultants in the members are denoted by the matrices $\mathbf{S}$ and $\mathbf{s}_i$ as follows:

$$\mathbf{S} = \begin{bmatrix} N \\ M_y \\ M_z \\ S_y \\ S_z \\ T \end{bmatrix}, \qquad \mathbf{s}_i = \begin{bmatrix} n_i \\ m_{yi} \\ m_{zi} \\ s_{yi} \\ s_{zi} \\ t_i \end{bmatrix} . \qquad (3.68a, b)$$

Finally the flexibility properties of a member are denoted by

$$\boldsymbol{\phi} = \begin{bmatrix} \dfrac{1}{EA} & 0 & 0 & 0 & 0 & 0 \\ 0 & \dfrac{1}{EI_y} & 0 & 0 & 0 & 0 \\ 0 & 0 & \dfrac{1}{EI_z} & 0 & 0 & 0 \\ 0 & 0 & 0 & \dfrac{k_y}{GA} & 0 & 0 \\ 0 & 0 & 0 & 0 & \dfrac{k_z}{GA} & 0 \\ 0 & 0 & 0 & 0 & 0 & \dfrac{1}{GJ} \end{bmatrix} . \qquad (3.69)$$

With the above notation, the equations and results previously expressed explicitly take the following forms.

1. *The equation of virtual work (3.55)*

$$\mathbf{P}^{*T}\mathbf{d} + \mathbf{R}^{*T}\mathbf{d}_s = \oint (\mathbf{S}^{*T}\boldsymbol{\phi}\mathbf{S}) \, dx + \oint N^*\alpha\Delta\theta \, dx . \qquad (3.70)$$

2. *The displacement d_i (3.57)*

$$d_i = \oint (\mathbf{s}_i^T\boldsymbol{\phi}\mathbf{S}) \, dx + \oint n_i\alpha\Delta\theta \, dx - \mathbf{r}_i^T\mathbf{d}_s . \qquad (3.71)$$

3. *The flexibility coefficient f_{ij} (3.67)*

$$f_{ij} = \oint (\mathbf{s}_i^T\boldsymbol{\phi}\mathbf{s}_j) \, dx . \qquad (3.72)$$

Equations (3.71) and (3.72) give particular displacements and particular flexibility coefficients for a general space framework. However, it is also useful

to be able to produce concise expressions for the *complete* set of displacements
d and the flexibility matrix **f**. For this purpose we introduce the further notation

$$
\mathbf{s} = \begin{bmatrix} \mathbf{s}_1^T \\ \mathbf{s}_2^T \\ \vdots \\ \mathbf{s}_N^T \end{bmatrix}, \quad
\mathbf{n} = \begin{bmatrix} n_1 \\ n_2 \\ \vdots \\ n_N \end{bmatrix}, \quad
\mathbf{r} = \begin{bmatrix} \mathbf{r}_1^T \\ \mathbf{r}_2^T \\ \vdots \\ \mathbf{r}_N^T \end{bmatrix}. \qquad (3.73a,b,c)
$$

Then

$$
\mathbf{d} = \oint (\mathbf{s}\boldsymbol{\phi}\mathbf{S} + \mathbf{n}\alpha\Delta\theta)\, dx - \mathbf{r}\mathbf{d}_s \qquad (3.74)
$$

$$
\mathbf{f} = \oint (\mathbf{s}\boldsymbol{\phi}\mathbf{s}^T)\, dx . \qquad (3.75)
$$

In order to help his understanding the above expressions, the reader may find
it useful to write out **s**, **n** and **r** in full for $N = S = 2$, and develop the explicit
expressions for **d** and **f** by carrying out the appropriate matrix multiplications.

3.15 PROBLEMS

Young's moduli are as follows:
$E(\text{steel}) = 200.0\ \text{GN/m}^2$, $E(\text{concrete}) = 25.0\ \text{GN/m}^3$, $E(\text{aluminium alloy}) = 70.0\ \text{GN/m}^2$.

3.1   The members of the pin-jointed framework in Fig. 3.28 are composed of
uniform section circular steel tubing. The external diameter of the tubing
is 50.0 mm and the wall thickness is 3.0 mm. Calculate the horizontal and
vertical components of the deflection of joint 1 due to the vertical force of
20.0 kN acting as shown.

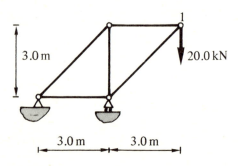

Fig. 3.28

3.2   A travelling crane truss is shown in Fig. 3.29. All the members are of the same material and have the same cross-sectional area. The inclined members are at $30°$ to the horizontal. A vertical force is applied at joint 1. What is the ratio of the vertical displacement d_2 of joint 2 to the vertical displacement d_1 of joint 1?

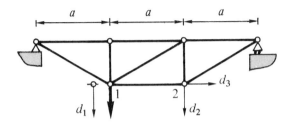

Fig. 3.29

3.3   In the pin-jointed framework shown in Fig. 3.30, all the members are of the same material of Young's modulus E and have the same cross-sectional area A. A force P is applied at joint 1 at an angle α to the vertical. Find the value of α for which the deflection of joint 1 is at $45°$ to the vertical as shown in the figure. What then is the deflection of joint 1 in terms of P, E, A and a?

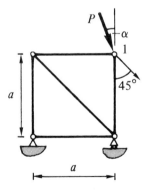

Fig. 3.30

3.4   The steel pin-jointed bridge truss shown in Fig. 3.31 is constructed so that the bottom chord is shaded by a road deck and sidewalks. What will be the vertical deflection of the centre of the bottom chord if the remaining structure undergoes a temperature rise of $20°C$? It is desired to construct the truss so that it has a slight upward camber when all the members are at the same temperature and are unstressed. This is to be achieved by increasing the length of each of the members a, b, c and d by Δ_C from the nominal design length. Find Δ_C if the centre of the lower chord is to be 0.1 m above the level of the supports (α(steel) $= 1.25 \times 10^{-5}/^{\circ}$C.)

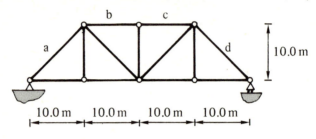

Fig. 3.31

3.5   The cross-sectional areas of the members of the steel crane structure shown in Fig. 2.74(f) are such that when loaded by the 40.0 kN force, all the tension members are stressed to 120.0 MN/m^2, and all the compression members to 60.0 MN/m^2. Calculate the vertical deflection of the point of application of the force.

3.6   In the derrick shown in Fig. 2.74(g), member a is a 7.5 mm diameter steel wire, and members b and c are aluminium alloy tubes, 30.0 mm in external diameter and 2.0 mm in wall thickness. Calculate the deflections in the x and y directions, of the point of application of the 5.0 kN force.

3.7   A uniform-section simply supported beam is subjected to the three loading cases shown in Figs. 3.32(a), (b) and (c). For each case calculate the vertical deflection at the centre of the beam in terms of P, E, I and l.

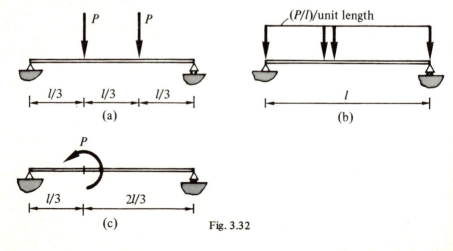

(a)   (b)

(c)   Fig. 3.32

3.8   The pin-jointed framework shown in Fig. 3.33 is loaded by a vertical force
      P at a point half-way between joints 1 and 2. Show that the vertical deflec-
      tion d_1 of joint 2 is given by

$$d_1 = \frac{Pa}{2E_aA_a} + \frac{\sqrt{2}Pa}{E_bA_b}$$

and that the vertical deflection d_2 of a point in member at at $a/4$ from joint
2, is given by

$$d_2 = \frac{3Pa}{8E_aA_a} + \frac{3\,Pa}{2\sqrt{2}E_bA_b} + \frac{11}{768}\frac{Pa^3}{E_aI_a}\;.$$

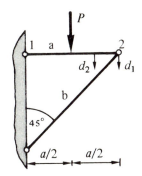

Fig. 3.33

3.9   The rigid-jointed plane framework shown in Fig. 3.34, is subjected to a
      uniformly distributed vertical force of p/unit length as shown. The two
      members of the framework are of the same material and have the same
      cross-section. Neglecting axial effects, calculate the horizontal displacement
      of the roller support in terms of p, E, I and a.

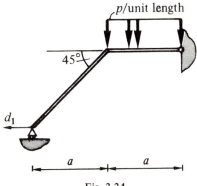

Fig. 3.34

3.10 The tower crane shown in Fig. 3.35 is made up of steel members and has the overall dimensions shown. The equivalent second moments of area of the column and beam structures about horizontal axes are 20.0×10^{-3} m^4 and 15.0×10^{-3} m^4 respectively. A load of 15.0 Mg is suspended from the beam at a point 5.0 m from the centre-line of the column and 10.0 m from the ground. It is then slowly moved outwards along the beam to a point 15.0 m from the centre-line of the column. Obtain an expression for the trajectory of the load in space. Why are the axial flexibilities of the column and beam structures irrelevant to determining this trajectory?

(Southampton University)

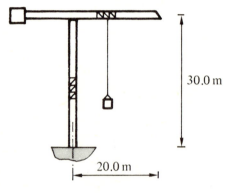

30.0 m

20.0 m

Fig. 3.35

3.11 Part of a hangar structure takes the form of a uniform-section steel beam structure supported on a concrete cantilever column by means of a pinned bearing. The beam is restrained by a steel tie as shown in Fig. 3.36. The relevant sectional properties of the members are as follows:
(i)   second moment of area of the beam about a horizontal axis: 0.2 m^4,
(ii) cross-sectional area of the column: 0.5 m^2,
(iii) cross-sectional area of the tie: 0.01 m^2.

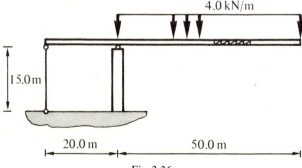

4.0 kN/m

15.0 m

20.0 m   50.0 m

Fig. 3.36

Calculate the vertical deflection of the tip of the beam due to the uniformly distributed force of 4.0 kN/m.

(Southampton University)

3.12 The slender mast shown in Fig. 3.37 has the uniform properties E, A and I throughout its length. It is pinned to a rigid foundation at its lower end and is stayed by an inclined cable as shown. Neglecting axial effects in both the mast and the cable, show that the horizontal deflection d_1 of the top of the mast due to P is given by Pa^3/EI. In fact the flexibility of the cable is non-linear and its extension Δ is related to its axial force N by the relation

$$\Delta = (\alpha N - \beta N^2)$$

where α and β are constants. Show that including all axial effects in the calculation gives the following expression for d_1:

$$d_1 = 9.0P(\alpha - 3.0\beta P) + (13.5\,Pa/EA) + Pa^3/EI\ .$$

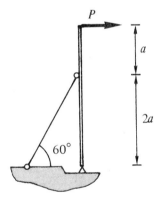

Fig. 3.37

3.13 The support for the street sign described in Problem 2.5 (Fig. 2.70), is composed of steel members of hollow box section. The section is a rectangle of sides 300.0 mm by 150.0 mm, oriented to give maximum resistance to in-plane bending. The wall thickness is uniform and can be assumed to be small compared with the overall dimensions of the section. What is the minimum allowable value for the wall thickness if the rotation of the sign due to self-weight is to be restricted to 0.02 rads?

(Southampton University)

3.14 The semicircular three-pinned arch shown in Fig. 3.38 has the uniform properties E, α and I. It carries a vertical force of P at its centre. Show that if axial effects are neglected, the vertical deflection d_1 of P is given by

$$d_1 = \frac{Pr^3}{EI}\left(\frac{\pi}{2} - \frac{3}{2}\right) \ .$$

The arch is subjected to a temperature rise of $\Delta\theta\,°$. Show that the additional deflection of P is upwards, and is given by

$$d_1 = -2\alpha r \Delta\theta \ .$$

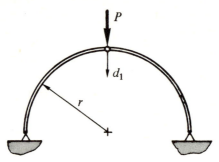

Fig. 3.38

3.15 The semicircular cantilever described in Problem 2.8 (Fig. 2.73) is of uniform material and cross-section. Obtain expressions in terms of E, G, I and J, for the vertical displacement of the free end, and for the rotation of the free end about the z axis.

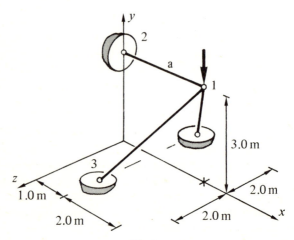

Fig. 3.39

3.16 The space framework in Fig. 3.39 is symmetrical about the x axis and member a is horizontal. It is subjected to a vertical force at joint 1 as shown. This force causes settlements at two of the supports; a horizontal settlement of 5.0 mm in the x-direction at support 2, and a downward vertical settlement of 7.5 mm at support 3. Using the virtual work equation, calculate the resulting vertical displacement of joint 1.

3.17 In the travelling crane truss described in Problem 3.2 (Fig. 3.29), the member properties are denoted by E and A. A third displacement d_3 is specified in addition to d_1 and d_2, this being the horizontal displacement of joint 2 shown in the figure. Calculate the flexibility matrix of the truss with respect to the three displacements.

APPENDIX A3.1 THE EQUATION OF VIRTUAL WORK

A3.1.1 Introduction
This appendix summarises the derivation of the equation of virtual work. The derivation is considerably simplified by the indicial notation discussed in Section A3.1.2. It employs equilibrium and compatibility conditions from the theory of continuum mechanics. These are summarised in Section A3.1.3. It also employs several well-known mathematical results including the divergence theorem – an important result in the theory of vector mechanics, summarised in Section A3.1.4. The derivation of the equation is given in Section A3.1.5.

A3.1.2 Indicial Notation
Using indicial notation the x, y and z coordinates of a defining coordinate system are respectively written as x_1, x_2 and x_3, and are denoted by the single indicial expression x_i. In this and all subsequent expressions, indices such as i. j, k etc., are assumed to range through the integers 1, 2 and 3. The components of a vector P are denoted by P_i, where $P_1 \equiv P_x$, $P_2 \equiv P_y$ and $P_3 \equiv P_z$. The stress components are denoted by σ_{ij}, where $\sigma_{11} \equiv \sigma_{xx}$, $\sigma_{22} \equiv \sigma_{yy}$, etc., and $\sigma_{12} \equiv \tau_{xy}$ etc. The strain components are denoted by ϵ_{ij}, where $\epsilon_{11} \equiv \epsilon_{xx}$, $\epsilon_{22} \equiv \epsilon_{yy}$ etc., *but* where $\epsilon_{12} \equiv \frac{1}{2}\gamma_{xy}$, etc. ϵ_{12} is called the **mathematical shear strain** to distinguish it from the engineering shear strain γ_{xy}. Its introduction simplifies the strain-displacement relations given in (A3.1.1) below.

Indicial expressions can be composed of the product of several subscripted variables. Thus $p_i d_j$ represents the nine terms $p_1 d_1, p_1 d_2, p_1 d_3, \ldots, p_3 d_3$.

An equation between indicial expressions represents a set of equations between individual terms as the indices range from 1 to 3. Thus $A_{ij} = B_i C_j$ for example, represents the nine equations $A_{11} = B_1 C_1$, $A_{12} = B_1 C_2$, $A_{13} = B_1 C_3$, $\ldots, A_{33} = B_3 C_3$.

A comma before a subscript represents partial differentiation of a variable with respect to the coordinate carrying that subscript. Thus $P_{1,2} \equiv \partial P_1/\partial x_2 \equiv \partial P_x/\partial y$.

If an indicial expression contains an index that is repeated, the terms of the expression are assumed to be summed as that index ranges from 1 to 3. Thus $P_i d_i$ represents the single term $(P_1 d_1 + P_2 d_2 + P_3 d_3)$. $P_i d_i$ is called the **inner product** of P_i with d_j. An inner product of σ_{ij} with d_k could take the form $\sigma_{ij} d_i$, representing the three terms

$$(\sigma_{11} d_1 + \sigma_{21} d_2 + \sigma_{31} d_3)$$

$$(\sigma_{12} d_1 + \sigma_{22} d_2 + \sigma_{32} d_3)$$

$$(\sigma_{13} d_1 + \sigma_{23} d_2 + \sigma_{33} d_3) \ .$$

In the expression $\sigma_{ij} d_i$, j is called a **free index** and generates the three terms as it ranges from 1 to 3. i is called a **dummy index**, since if it were replaced by k say, the meaning of the expression would be unaltered.

As an example of indicial notation, note that

$$\sigma_{ij,j} + f_i = 0$$

represents the three equations

$$\sigma_{i1,1} + \sigma_{i2,2} + \sigma_{i3,3} + f_i = 0 \ .$$

In the explicit notation of the main text, these take the form

$$\frac{\partial \sigma_{xx}}{\partial x} + \frac{\partial \tau_{xy}}{\partial y} + \frac{\partial \tau_{xz}}{\partial z} + f_x = 0$$

$$\frac{\partial \tau_{yx}}{\partial x} + \frac{\partial \sigma_{yy}}{\partial y} + \frac{\partial \tau_{yz}}{\partial z} + f_y = 0$$

$$\frac{\partial \tau_{zx}}{\partial x} + \frac{\partial \tau_{zy}}{\partial y} + \frac{\partial \sigma_{zz}}{\partial z} + f_z = 0 \ .$$

A3.1.3 Basic Results from the Theory of Continuum Mechanics [3.4], [3.5]

1. *The strain-displacement relations*

If d represents a continuous vector field of differentially small displacements throughout a body, as in Fig. A3.1.1, the strain components are given by

$$\epsilon_{ij} = \tfrac{1}{2}(d_{i,j} + d_{j,i}) \ . \tag{A3.1.1}$$

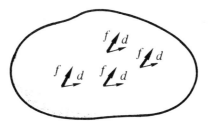

Fig. A3.1.1

2. *Internal equilibrium*
Suppose a body is subjected to a vector field of body forces f/unit volume, throughout its volume as in Fig. A3.1.1. These forces could be due for example to the Earth's gravity or to a magnetic field. Although they act at internal points, body forces are in fact *external forces* on the body, since they are not self-equilibrating and are caused by external agencies which are themselves subject to equal and opposite net reactions. By considering the linear equilibrium in the coordinate directions of the differentially small parallelepiped of material within the body shown in Fig. A3.1.2(a), and taking variations in the stresses into account, the following **internal equilibrium equations** are obtained

$$\sigma_{ij,j} + f_i = 0 \ . \tag{A3.1.2}$$

Further, the rotational equilibrium of the parallelepiped requires that the stress components are symmetric in i and j. Thus

$$\sigma_{ij} = \sigma_{ji} \ , \tag{A3.1.3}$$

— a result quoted in the main text in Section 1.6.

3. *Surface equilibrium*
Suppose a body is subjected to a vector field of surface forces p/unit area, over the surface. Again these are *external* forces. By considering the linear equilibrium in the coordinate directions of the differentially small tetrahedron of material at the surface of the body shown in Fig. A3.1.2(b), the following **surface equilibrium equations** are obtained

$$P_i = \sigma_{ij} n_j \ . \tag{A3.1.4}$$

σ_{ij} are the internal stress components at the surface. n_j are the components of the outward unit normal n to the element of body surface dS forming one side of the tetrahedron, as shown in the figure.

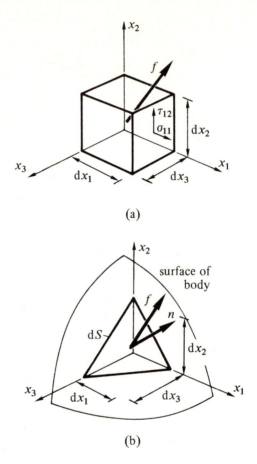

(a)

(b)

Fig. A3.1.2

A3.1.4 Basic Mathematical Results [3.6]

1. *Partial derivative of a product*

The partial derivative of the product of two variables $(P_i d_j)$ say, is related to the partial derivatives of the individual variables as follows:

$$(P_i d_j)_{,k} = P_i d_{j,k} + P_{i,k} d_j \ . \tag{A3.1.5}$$

2. *Vectors*

When describing physical systems, any indicial expression containing a single free variable such as $\sigma_{ij} d_i$ say, behaves as a vector. In particular it obeys the divergence theorem.

3. *The divergence theorem*
Consider a continuous vector field A say, defined within the volume and on the surface of a body as shown in Fig. A3.1.3. An important result in the theory of vector mechanics relates the volume and surface integrals of A as follows:

$$\int_V A_{i,i}\, \mathrm{d}V + \int_S A_i n_i\, \mathrm{d}S \tag{A3.1.6}$$

where n_i are again the components of the outward unit normal to the surface element $\mathrm{d}S$. $A_{i,i}$ is the **divergence** of A, and the result (A3.1.6) is called the **divergence theorem**. It is central to the derivation of the equation of virtual work discussed below.

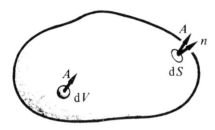

Fig. A3.1.3

A3.1.5 Equation of Virtual Work [3.3], [3.7]
Suppose the body shown in Fig. A3.1.4 is subjected to body forces f^* and surface forces p^*, giving rise to the internal stresses σ_{ij}^*. This is called the **equilibrium force system**. Then from (A3.1.2)

$$\sigma_{ij,j}^* + f_i^* = 0 \ . \tag{A3.1.7}$$

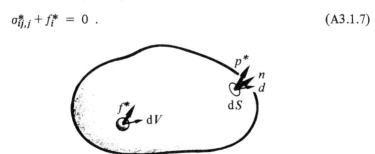

Fig. A3.1.4

Consider an *entirely independent* set of differentially small displacements d forming a continuous field of displacements throughout the body. This is called the **compatible displacement system**. Taking the inner product of d_i with the terms in (A3.1.7) and integrating through the volume then gives

$$\int_V \sigma^*_{ij,j} d_i \, dV + \int_V f^*_i d_i \, dV = 0 \ . \tag{A3.1.8}$$

The first term in (A3.1.8) is developed as follows:

(i) Using (A3.1.5)

$$\int_V \sigma^*_{ij,j} d_i \, dV = \int_V (\sigma^*_{ij} d_i)_{,j} \, dV - \int_V \sigma^*_{ij} d_{i,j} \, dV \ . \tag{A3.1.9}$$

(ii) Using the divergence theorem and the surface equilibrium equations:

$$\int_V (\sigma^*_{ij} d_i)_{,j} \, dV = \int_S \sigma^*_{ij} d_i \, n_j \, dS = \int_S P^*_i d_i \, dS \ . \tag{A3.1.10}$$

(iii) Using the symmetry of the stress components and the strain-displacement relations

$$\int_V \sigma^*_{ij} d_{i,j} \, dV = \int_V \tfrac{1}{2}\sigma^*_{ij}(d_{i,j} + d_{j,i}) \, dV = \int_V \sigma^*_{ij} \epsilon_{ij} \, dV \ . \tag{A3.1.11}$$

Whence substituting back into (A3.1.8) gives

$$\int_S P^*_i d_i \, dS + \int_V f^*_i d_i \, dV = \int_V \sigma^*_{ij} \epsilon_{ij} \, dV \ . \tag{A3.1.12}$$

(A3.1.12) is the general form of the equation of virtual work. The left-hand side represents the work done by the external forces of the equilibrium force system if they were to move through the displacements of the compatible displacement system — the **external virtual work**. The right-hand side represents the work done by the stresses of the equilibrium force system if they were to move through the strains of the compatible displacement system — the **internal virtual work**. In the explicit notation of the main text $\int_V \sigma^*_{ij} \epsilon_{ij} \, dV$ is given by

$$\int_V (\sigma^*_{xx} \epsilon_{xx} + \sigma^*_{yy} \epsilon_{yy} + \sigma^*_{zz} \epsilon_{zz} + \tau^*_{xy} \gamma_{xy} + \tau^*_{yz} \gamma_{yz} + \tau^*_{zx} \gamma_{zx}) \, dV \ ,$$

— an expression identical to that on the right-hand side of (3.1).

A3.1.6 Application to Frameworks

For the particular case of frameworks subjected only to concentrated forces discussed in Section 3.2, the following changes are necessary in the expression for the external virtual work:

1. *Body forces*
The body forces are zero and the second term in the expression for the external virtual work disappears.

2. *Concentrated forces*
In the presence of concentrated forces, most of the surface of the body is unloaded. A concentrated direct force P^* can be regarded as a very large distributed surface force $(P^*/\Delta S)$ acting over a small element of the surface area ΔS, as in Fig. A3.1.5(a). As ΔS decreases in size, the surface integral tends to $P_i^* d_{i'}$, where d is the linear displacement of the point of application of P^*. A concentrated couple P^* can be regarded as two equal and opposite distributed surface forces $P^*/(\Delta S\Delta l)$ acting over small surface area elements ΔS, separated by Δl, as in Fig. A3.1.5(b). As ΔS and Δl decrease in size, the surface integral again tends to $P_i^* d_i$, where d is the rotation at the point of application of P^*.

For convenience in structural analysis, $P_i^* d_i$ can also be expressed as P^*d if d is specifically defined as a displacement or rotation component 'corresponding' to the force P^* as in Section 1.11.1. Whence *employing the suffix notation of the main text*, if the loading on the framework comprises N concentrated forces $P_i^* (i=1,N)$, the external virtual work becomes $\sum_i (P_i^* d_i)$, an expression identical to that on the left-hand side of (3.1).

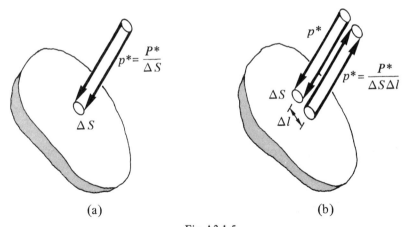

Fig. A3.1.5

3. *Articulation*
Finally note that the equation of virtual work applies without modification to a framework containing internal articulation in the form of hinges or pins. This is because the equation for the complete framework can be derived by summing the equations for the individual sub-sections joined by the hinges. From the point of view of these sub-sections, the reactions at the hinges are external

forces. However, since the reactions are self-equilibrating, their external virtual work disappears in the summation. The equation of virtual work for the complete framework is therefore unchanged.

REFERENCES

[3.1] (1966), *Steel Designers' Manual,* 3rd edn, Crosby-Lockwood, London, pp. 148–160.

[3.2] Bisplinghoff, R. L., Mar, J. W., and Pian, T.H.H. (1965), *Statics of Deformable Solids,* Addison-Wesley, Reading, Mass., Section 9.4.

[3.3] McCracken, D. D., and Dorn, W. S. (1966), *Numerical Methods and Fortran Programming,* Wiley International, New York, p. 172.

[3.4] Bisplinghoff, R. L., Mar, J. W., and Pian. T. H. H., op. cit., Chapters 5 and 6.

[3.5] Timoshenko, S., and Goodier, J. N. (1951), *Theory of Elasticity,* 2nd edn, McGraw-Hill, New York, Chapters 1 and 9.

[3.6] Sokolnikoff, I. S. (1964), *Tensor Analysis,* 2nd edn, John Wiley, New York, Sections 25 and 92.

[3.7] Sokolnikoff, I. S., op. cit., Section 117.

CHAPTER 4

Flexibility Method

4.1 INTRODUCTION

We have seen in Chapter 2 that the stress resultants in certain relatively simple structures can be calculated by considering only static equilibrium. However, for most structures, equilibrium alone provides insufficient equations for the unknowns and extra equations have to be found by considering the **compatibility** of the structures under the loading. Thus in the case of the plane pin-jointed framework shown in Fig. 4.1(a), the number of unknowns, $(M + R)$ in the notation of Section 2.8.5, is 9, and the number of equilibrium equations $2J$, is 8. A further equation has to be found by considering how the deformed members fit together under the loading.

As an introduction to the use of compatibility in providing extra equations for structural analysis, we shall consider an explicit procedure for analysing the framework in Fig. 4.1.

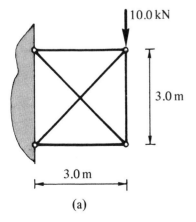

(a)

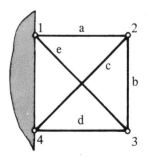

Cross-sectional areas:
a, b, e = 500.0 mm²
c, d   = 750.0 mm²

(b)

Fig. 4.1

Example 4.1

The framework in Fig. 4.1(a) is composed of steel members. It is given the joint numbering and member lettering shown in Fig. 4.1(b). The member cross-sectional areas are then as indicated. We first consider the framework with member e disconnected at joint 3 as shown in Fig. 4.2(a). This disconnection, called a **release**, reduces the axial force in e to zero, and the axial forces in the remaining framework, which is now statically determinate, can be calculated. The structure derived from the original structure by making such a release, is called a **primary structure**. For clarity, we shall call the released end of e in the primary structure, joint 5.

Consider now the deformation of the primary structure due to the 10.0 kN force. Joint 5 remains stationary, while joint 3 moves through the total displacement d shown in Fig. 4.2(b). We shall concentrate on the component of d in the direction of member e, which we shall call d_1, say. This can be regarded as the lack of fit of the member caused by the loading. Since the primary structure is statically determinate, we can calculate d_1 by the virtual work method discussed in Chapter 3. Thus we recall that the equation (3.16) for the displacement d_i of the pin-jointed framework is

$$d_i = \sum_{\mathrm{m}} \{n_i \Delta\}_{\mathrm{m}} \qquad (4.1)$$

where Δ is given in Section 3.3.1 as

$$\Delta = \Delta_E + \Delta_\Theta = \left(\frac{Nl}{EA} + \alpha l \Delta\theta \right). \qquad (4.2)$$

In this first example we shall assume that the temperature changes, $\Delta\theta$ are zero.

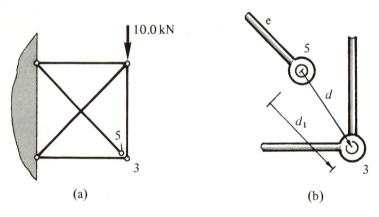

(a)   (b)

Fig. 4.2

In calculating d_1, N are the member forces due to the external loading P_i on the primary structure, which in this case is composed of just the single 10.0 kN force. In the following discussion, it is helpful for clarity to distinguish these member forces by the subscript P. They are then shown in Fig. 4.3(a). The corresponding member extensions are called Δ_P. n_1 are the member forces due to the unit force $P_1^* = 1.0$ acting on the primary structure in the direction of d_1. These are shown in Fig. 4.3(b). The calculation for d_1 is then carried out in Table 4.1, and d_1 is shown to be 0.612 mm.

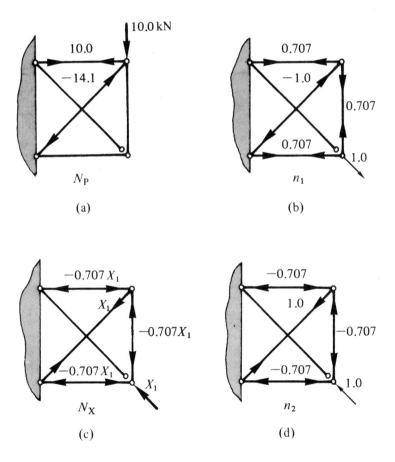

Fig. 4.3

We next imagine a process for converting the primary structure back into the original structure. This can be done by applying equal and opposite, self-equilibrating forces X_1 say, to joints 3 and 5 as in Fig. 4.4. Their function is to draw joint 3 back by d_2 say, in the direction of member e, and extend member e by d_3 say, so that the final positions of joints 3 and 5 coincide. This will occur when

$$d_2 + d_3 = d_1 . \tag{4.3}$$

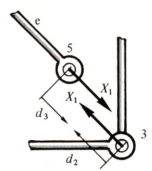

Fig. 4.4

The unknown forces X_1 are called **release forces**, and (4.3) is called the **compatibility equation**. d_2 and d_3 are again calculated from (4.1) and (4.2). In the case of d_2, N are the member forces due to the force X_1. It is helpful for clarity to distinguish these forces by the subscript X. They are shown in Fig. 4.3(c). The corresponding member extensions are called Δ_X. n_2 are the member forces due to the unit force $P_2^* = 1.0$ in the direction of d_2. These are shown in Fig. 4.3(d). The calculation for d_2 is then carried out in second part of Table 4.1 and d_2 is shown to be $68.3 \times 10^{-9} X_1$ m. In the case of d_3, we simply have to consider the single member e for which, $N_X = X_1$ and $n_3 = 1.0$. Thus

$$d_3 = \left\{ 1.0 \frac{X_1 l}{EA} \right\}_e = 42.4 \times 10^{-9} X_1 \text{ m} .$$

The compatibility equation (4.3) then gives

$$(68.3 \times 10^{-9} + 42.4 \times 10^{-9}) X_1 = 0.612 \times 10^{-3}$$

and

$$X_1 = 5.53 \text{ kN} .$$

The complete framework in Fig. 4.1 is identically equivalent to the primary structure subject to the 10.0 kN load and the release forces X_1. The required axial forces are therefore the same as the axial forces in the primary structure, and these are calculated by summing N_P and N_X for each member, the latter now being calculable since X_1 is known. The calculation is included in Table 4.1 giving the final forces in members a, b, c, and d. The force in member e is obviously equal to X_1, that is equal to 5.53 kN. The forces in the solved pin-jointed framework are then shown in Fig. 4.5.

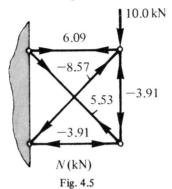

N (kN)

Fig. 4.5

Table 4.1
Calculations for Example 4.1

Member		a	b	c	d	Σ m
l		3.0	3.0	4.24	3.00	
A	10^{-3}	0.5	0.5	0.75	0.75	
N_P	10^3	10.00	0	−14.14	0	
Δ_P	10^{-3}	0.3	0	−0.4	0	
n_1		0.707	0.707	−1.0	0.707	
$n_1\Delta_P$	10^{-3}	0.212	0	0.4	0	0.612 $(= d_1)$
N_X		$-0.707X_1$	$-0.707X_1$	$+X_1$	$-0.707\,X_1$	
Δ_X	10^{-9}	$-21.2\,X_1$	$-21.2\,X_1$	$28.3\,X_1$	$-14.1\,X_1$	
n_2		−0.707	−0.707	−1.0	−0.707	
$n_2\Delta_X$	10^{-9}	$15.0\,X_1$	$15.0\,X_1$	$28.3\,X_1$	$10.0\,X_1$	$68.3\,X_1\,(= d_2)$
N_X	10^3	−3.91	−3.91	5.53	−3.91	
$(N_P + N_X)$	10^3	6.09	−3.91	−8.61	−3.91	

This method for carrying out the analysis of a statically indeterminate structure is called the **flexibility method,** because in determining the displacements of the primary structure due to the various forces, we are again finding its flexibility. We shall discuss a more formal presentation of the flexibility method in the next section.

4.2 FLEXIBILITY METHOD FOR THE SOLUTION OF PIN-JOINTED FRAMEWORKS

4.2.1 Framework with a Single Redundancy

The method described in Section 4.1 was presented in a way that emphasised the use of compatibility to provide an extra equation for the forces in a simple pin-jointed framework. Here we consider the same method, but in a form that is appropriate for analysing more complicated cases.

Example 4.2
Consider again the framework shown in Fig. 4.1. This time we produce the primary structure by making a cut anywhere in member e as in Fig. 4.6. This cut is again called a **release,** which we now define more precisely as that which destroys a single stress resultant at a point in a framework. In this case, the release destroys the axial force N_e. The primary structure, which will be taken to *include* the member e, is statically determinate, since N_e is zero, and all the other axial forces can be obtained by the method of joints. The axial forces N_P are then the same as those calculated before and shown in Fig. 4.3(a).

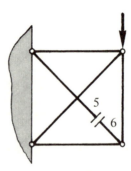

Fig. 4.6

We next consider the relative displacements of the cut surfaces at the release, which for clarity we call joints 5 and 6 of the primary structure. Suppose, owing to the loading, that 5 moves d_1 and 6 moves d_2 as shown in the expanded view in Fig. 4.7(a). Then the total **relative displacement** of the two surfaces is $d_1 + d_2$. We denote this relative displacement by u_1 say. The object of the analysis is to obtain the **initial relative displacement** u_{P1}, due to the applied loading acting by itself, and the relative displacement u_{X1}, due to the two self-equilibrating release forces X_1 acting on the cut surfaces of the release, as in Fig. 4.7(b). Compatibility at the release then requires that the final relative displacement of the cut surfaces is zero, that is

$$u_{X1} + u_{P1} = 0 . \tag{4.4}$$

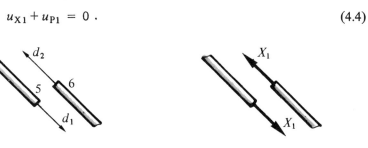

(a)   (b)

Fig. 4.7

The relative displacement u_1 for any particular loading can be calculated from the virtual work equation, by using as the equilibrium force system, self-equilibrating unit forces acting *simultaneously* on the cut surfaces as shown in Fig. 4.8(a). The corresponding internal forces n_1 are calculated for the statically determinate primary structure and are shown in Fig. 4.8(b). The external virtual work is then given by

$$W_E^* = 1.0 \, d_1 + 1.0 \, d_2 = u_1 . \tag{4.5}$$

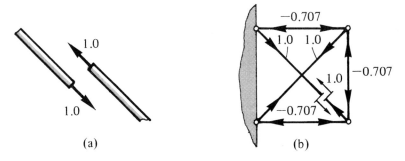

(a)   (b)

Fig. 4.8

The internal virtual work is the usual expression for pin-jointed frameworks

$$W_I^* = \sum_m \{n_1 \Delta\}_m \tag{4.6}$$

where Δ is calculated using the member forces N corresponding to the particular loading. The equation of virtual work then leads to the deflection equation

$$u_1 = \sum_m \{n_1 \Delta\}_m \tag{4.7}$$

where

$$\Delta = \left(\frac{Nl}{EA} + \alpha l \Delta\theta\right) . \tag{4.8}$$

The initial relative displacement u_{P1} is calculated from (4.7) using $\Delta = \Delta_P$, where, again ignoring temperature changes,

$$\Delta_P = \frac{N_P l}{EA} . \tag{4.9}$$

The calculation is carried out in the first part of Table 4.2, member e being included in the table. Note that wherever the release is made in e, N_P and n_1 will be the same throughout the length of e. The contribution of e to the internal virtual work is therefore given by $\{n_1 N_P l/EA\}_e$ and e is treated in the table in the same way as the other members. u_{P1} comes out to be -0.612 mm. The minus sign indicates that the release opens under the 10.0 kN force and this is consistent with the positive value obtained for d_1 in Example 4.1.

In order to be able to generalise the flexibility method to structures needing more than one release, the calculation for the relative displacement u_{X1} due to the release forces X_1, is found in terms of the flexibility of the primary structure f_{11} *at the release*. This by analogy with the definition of the flexibility coefficients given in Chapter 3, Section 3.13, is defined as the relative displacement u_1 due to *unit* self-equilibrating forces acting on the cut surfaces of the release. If then f_{11} is known, we simply have

$$u_{X1} = f_{11} X_1 \tag{4.10}$$

and the compatibility equation (4.4) changes to

$$f_{11} X_1 + u_{P1} = 0 . \tag{4.11}$$

In calculating f_{11}, the member extensions Δ in (4.7) are those due to the unit release forces. N in (4.8) are the corresponding axial stress resultants which have already been calculated as n_1. Thus

$$f_{11} = \sum_m \{n_1 \Delta\}_m = \sum_m \left\{\frac{n_1^2 l}{EA}\right\}_m . \tag{4.12}$$

f_{11} is calculated in the second part of Table 4.2, and it comes out to be 110.7×10^{-9} m/N. Thus rearranging the compatibility equation we obtain

$$X_1 = \frac{-u_{P1}}{f_{11}} = \frac{0.612 \times 10^{-3}}{110.7 \times 10^{-9}} = 5.53 \text{ kN}$$

giving the same value for the release forces as that obtained in Example 4.1.

Finally we note that the axial forces due to X_1 are simply $n_1 X_1$ and that the total axial forces in the members are therefore given by

$$N = N_P + n_1 X_1 . \tag{4.13}$$

These forces are evaluated in Table 4.2 and agree with the previous results.

Table 4.2
Calculations for Example 4.2

Member		a	b	c	d	e	Σ m
l		3.0	3.0	4.24	3.0	4.24	
A	10^{-3}	0.5	0.5	0.75	0.75	0.5	
N_P	10^3	10.0	0	-14.14	0	0	
Δ_P	10^{-3}	0.3	0	-0.4	0	0	
n_1		-0.707	-0.707	1.0	-0.707	1.0	
$n_1 \Delta_P$	10^{-3}	-0.212	0	-0.4	0	0	$-0.612 (= u_{P1})$
$\dfrac{n_1^2 l}{EA}$	10^{-9}	15.0	15.0	28.3	10.0	42.4	$110.7 (= f_{11})$
$n_1 X_1$	10^3	-3.91	-3.91	5.53	-3.91	5.53	
$(N_P + n_1 X_1) \, 10^3$		6.09	-3.91	-8.61	-3.91	5.53	

In Example 4.2 we have selected a release in a particular member. In fact, selecting a release to form a primary structure can usually be done in several ways, and in the present framework, any of the five members could have been cut. We illustrate that this does indeed lead to a consistent solution by recalculating the example with a release in member a.

Example 4.3

The pin-jointed framework with the release in a is shown in Fig. 4.9(a). The axial force systems N_P and n_1 are calculated by the method of joints and are shown in Figs. 4.9(a) and (b). The initial relative displacement and the flexibility are then calculated in Table 4.3. Whence

$$X_1 = \frac{-u_{P1}}{f_{11}} = \frac{1.348 \times 10^{-3}}{221.4 \times 10^{-9}} = 6.09 \text{ kN} .$$

The corresponding total axial forces are also calculated in Table 4.3, and can be seen to agree with the previous solution.

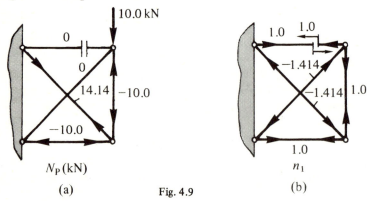

(a)   Fig. 4.9   (b)

Table 4.3

Calculations for Example 4.3

Member		a	b	c	d	e	Σ m
N_P	10^3	0	−10.0	0	−10.0	14.14	
Δ_P	10^{-3}	0	−0.3	0	−0.2	0.6	
n_1		1.0	1.0	−1.414	1.0	−1.414	
$n_1\Delta_P$	10^{-3}	0	−0.3	0	−0.2	−0.848	−1.348 ($= u_{P1}$)
$\dfrac{n_1^2 l}{EA}$	10^{-9}	30.0	30.0	56.6	20.0	84.8	221.4 ($= f_{11}$)
$n_1 X_1$	10^3	6.09	6.09	−8.61	6.09	−8.61	
$(N_P + n_1 X_1)$	10^3	6.09	−3.91	−8.61	−3.91	5.53	

In order to simplify the generalisation of the flexibility method to the solution of multiply-redundant frameworks, further observations can be made about the self-equilibrating forces X_1 and the relative release displacement u_1. Firstly, it should be emphasised that they they are *external* forces and displacements associated with the primary structure. Then, from the point of view of the virtual work equation, X_1 and u_1 can be regarded as a single force and its 'corresponding' displacement, since the work done by X_1 at the release is $X_1 (d_1 + d_2) = X_1 u_1$. The virtual work equation for the primary structure is therefore

$$\sum_i (P_i^* d_i) + X_1^* u_1 = \sum_m \{N^* \Delta\}_m \ . \tag{4.14}$$

Thus for simplicity in the remainder of this text, we shall refer to the system of two self-equilibrating forces as the *single* **release force**, X_1. It is also worth noting that the positive direction chosen for the release force and for the relative displacement, is quite arbitrary. Thus if X_1 were chosen as two compressive forces on the cut surfaces of the release in Fig. 4.7(b), u_1 would be the relative movement apart of the surfaces. The actual direction of X_1 is determined by its sign in the solution.

4.2.2 Temperature Changes
In statically indeterminate structures, temperature changes alter the internal forces. This is because enforcing the compatibility of members that have expanded or contracted cannot be done without changing the forces in them. We illustrate this by recalculating the framework problem in Fig. 4.1, but with one of the members subject to a temperature rise.

Example 4.4
Suppose member a in the framework shown in Fig. 4.1 is subject to a temperature rise of $20°C$. Choosing the same primary structure as in Example 4.2, the extension of member a given by (4.9) is then increased by $\{\alpha l \Delta \theta\}_a (= 1.25 \times 10^{-5} \times 3.0 \times 20.0 = 0.75 \times 10^{-3} m)$. The effect of this is to change the initial relative displacement u_{P1}, and its new value is calculated in Table 4.4. Whence

$$X_1 = \frac{-u_{P1}}{f_{11}} = \frac{1.142 \times 10^{-3}}{110.7 \times 10^{-9}} = 10.32 \text{ kN} \ .$$

The corresponding total axial forces are also calculated in the table and when compared with those in Table 4.2, it is apparent that there is a large reduction in the tensile force in a. This is caused by the rest of the framework restraining its expansion. There are correspondingly large changes in the axial forces in the other members, notably increases in the compressive forces in b and d.

Table 4.4
Calculations for Example 4.4

Member		a	b	c	d	e	Σ m
N_P	10^3	10.0	0	-14.14	0	0	
$\Delta_P = \Delta_{EP}$ $+ \Delta_{\Theta P}$	10^{-3}	$+\begin{array}{c}0.3\\0.75\end{array}$	0	-0.4	0	0	
n_1		-0.707	-0.707	1.0	-0.707	1.0	
$n_1\Delta_P$	10^{-3}	-0.742	0	-0.4	0	0	$-1.142 \, (= u_{P1})$
$\dfrac{n_1^2 l}{EA}$	10^{-9}	15.0	15.0	28.3	10.0	42.4	$110.7 \, (= f_{11})$
$n_1 X_1$	10^3	-7.29	-7.29	10.32	-7.29	10.32	
$(N_P + n_1 X_1) \, 10^3$		2.71	-7.29	-3.82	-7.29	10.32	

It is important to note that temperature stresses like the above, are not generated in a statically *determinate* framework. This is because such a framework can deform freely to accommodate the expansion and contraction of its members.

4.2.3 Prestress
It is possible for a statically indeterminate structure to be in a state of internal stress before external loading is applied. This internal stress, called **prestress**, can for example, be brought about by construction errors, where the lengths of members are different from their nominal design lengths. To see how this arises, let us consider a hypothetical construction process for the framework in Fig. 4.1. It is possible to reach the stage shown in Fig. 4.10, when two structural frameworks comprising members a and c, and d and e have been formed and joints 2 and 3 are fixed in space. If member b is then found to be too long, inserting it into the framework will entail jacking joints 2 and 3 apart; a process generating axial forces in the entire framework and leaving a compressive force in b when the jacks are removed. Such prestress can be calculated by considering members to have extra extensions Δ_C, when finding the initial relative displacement in the primary structure. Again we illustrate this by means of the framework in Fig. 4.1.

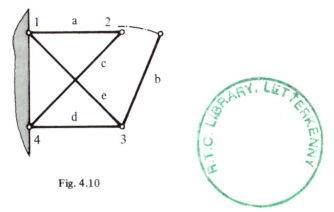

Fig. 4.10

Example 4.5

Consider the pin-jointed framework in Fig. 4.1, with member b made 5.0 mm too long. u_{P1} is calculated with this extra extension included in Δ_{Pb}, as in Table 4.5. Thus $u_{P1} = -4.15$ mm, f_{11} again is 110.7×10^{-9} m/N, and therefore $X_1 = 37.49$ kN. The total axial forces are as shown in the table. Again, comparing with those in Table 4.2, we note that this fairly large construction error has completely changed the internal force system in the framework and the compressive forces needed to enforce compatibility dominate the final state of internal stress.

Table 4.5

Calculations for Example 4.5

Member		a	b	c	d	e	Σ m
Δ_P	10^{-3}	0.3	5.0	−0.4	0	0	
n_1		−0.707	−0.707	1.0	−0.707	1.0	
$n_1\Delta_P$	10^{-3}	−0.212	−3.54	−0.4	0	0	−4.15 $(= u_{P1})$
$n_1 X_1$	10^3	−26.51	−26.51	37.49	−26.51	37.49	
$(N_0 + n_1 X_1)\,10^3$		−16.51	−26.51	23.35	−26.51	37.49	

Prestress can be introduced deliberately into a structure in order to control the final internal forces. The most important practical example of this is prestressed concrete construction, where large compressive forces are introduced into concrete beams in order to eliminate the tensile bending stresses. In our framework example, it might be desirable say, to reduce the compressive force in member b to avoid buckling problems. This we could do by deliberately making b too *short*.

4.2.4 Multiply-redundant Pin-jointed Frameworks

Hitherto we have considered a framework that can be analysed by the flexibility method, by making a single release to form a primary structure. Such a framework is said to be **singly-redundant**. In more complicated frameworks, several releases are often necessary, and such frameworks are said to be **multiply-redundant**. The flexibility method can easily be generalised to deal with multiply-redundant frameworks, and to illustrate the extended method, it is sufficient only to consider the problem of a framework needing two releases. Such a framework is shown in Fig. 4.11, and we shall explain its solution using the same sequence of ideas as in Example 4.2.

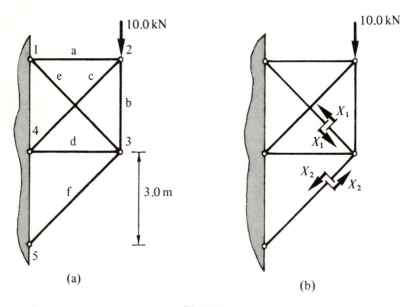

Fig. 4.11

Example 4.6

Consider the framework in Fig. 4.11(a) which is the same as the framework in the previous example, but with the addition of the extra supporting member f. f is again made of steel and is of cross-sectional area equal to 750.0 mm². We now have six support reactions and six members in the framework, so that the number of unknowns $(M + R)$ is 12. The number of equilibrium equations $(2J)$ is 10. Two compatibility equations have therefore to be found. A statically determinate primary structure can be formed by introducing two releases into the structure, in, say, members e and f, as shown in Fig. 4.11(b). These will be called releases 1 and 2 respectively. All parameters associated with the releases will then be distinguished by the subscripts 1 and 2. To simplify the discussion, we shall also introduce the matrix notation

$$\mathbf{u} = \begin{bmatrix} u_1 \\ u_2 \end{bmatrix}, \quad \mathbf{X} = \begin{bmatrix} X_1 \\ X_2 \end{bmatrix} \quad \text{etc.} \tag{4.15}$$

The primary structure is subject to the external force of 10.0 kN and the two unknown release forces X_1 and X_2. The corresponding relative displacements at the releases are u_1 and u_2. The object of the analysis is to obtain the initial relative displacements $\mathbf{u}_P$ due to the applied load acting by itself, and the relative displacements $\mathbf{u}_X$ due to the release forces. Compatibility of the final structure is then expressed by the *two* conditions that $u_1 = 0$ and $u_2 = 0$. The two compatibility equations in matrix form are thus

$$\mathbf{u}_X + \mathbf{u}_P = 0 \ . \tag{4.16}$$

In order to evaluate the initial relative displacements, we determine the axial forces N_P due to the 10.0 kN force on the primary structure, and the axial forces n_1 and n_2 due to self-equilibrating unit forces acting at the two releases in turn. These are shown in Figs. 4.12(a), (b), and (c). Then from the virtual work equation (4.14)

$$\mathbf{u}_P = \sum_m \{\mathbf{n}\Delta_P\}_m \tag{4.17}$$

where $\mathbf{n}$ is the column matrix of n_1 and n_2. The calculation for $\mathbf{u}_P$ for this particular example is carried out in Table 4.6 (p. 212) and

$$\mathbf{u}_P = \begin{bmatrix} -0.612 \\ 0.612 \end{bmatrix} \text{mm} \ .$$

The relative displacements $\mathbf{u}_X$ are found by determining the flexibility matrix $\mathbf{f}$ of the primary structure, corresponding to the two *release* forces $\mathbf{X}$. Recalling the definition of $\mathbf{f}$ discussed in Chapter 3, Section 3.13, in this context it is the matrix of flexibility coefficients f_{ij}, where

$$f_{ij} = u_i [X_j = 1, \ X_k = 0, \ k \neq j] \ . \tag{4.18}$$

In the present case of just two releases, f_{12} say, is the relative displacement of release 1 due to a unit release force at release 2, with all other forces on the structure zero. If then f is known, we have

$$\mathbf{u}_X = \mathbf{f} \mathbf{X} \tag{4.19}$$

a relation derived in a similar manner to (3.66). The compatibility equation (4.16) then changes to

$$\mathbf{f} \mathbf{X} + \mathbf{u}_P = 0 \ . \tag{4.20}$$

For joint loaded pin-jointed frameworks, equation (3.67) for f_{ij} reduces to

$$f_{ij} = \sum_m \left\{ n_i n_j \frac{l}{EA} \right\}_m \tag{4.21}$$

and therefore using the notation n defined above, f is given by

$$\mathbf{f} = \sum_m \left\{ \mathbf{n} \mathbf{n}^T \frac{l}{EA} \right\}_m \ . \tag{4.22}$$

The calculation of the flexibility coefficients for the present framework is shown in the second part of Table 4.6. Whence

$$\mathbf{f} = \begin{bmatrix} 110.7 & -48.3 \\ -48.3 & 96.6 \end{bmatrix} \times 10^{-9} \text{ m/N} \ .$$

The matrix compatibility equation (4.20) represents two simultaneous equations in the unknowns $\mathbf{X}$. Their solution can be expressed formally as

$$\mathbf{X} = -\mathbf{f}^{-1} \mathbf{u}_P \ . \tag{4.23}$$

They can be solved by hand or by any standard numerical procedure, to give

$$\mathbf{X} = \begin{bmatrix} 3.54 \\ -4.57 \end{bmatrix} \text{ kN} \ .$$

Finally the axial forces due to the two release forces acting together are $n_1 X_1 + n_2 X_2 = \mathbf{n}^T \mathbf{X}$. Thus the total axial forces in the member are given by

$$\mathbf{N} = \mathbf{N}_P + \mathbf{n}^T \mathbf{X} \ . \tag{4.24}$$

These forces are evaluated in the final part of Table 4.6.

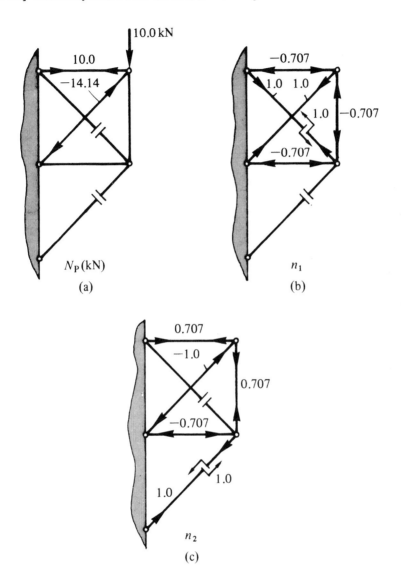

Fig. 4.12

The above presentation of the flexibility method is applicable to a pin-jointed framework requiring any number of releases. It is simply necessary to suitably increase the sizes of the matrices, u_P, X, n and f.

Table 4.6
Calculations for Example 4.6

Member		a	b	c	d	e	f	Σ_m
l		3.0	3.0	4.24	3.0	4.24	4.24	
A	10^{-3}	0.5	0.5	0.75	0.75	0.5	0.75	
N_P	10^3	10.0	0	−14.14	0	0	0	
Δ_P	10^{-3}	0.3	0	−0.4	0	0	0	
n_1		−0.707	−0.707	1.0	−0.707	1.0	0	
n_2		0.707	0.707	−1.0	−0.707	0	1.0	
$n_1\Delta_P$	10^{-3}	−0.212	0	−0.4	0	0	0	$-0.612 (= u_{P1})$
$n_2\Delta_P$	10^{-3}	0.212	0	0.4	0	0	0	$0.612 (= u_{P2})$
$\dfrac{n_1^2 l}{EA}$	10^{-9}	15.0	15.0	28.3	10.0	42.4	0	$110.7 (= f_{11})$
$\dfrac{n_1 n_2 l}{EA}$	10^{-9}	−15.0	−15.0	−28.3	10.0	0	0	$-48.3 (= f_{12})$
$\dfrac{n_2^2 l}{EA}$	10^{-9}	15.0	15.0	28.3	10.0	0	28.3	$96.6 (= f_{22})$
$n_1 X_1$	10^3	−2.50	−2.50	3.54	−2.50	3.54	0	
$n_2 X_2$	10^3	−3.23	−3.23	4.57	3.23	0	−4.57	
N	10^3	4.27	−5.73	−6.04	0.73	3.54	−4.57	

4.2.5 External Releases

The releases discussed in the previous sections, each destroying an internal stress resultant at a point in the framework, are **internal releases**. We can, however, produce primary structures by destroying support reactions. Each reaction destroyed is called an **external release**. The corresponding release force X_i, is then a single external force applied at the support, and takes the place of the reaction. The relative displacement at the release u_i is the displacement of the structure relative to the support. Compatibility at the support then requires that $u_i = 0$. Using this notation, exactly the same equations apply for finding X as for finding the internal release forces, namely (4.17), (4.22) and (4.23).

Example 4.7

We recalculate the doubly-redundant framework in Example 4.6, but this time taking as release 2, the external release formed by destroying the vertical reaction at joint 4. The primary structure is shown in Fig. 4.13 with the release force X_2 as shown. u_2 is the vertical deflection of 4. N_P, n_1 and n_2 are shown in Figs. 4.14(a), (b) and (c). u_P and f are then calculated in Table 4.7 as

$$\mathbf{u_P} = \begin{bmatrix} -0.682 \\ -1.066 \end{bmatrix} \text{ mm,} \quad \mathbf{f} = \begin{bmatrix} 110.7 & 68.3 \\ 68.3 & 193.0 \end{bmatrix} \times 10^{-9} \text{ m/N .}$$

Thus $$\mathbf{X} = -\mathbf{f}^{-1}\mathbf{u_P} = \begin{bmatrix} 3.54 \\ 4.27 \end{bmatrix} \text{ kN .}$$

The final total internal forces are calculated at the end of the table, and are the same as those obtained in Example 4.6.

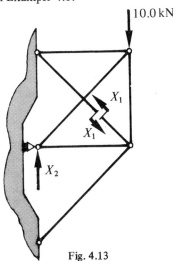

10.0 kN

Fig. 4.13

Table 4.7
Calculations for Example 4.7

Member		a	b	c	d	e	f	$\sum_m$
l		3.0	3.0	4.24	3.0	4.24	4.24	
A	10^{-3}	0.5	0.5	0.75	0.75	0.5	0.75	
N_P	10^3	0	−10.0	0	10.0	0	−14.14	
Δ_P	10^{-3}	0	−0.3	0	0.2	0	−0.4	
n_1		0	0	0	−1.414	1.0	1.0	
n_2		1.0	1.0	−1.414	−1.0	0	1.414	
$n_1\Delta_P$	10^{-3}	0	0	0	−0.282	0	−0.4	−0.682 (= u
$n_2\Delta_P$	10^{-3}	0	−0.3	0	−0.2	0	−0.566	−1.066 (= u
$\dfrac{n_1^2 l}{EA}$	10^{-9}	0	0	0	40.0	42.4	28.27	110.7 (= f_1
$\dfrac{n_1 n_2 l}{EA}$	10^{-9}	0	0	0	28.28	0	40.00	68.3 (= f_{12}
$\dfrac{n_2^2 l}{EA}$	10^{-9}	30.0	30.0	56.52	20.0	0	56.52	193.0 (= f_2
$n_1 X_1$	10^3	0	0	0	−5.00	3.54	3.54	
$n_2 X_2$	10^3	4.27	4.27	−6.04	−4.27	0	6.04	
N	10^3	4.27	−5.73	−6.04	0.73	3.54	−4.57	

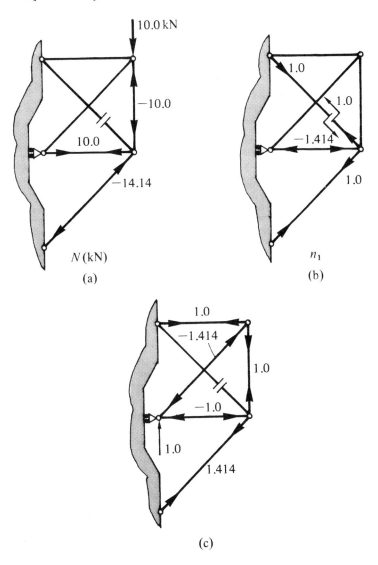

N (kN)

(a)

n_1

(b)

(c)

Fig. 4.14

4.2.6 Selecting Releases

In Example 4.6 we indicated how the formula for the statical determinacy of a pin-jointed framework, derived in Section 2.8.5, may be extended to determine the number of releases required in an indeterminate framework. Thus in a plane framework, if M is the number of members, R the number of support reactions and J the number of joints, the number of releases N is given by

$$N = (M + R - 2J) . \tag{4.25}$$

N is called the **degree of redundancy of the framework**. In a space framework, where three equilibrium equations are available at each joint,

$$N = (M + R - 3J) . \tag{4.26}$$

Perhaps we should note that the flexibility method is identical whether applied to a space framework or a plane framework. The only difficulty with the former arises in calculating N_P and n in the primary structure, and they will usually have to be found using the method of tension coefficients described in Section 2.8.3.

Equations (4.25) and (4.26) tell us how many releases to make in a pin-jointed framework; they do not tell us where to make them. In fact, as indicated in Examples 4.2 and 4.3, we usually have a choice as to where the releases are made. What has to be avoided is producing a primary structure that is a mechanism. Thus for the truss bridge shown in Fig. 4.15(a) which requires two releases, we cannot make cuts in the diagonals of just one panel as shown. Allowable release systems for the bridge are shown in Figs. 4.15(b), (c) and (d).

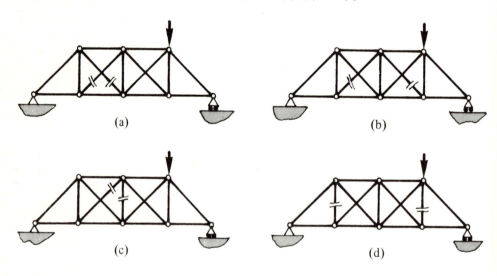

Fig. 4.15

Selecting releases is a structure-oriented problem. By this is meant that it is difficult to devise a simple set of rules for selecting releases that will work for any framework. This is the great weakness of the flexibility method, because it is therefore difficult to use it for programming a computer to solve the general case.

The strength of the method is that quite complicated structures can be solved with few unknowns. Thus for the truss bridge in Fig. 4.15, the determination of the release forces involves solving just two simultaneous equations. In contrast, the computer-oriented stiffness method to be discussed in Chapters 5 and 6, would use 16 unknowns for this problem. The selection of releases in plane frameworks at least, can be carried out by inspection by a competent engineer. Thus the flexibility method is very useful for hand analysis and particularly for preliminary design when complicated structures can often be approximated to simple structures with few redundancies.

4.3 FRAMEWORKS WITH MEMBERS IN BENDING

4.3.1 Tied Cantilever
As an introduction to the problem of solving statically indeterminate frameworks with members in bending, we consider the simple case of a tied cantilever.

Example 4.8
A tied cantilever is loaded by the single concentrated 10.0 kN force P at its centre as in Fig. 4.16(a). Both members are of steel. The pin-ended tie has $A_a = 500.0$ mm^2. The cantilever has $A_b = 4.0 \times 10^3$ mm^2 and $I_b = 30.0 \times 10^6$ mm^4. Since the cantilever without the tie is statically determinate, we can produce a primary structure by forming a release in the tie as shown in Fig. 4.16(b). As before, compatibility requires that the relative displacement u_1 at the release is zero, when the primary structure is loaded by P and the release

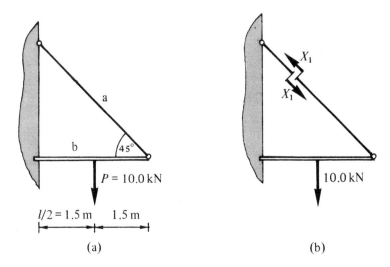

Fig. 4.16

force X_1. In this example, the cantilever is subject to both axial forces and bending. Thus the appropriate form of the virtual work equation is

$$\sum_i (P_i^* d_i) + X_1^* u_1 = \oint \left(N^* \left(\frac{N}{EA} + \alpha \Delta \theta \right) + \frac{M^* M}{EI} \right) \, dx \; .$$

(4.27)

In this present example we again ignore temperature changes.

If therefore N_P and M_P are the stress resultants in the primary structure due to P, and n_1 and m_1 are the stress resultants due to unit self-equilibrating forces on the cut surfaces at the release we have

$$u_{P1} = \oint \left(\frac{n_1 N_P}{EA} + \frac{m_1 M_P}{EI} \right) \, dx$$

(4.28)

and

$$f_{11} = \oint \left(\frac{n_1^2}{EA} + \frac{m_1^2}{EI} \right) \, dx \; .$$

(4.29)

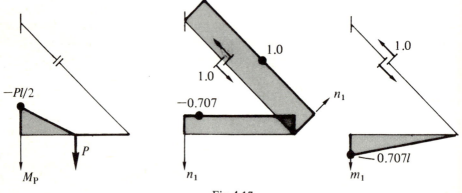

Fig. 4.17

The stress resultant diagrams calculated by considering equilibrium, are shown in Fig. 4.17. (Note that $N_P = 0$.) Thus carrying out the integrations we have

$$u_{P1} = (-0.0736 \, Pl^3 / EI_b) = -3.31 \times 10^{-3} \text{ m}$$

$$f_{11} = \left(\frac{1.414 l}{EA_a} \right) + \left(\frac{0.5 l}{EA_b} \right) + \left(\frac{0.1667 l^3}{EI_b} \right)$$

$$= 42.42 \times 10^{-9} + 1.875 \times 10^{-9} + 750.15 \times 10^{-9} =$$

$$= 794.45 \times 10^{-9} \text{ m/N} \; .$$

Thus

$$X_1 = \frac{-u_{P1}}{f_{11}} = 4.17 \, \text{kN} \;.$$

The final stress resultant diagrams are determined by summing N_P and $n_1 X_1$ and M_P and $m_1 X_1$. These are shown in Figs. 4.18(a) and (b).

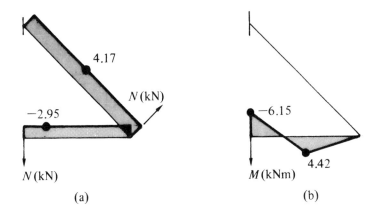

(a)   (b)

Fig. 4.18

4.3.2 Internal Moment Release

Hitherto we have considered structures for which the internal releases have been equivalent to cuts at points in the member. Other types of internal release are possible, and we consider next a release which is equivalent to a hinge at a point. This destroys the bending moment at the point, and is called an **internal moment release**.

Example 4.9
Suppose we wish to analyse the pin-footed portal frame shown in Fig. 4.19. All the members are uniform and have the same values for E and I. Unlike the framework in Example 2.18, this is not a statically determinate structure, because there now exists a horizontal restraint at the right-hand foot. However, we do know that a three-pinned portal frame is statically determinate, so one way of producing a primary structure in this case is to insert a horizontal hinge at any convenient point in the framework. For ease of analysis, we shall insert the hinge at joint 3, thus destroying the bending moment at the junction of b and c. The bending moments M_P due to the forces on the primary structure, have been calculated in Example 2.22. For convenience, they are reproduced in Fig. 4.20.

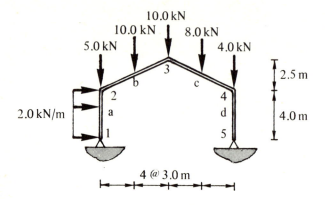

Fig. 4.19

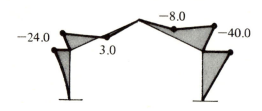

M_P (kNm)

Fig. 4.20

The introduction of the hinge at 3 destroys the compatibility of the original framework since the angle between members b and c can now change. The corresponding relative displacement u_1 at the release is therefore taken to be the relative rotation of the two members. Thus in Fig. 4.21(a), u_1 is equal to the sum of the rotations d_1 and d_2. Compatibility then requires that $u_1 = 0$. The release force system X_1 needed to produce compatibility now consists of two self-equilibrating couples acting on the ends of members b and c on either side of the hinge as in Fig. 4.21(b). It is then evident that the work done by X_1 in moving through the rotations is $X_1(d_1 + d_2) = X_1 u_1$, so that u_1 is again the 'corresponding' displacement to X_1 in the virtual work equation. All the previous equations then apply, so that if m_1 are the bending moments due to unit self-equilibrating couples acting on either side of the hinge, and we neglect axial forces,

$$u_{P1} = \oint \left(\frac{m_1 M_P}{EI} \right) \, dx \tag{4.30}$$

and

$$f_{11} = \oint \left(\frac{m_1^2}{EI} \right) \, dx \tag{4.31}$$

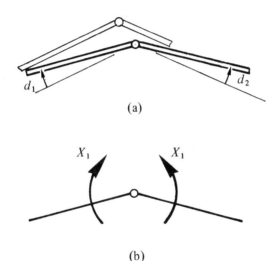

(a)

(b)

Fig. 4.21

The calculation of m_1 presents no difficulty, if we analyse the statically determinate three-pinned portal frame in the manner described in Section 2.7.3. Thus we first obtain the reactions R_1 to R_4 shown in the free-body diagram in Fig. 4.22(a) by satisfying the three global equilibrium equations, and one internal equilibrium equation for the structure to one side of the hinge. This leads to

$$\Sigma \rightarrow \quad R_1 + R_3 = 0$$

$$\Sigma \uparrow \quad R_2 + R_4 = 0$$

$$\Sigma_z \, \textcircled{1} \quad R_4 = 0$$

$$\Sigma_z \, \textcircled{3}_r \quad 1.0 + R_3 \times 6.5 + R_4 \times 6.0 = 0 \ .$$

Thus

$$\begin{bmatrix} R_1 \\ R_2 \\ R_3 \\ R_4 \end{bmatrix} = \begin{bmatrix} 0.154 \\ 0 \\ -0.154 \\ 0 \end{bmatrix}$$

and the corresponding bending moment diagram for m_1 is constructed without difficulty. This is shown in Fig. 4.22(b).

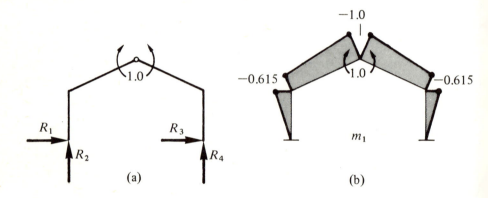

Fig. 4.22

The calculation of u_{P1} and f_{11} is carried out in Table 4.8, using the standard product integrals. Thus

$$u_{P1} = (133.02 \times 10^3/EI) \text{ rad}, \quad f_{11} = (9.651/EI) \text{ rad/Nm} .$$

Whence the release couples X_1 are given by

$$X_1 = \frac{-u_{P1}}{f_{11}} = \frac{-133.02 \times 10^3}{9.651} = -13.78 \text{ kNm} .$$

Note that when carrying out this analysis of a framework for its internal forces, it is not necessary to know the absolute values of E and I for the members. In this case, with E and I the same for all the members, the product (EI) cancels out of the compatibility equation, and need not be considered.

The final bending moments are found by summing M_P and $m_1 X_1$. This is accomplished in Table 4.8, by working out spot values and interpolating between them. The resulting bending moment diagram for the framework is shown in Fig. 4.23.

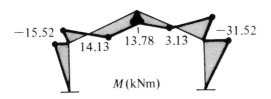

-15.52 14.13 13.78 3.13 -31.52

M (kNm)

Fig. 4.23

Table 4.8

Calculations for Example 4.9

member	a	b	c	d	Σ m
l	4.0	6.5	6.5	4.0	
M_P 10^3	-24.0 ... $+$ ... 4.0	-24.0 ... $+$ ... 15.0	-40.0 ... $+$ ... 12.0	-40.0	
m_1	-0.615	-1.00 ... -0.615	-1.00 ... -0.615	-0.615	
$\int m_1 M_P dx \times 10^3$	16.41	18.63	65.17	32.82	133.02
$\int m_1^2 dx$	0.505	4.321	4.321	0.505	9.651
$m_1 X_1$ 10^3	8.48	8.48 ... 13.78	13.78 ... 8.48	8.48	
M 10^3	-3.76 ... -15.52	-15.52 ... 14.13 13.78	-31.52 ... 13.78 3.13	-31.52	

4.3.3 External Moment Release

An **external moment release** destroys a couple reaction at a support. This can be illustrated by the problem of a rigid-jointed portal frame with encastered feet.

Example 4.10

Consider the portal frame shown in Fig. 4.24(a), which is the same as that analysed in the previous example except that the feet are fixed against rotation. A statically determinate primary structure in the form of the three-pinned portal frame, can be formed by introducing horizontal hinges at the crown and at the two feet, the latter being external moment releases. The releases are numbered as shown in Fig. 4.24(b). The release forces at the feet are the unknown external couples X_1 and X_3, and the corresponding relative displacements at the releases are the rotations u_1 and u_3 at the feet. At the crown, the release force X_2 and the relative rotation u_2 have the same meaning as in Example 4.9. Compatibility then requires that $\mathbf{u} = 0$.

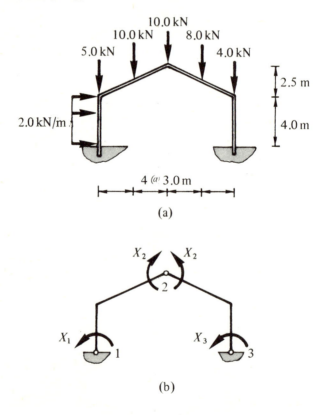

Fig. 4.24

The bending moments M_P due to the applied loading are as before. The bending moments m_1 and m_3 due to unit couples at the releases 1 and 3 are shown in Figs. 4.25(a) and (b). m_2, due to the unit couples at release 2, are the same as m_1 in Example 4.8. Then in matrix notation

$$\mathbf{u}_P = \oint \left(\frac{\mathbf{m} M_P}{EI} \right) dx \tag{4.32}$$

$$f_{ij} = \oint \left(\frac{m_i m_j}{EI} \right) dx \tag{4.33}$$

and

$$\mathbf{f} = \oint \left(\frac{\mathbf{m} \mathbf{m}^T}{EI} \right) dx \tag{4.34}$$

where $\mathbf{m}$ is the column matrix of m_1, m_2 and m_3.

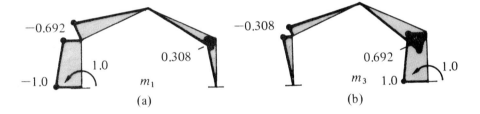

Fig. 4.25

The calculations are carried out in Table 4.9, which gives

$$\mathbf{u}_P = \begin{bmatrix} 11.18 \\ 133.02 \\ -93.38 \end{bmatrix} \times \frac{10^3}{EI} \text{ rad}, \quad \mathbf{f} = \begin{bmatrix} 4.27 & -1.66 & 1.90 \\ -1.66 & 9.65 & 1.66 \\ 1.90 & 1.66 & 4.27 \end{bmatrix} \times \frac{1}{EI} \text{ rad/Nm} \ .$$

Whence the release couples $\mathbf{X}$ are given by

$$\mathbf{X} = -\mathbf{f}^{-1} \mathbf{u}_P = \begin{bmatrix} -9.74 \\ -8.16 \\ 23.07 \end{bmatrix} \text{ kNm} \ .$$

Table 4.9

Calculations for Example 4.10

Member	a	b	c	d	Σ m
l	4.0	6.5	6.5	4.0	
$M_P\ 10^3$	−24.0 + 4.0	−24.0 + 15.0	−40.0 + 12.0	−40.0	
m_1	−1.0 −0.692	−0.692	0.308	0.308	
m_2	−0.615	−1.0 −0.615	−1.0 −0.615	−0.615	
m_3	−0.308	−0.308	0.692	0.692 1.0	
$\int m_1 M_P\,dx\,10^3$	29.13	19.13	−20.67	−16.41	11.18
$\int m_2 M_P\,dx\,10^3$	16.41	18.63	65.17	32.82	133.02
$\int m_3 M_P\,dx\,10^3$	8.21	8.50	−46.50	−63.59	−93.38
$\int m_1 m_1\,dx$	2.90	1.04	0.21	0.13	4.27
$\int m_1 m_2\,dx$	−0.98	−1.67	0.74	0.25	−1.66
$\int m_1 m_3\,dx$	0.49	0.46	0.46	0.49	1.90
$\int m_2 m_2\,dx$	0.50	4.32	4.32	0.50	9.65
$\int m_2 m_3\,dx$	−0.25	−0.74	1.67	0.98	1.66
$\int m_3 m_3\,dx$	0.13	0.21	1.04	2.90	4.27
$m_1 X_1\ 10^3$	9.74 6.74	6.74	−3.0	−3.0	
$m_2 X_2\ 10^3$	5.02	5.02 8.16	8.16 5.02	5.02	
$m_3 X_3\ 10^3$	−7.11	−7.11	15.96	15.96 23.07	
$M\ 10^3$	−19.35 −0.81 9.74	−19.35 9.41 8.16	−22.02 8.16 5.07	−22.02 23.07	

The final bending moment diagram is found by summing M_P and $m_1 X_1$, $m_2 X_2$ and $m_3 X_3$. Thus

$$M = M_P + m^T X .$$ (4.35)

The summation is carried out in the table, and the bending moment diagram is shown in Fig. 4.26.

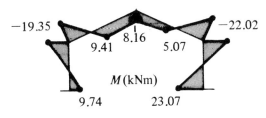

Fig. 4.26

4.4 CONTINUOUS BEAM

A frequent problem in structural design offices is the analysis of multispan continuous beams. This is because such beams are often considered as an approximation for multispan bridges. Similarly beams in building frames are often analysed as separate continuous beams, simply supported at the columns. Continuous beams are solved by the introduction of internal moment releases at the supports as in the following example.

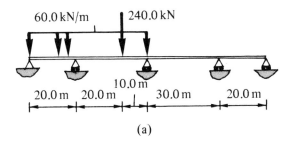

(a)

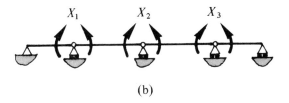

(b)

Fig. 4.27

Example 4.11
A 4-span bridge is idealised as the continuous beam, shown in Fig. 4.27(a). The loading on the bridge is chosen as typical of traffic live loading, being made up of a uniformly distributed load and a concentrated load as shown. We shall assume that the bridge is of uniform material properties, and that I is constant throughout. The statically determinate primary structure is formed by introducing hinges into the beam above the three internal supports, as in Fig. 4.27(b). Again, these are internal releases, destroying the bending moments at the support positions. The primary structure is then composed of four simply supported beams, and the release forces are the couples X_1, X_2 and X_3 acting on the ends of the beams at the three releases as shown. The corresponding relative release displacements **u** are the relative rotations of the ends of the beams.

The bending moments in the primary structure are easily calculated as the bending moments in the separate beams. Thus M_P due to the live loading is as shown in Fig. 4.28(a). Similarly the bending moments **m**, due to unit self-equilibrating couples at the releases are as shown in Figs. 4.28(b), (c) and (d).

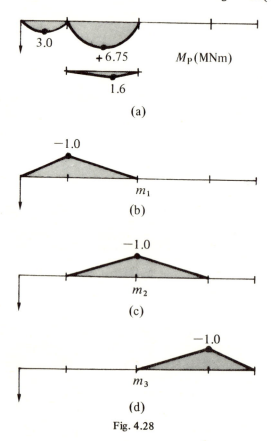

M_P(MNm)

(a)

(b)

(c)

(d)

Fig. 4.28

u_P and f, again given by (4.32) and (4.34) are calculated in Table 4.10. Thus

$$
u_P = \begin{bmatrix} -98.17 \\ -80.83 \\ 0 \end{bmatrix} \times \frac{10^6}{EI} \text{ rad}, \quad f = \begin{bmatrix} 16.67 & 5.0 & 0 \\ 5.0 & 20.0 & 5.0 \\ 0 & 5.0 & 16.67 \end{bmatrix} \times \frac{1}{EI} \text{ rad/Nm}
$$

Therefore

$$
X = -f^{-1}u_P = \begin{bmatrix} 4.98 \\ 3.02 \\ -0.91 \end{bmatrix} \times 10^6 \text{ Nm }.
$$

The final bending moments given by (4.35) are calculated at the end of the table. They are shown in Fig. 4.29.

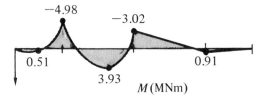

Fig. 4.29

The multispan bridge is frequently designed with members of varying section properties as in Fig. 4.30. In this case, the product integrals in (4.32) and (4.34) are evaluated numerically, in the manner described in Chapter 3, Section 3.9.

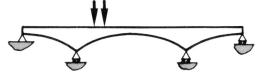

Fig. 4.30

Examination of Table 4.10 reveals a pattern in the calculations that makes the solution of a continuous beam a simple programming exercise. In fact, for this particular structural type, especially with members of varying section properties, the flexibility method has some advantage over the stiffness method for automatic computation. However, for conciseness, the computer programming of the flexibility method is not discussed in this text.

Table 4.10

Calculations for Example 4.11

Member	a	b	c	d	Σ m
l	20.0	30.0	30.0	20.0	
$M_P \ 10^6$	3.0	+ 6.75 / 1.6	0	0	
m_1	−1.0	−1.0	0	0	
m_2	0	−1.0	−1.0	0	
m_3	0	0	−1.0	−1.0	
$\int m_1 M_P dx\, 10^6$	−20.0	−78.17	0	0	−98.17
$\int m_2 M_P dx\, 10^6$	0	−80.83	0	0	−80.83
$\int m_3 M_P dx\, 10^6$	0	0	0	0	0
$\int m_1 m_1 dx$	6.67	10.0	0	0	16.67
$\int m_1 m_2 dx$	0	5.0	0	0	5.0
$\int m_1 m_3 dx$	0	0	0	0	0
$\int m_2 m_2 dx$	0	10.0	10.0	0	20.0
$\int m_2 m_3 dx$	0	0	5.0	0	5.0
$\int m_3 m_3 dx$	0	0	10.0	6.67	16.67
$m_1 X_1 \ 10^6$	−4.98	−4.98			
$m_2 X_2 \ 10^6$		−3.02	−3.02		
$m_3 X_3 \ 10^6$			0.91	0.91	
$M \ 10^6$	−4.98 / 0.51	−4.98 / −3.02 / 3.93	−3.02 / 0.91	0.91	

4.5 RELEASES

Up to now we have considered a limited number of release types suitable for forming primary structures in particular cases. In this section, we shall consider the complete range of possible release types.

4.5.1 Internal Release

We have defined an internal release as that which destroys a single stress resultant at a point in a framework. In a rigid-jointed space framework there are six stress resultants at any point and each can be destroyed separately by inserting a suitable hypothetical mechanism. Thus a hinge parallel to the member z_m direction destroys M_z, but is in theory capable of sustaining the axial force, the bending moment about the y_m axis, the two shear forces and the torque. Table 4.11 presents simplified diagrams of mechanisms for destroying other stress

Table 4.11

Types of release

released resultant	mechanism	release forces
N		
S_y		
M		
T		

resultants. The corresponding release forces needed to maintain compatibility are also shown. Note that the releases in the pin-jointed frameworks represented up to now by cuts, should strictly take the form of the axial force mechanism shown in the table.

4.5.2 External Release

An external release is that which destroys a single external reaction at a supported point of a framework. The reaction can either be a direct force or a couple, and the corresponding release force takes the place of the reaction.

4.5.3 Multiple Release

The problems discussed in the previous sections have been analysed by using one or more internal or external releases, each at separate points of the frame-work. However, it is possible to make several releases at a single point, and the most obvious example of this is to make a complete cut in the member at that point. Thus a cut in a member in a rigid-jointed plane framework simultaneously destroys N, S and M and is equivalent to three releases. Consider the cut in the beam member in Fig. 4.31. The release forces X_1, X_2 and X_3, comprising the two pairs of self-equilibrating direct forces and the pair of self-equilibrating couples shown in the expanded view, are then necessary to maintain compatibility at the cut. The three relative displacements of the cut surfaces are the horizontal relative displacement u_1, the vertical relative displacement u_2 and the relative rotation, u_3.

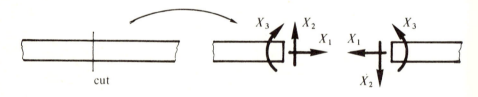

Fig. 4.31

Example 4.12
Let us consider again the problem of a fixed-footed portal frame discussed in Example 4.10, and shown in Fig. 4.24(a). This time we produce a primary structure by making a complete cut at the crown, which produces two separate statically determinate cantilevers, which are shown in expanded view in Fig. 4.32. The three release forces **X** are as shown.

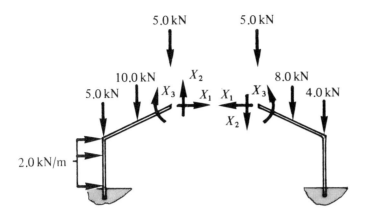

Fig. 4.32

The bending moments M_P due to the actual loading are shown in Fig.
4.33(a), where for symmetry the 10.0 kN force at the crown has been divided
into two 5.0 kN forces acting on the two cantilevers. This device is not essential,
since it would be equally satisfactory to take all the force to act on one or other
of the cantilevers. The bending moments **m** due to self-equilibrating unit forces
acting at each release in turn are given in Fig. 4.33(b), (c), and (d). $\mathbf{u}_P$ and **f**,
given by (4.32) and (4.34) are then calculated in Table 4.12. Thus

$$
\mathbf{u}_P = \begin{bmatrix} 2677.1 \\ -320.8 \\ 760.1 \end{bmatrix} \times \frac{10^3}{EI} \begin{matrix} \text{m} \\ \text{m} \\ \text{rad} \end{matrix} , \quad
\mathbf{f} = \begin{bmatrix} 199.75 & 0 & 52.25 \\ 0 & 444.0 & 0 \\ 52.25 & 0 & 21.0 \end{bmatrix} \times \frac{1}{EI} \begin{matrix} \text{m/N,} \\ \text{rad/Nm} \\ \text{etc.} \end{matrix}
$$

Therefore

$$
\mathbf{X} = -\mathbf{f}^{-1}\mathbf{u}_P = \begin{bmatrix} -11.27 \\ 0.72 \\ -8.16 \end{bmatrix} \begin{matrix} \text{kN} \\ \text{kN} \\ \text{kNm} \end{matrix} .
$$

The final bending moments calculated from (4.35) as in the table, come out
to be the same as those given in the solution of Example 4.10.

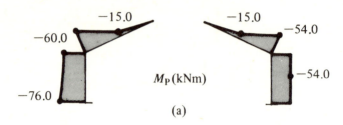

M_P(kNm)

(a)

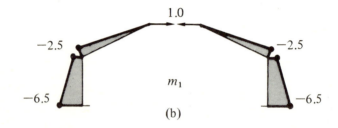

m_1

(b)

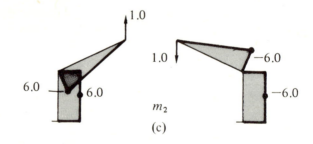

m_2

(c)

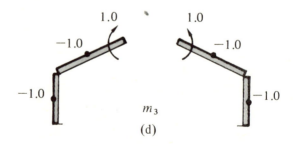

m_3

(d)

Fig. 4.33

Table 4.12
Calculations for Example 4.12

Member	a	b	c	d	Σ m
l	4.0	6.5	6.5	4.0	
M_P 10^3	−76.0 −60.0 4.0	−60.0 15.0	−54.0 12.0	−54.0	
m_1	−6.5 −2.5	−2.5	−2.5	−2.5 −6.5	
m_2	6.0	6.0	−6.0	−6.0	
m_3	−1.0	−1.0	−1.0	−1.0	
$\int m_1 M_P \mathrm{d}x 10^3$	1197.3	264.1	243.8	972.0	2677.1
$\int m_2 M_P \mathrm{d}x 10^3$	−1568.0	−633.8	585.0	1296.0	−320.8
$\int m_3 M_P \mathrm{d}x 10^3$	261.3	146.3	136.5	216.0	760.1
$\int m_1 m_1 \mathrm{d}x$	86.33	13.54	13.54	86.33	199.75
$\int m_1 m_2 \mathrm{d}x$	−108.0	−32.5	32.5	108.0	0
$\int m_1 m_3 \mathrm{d}x$	18.0	8.13	8.13	18.0	52.25
$\int m_2 m_2 \mathrm{d}x$	144.0	78.0	78.0	144.0	444.0
$\int m_2 m_3 \mathrm{d}x$	−24.0	−19.5	19.5	24.0	0
$\int m_3 m_3 \mathrm{d}x$	4.0	6.5	6.5	4.0	21.0
$m_1 X_1$ 10^3	28.18 73.26	28.18	28.18	28.18 73.26	
$m_2 X_2$ 10^3	4.32	4.32	−4.32	−4.32	
$m_3 X_3$ 10^3	8.16	8.16	8.16	8.16	
M 10^3	−19.34 −0.8 9.74	−19.34 9.41 8.16	−21.98 8.16 5.09	−21.98 23.10	

Example 4.12 illustrates the use of an internal multiple release. An external multiple release destroys several reactions simultaneously at one support. Thus the portal frame in Fig. 4.24(a) can be made statically determinate by making a complete cut at the right-hand foot. We then have a primary structure comprising a single cranked cantilever as shown in Fig. 4.34. The release forces X shown in the figure take the place of the three reactions at the support. u_P and f are again given by (4.32) and (4.34) where m now correspond to unit external forces at the support.

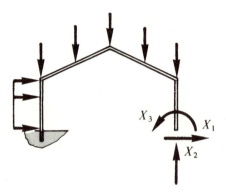

Fig. 4.34

4.6 DEGREE OF REDUNDANCY AND SELECTION OF RELEASES

4.6.1 Degree of Redundancy
We have seen in Examples 4.10 and 4.12 that the number of releases required to form a primary structure for the fixed-footed portal frame, is the same for both solutions. This number of releases is unique for any structure and is called its **degree of redundancy**. Simple rules for determining the degree of redundancy of pin-jointed frameworks have been presented in Section 4.2.6. Here we consider a method for determining the degree of redundancy of any framework.

The method is based on the concept of the **cantilever**, being a structure that is singly-connected and supported at a single fully encastered support. By singly-connected, we mean that the structure does not contain closed rings of members. Thus the cantilever is also very simply described as a structural 'tree'. An example is shown in Fig. 4.35(a). We have seen in Section 2.6.5 that a cantilever is statically determinate, because we can make a cut at any point and find the stress resultants by considering the equilibrium of that part of the structure remote from the root as in Fig. 4.35(b).

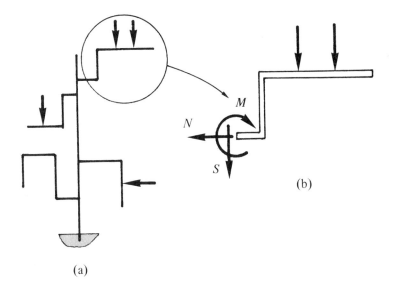

(b)

(a)

Fig. 4.35

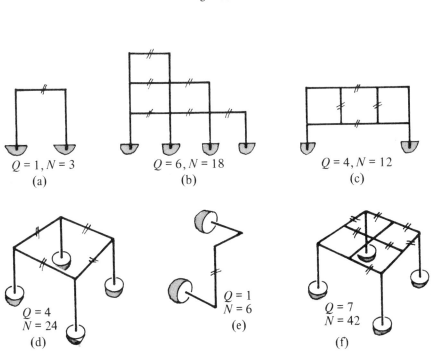

$Q = 1, N = 3$
(a)

$Q = 6, N = 18$
(b)

$Q = 4, N = 12$
(c)

$Q = 4$
$N = 24$
(d)

$Q = 1$
$N = 6$
(e)

$Q = 7$
$N = 42$
(f)

Fig. 4.36

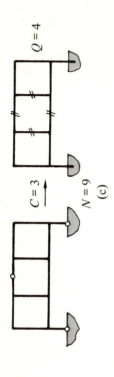

$Q = 4$

$C = 3$

$N = 9$

(c)

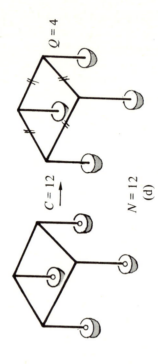

$Q = 4$

$C = 12$

$N = 12$

(d)

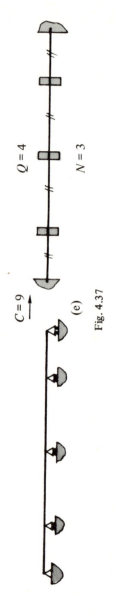

$Q = 4$

$C = 9$

$N = 3$

(e)

Fig. 4.37

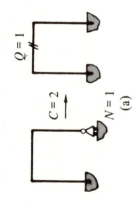

$Q = 1$

$C = 2$

$N = 1$

(a)

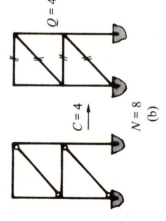

$Q = 4$

$C = 4$

$N = 8$

(b)

Consider first a completely rigid-jointed framework, fully encastered at its supports. It is always possible to reduce such a structure to a series of cantilevers, by making a certain number of cuts in the members, Q say. Each cut represents a multiple release, and in a plane framework, the corresponding total number of releases N is $3Q$, and in a space framework $6Q$. N then is the required degree of redundancy. Examples of this procedure applied to various plane and space frameworks are shown in Fig. 4.36. The selection of cuts can be done by inspection, although it often requires careful thought in the case of space frameworks.

Next consider a framework containing internal hinges or pins, and with partially fixed supports. We can produce a completely rigid-jointed framework by introducing a certain number of **constraints** at the hinges and supports, C say. The definition of a constraint is exactly opposite to the definition of a release, being that which creates a stress resultant at a point in a framework, or that which creates a reaction at a support. If we then produce a set of cantilevers from the rigid-jointed framework using Q cuts as above, the total number of releases required for the rigid-jointed framework is either $3Q$ or $6Q$. We therefore deduce that the number of releases N required for the actual framework is given by

$$N = 3Q - C \qquad (4.36)$$

for plane frameworks, or

$$N = 6Q - C \qquad (4.37)$$

for space frameworks. Examples of the use of this formula for various plane and space frameworks are shown in Fig. 4.37. In Fig. 4.37(e) we determine the degree of redundancy of a continuous beam from (4.36). However, it is clear that in this, as in many cases, the number of releases can be determined more easily by direct inspection.

4.6.2 Selection of Releases

The process of selecting releases to produce primary structures is difficult to generalise and therefore difficult to program. The selection of a particular set of releases also influences the efficiency of the solution. This latter point can be illustrated by the continuous beam example. Thus we can select releases comprising hinges at the supports as in Example 4.11. We then see in Table 4.10 that the bending moments $\mathbf{m}$, are localised to the spans adjacent to the releases, and this leads to a strongly banded flexibility matrix $\mathbf{f}$, since $f_{ij} = 0$ when $|i - j| > 1$. A bridge with many spans would have a flexibility matrix populated as follows:

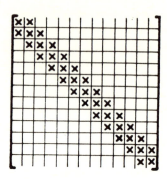

It will be shown in Chapter 6 that this strong banding can be utilised in numerical solving routines, reducing the required computer storage and solving time. If, however, we select releases for the continuous beam, in the form of three external releases destroying the support reactions as shown in Fig. 4.38, the primary structure is composed of a single simply supported beam. The bending

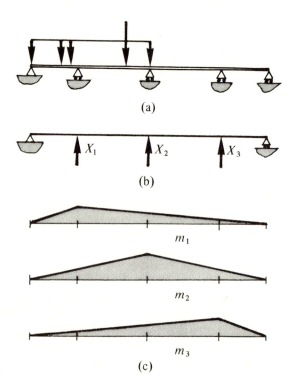

(a)

(b)

(c)

Fig. 4.38

moments **m**, corresponding to unit vertical forces at the releases, are then distributed along the beam as shown and all the coefficients f_{ij} take non-zero values. We therefore lose the banded property of the matrix. Analysing an unbanded matrix is not only expensive in storage and time, it can also be error prone, for since the matrix is less well conditioned, numerical procedures tend to be unstable owing to rounding errors. This alternative set of releases is therefore inefficient. A similar problem arises in the analysis of building frames. If the single bay multistorey portal frame in Fig. 4.39(a) is analysed by making cuts in all the beams to produce two cantilevers, the unit stress resultant diagrams are not localised. The two representative diagrams corresponding to m_2 and m_{11}, shown in Figs. 4.39(b) and (c), indicate that the bending moments due to unit forces in the upper storeys interact with the bending moments due to those in the lower storeys, and that the flexibility matrix is therefore fully populated. If, however, the cuts are made at each storey in one of the columns, producing a single complicated cantilever as in Fig. 4.39(d), the unit bending moment

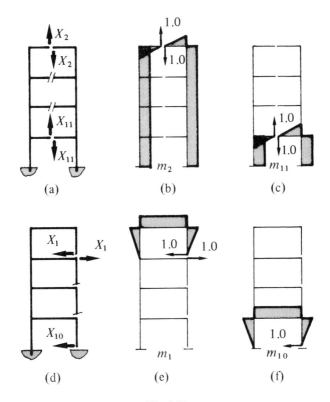

Fig. 4.39

diagrams shown in Figs. 4.39(e) and (f) are localised and the flexibility matrix is strongly banded. Extending this latter procedure, an appropriate release selection for a multistorey, multibay building frame is shown in Fig. 4.40.

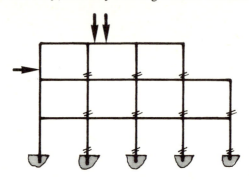

Fig. 4.40

A numerical procedure called the Rank Force Method has been proposed by Robinson [4.1] for selecting optimal release systems automatically. The procedure, however, is time consuming and not competitive with the stiffness method to be discussed in the next two chapters. Computer programs can be devised for special structures such as continuous beams or building frames. However, in this text, the flexibility method is presented essentially as a useful method for solving by hand, fairly complex structures having few degrees of redundancy. This is illustrated in the case of an important structural form – the tied arch, which we shall discuss in the next section.

4.7 CURVED TIED ARCH

In Section 3.10, we determined the deflections of a pin-footed curved arch bridge, with one foot free to move horizontally. These deflections, of the order of metres, would be unacceptable in a practical structure. One way to limit the deflections is to restrain the horizontal movement of the foot by attaching it to a pinned support mounted on a rigid foundation. However, in the case of an arch bridge forming part of a viaduct as in Fig. 4.41, such a rigid foundation is very difficult to produce practically. The alternative way of restraining the foot is to introduce a pin-ended tie between the two feet of the arch at deck level. Such a tie might, for example, take the form of high tensile steel cables. The tie is an extra member inserted into the framework making the structure statically indeterminate. The degree of redundancy is equal to one, since a primary structure can obviously be formed by making a cut in the tie.

Fig. 4.41

Example 4.13
A tied arch is shown in Fig. 4.42(a). It is subject to the same loading as the
arches considered in Examples 2.19 and 3.9 and the properties of the arch are
$E = 30.0$ GN/m^2 and $I = 1.34$ m^4. We shall assume that the tie is composed of
eight, 25 mm diameter steel cables, giving a total area $A = 3.926 \times 10^{-3}$ m^2. The
primary structure is shown in Fig. 4.42(b). The equations for u_{P1} and f_{11} are the
same as (4.28) and (4.29) which we repeat here for convenience. Thus

$$u_{P1} = \oint \left(\frac{n_1 N_P}{EA} + \frac{m_1 M_P}{EI} \right) dx \qquad (4.28)*$$

$$f_{11} = \oint \left(\frac{n_1^2}{EA} + \frac{m_1^2}{EI} \right) dx \ . \qquad (4.29)*$$

In the arch, bending actions predominate, while in the tie, only axial forces exist.
Thus in the above equations, only the one appropriate term will be calculated
for each member.

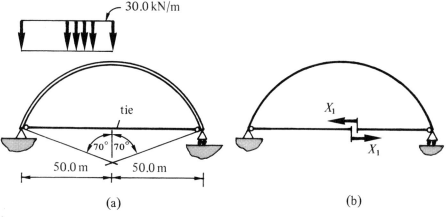

(a)   (b)

Fig. 4.42

The bending moments M_P in the arch, caused by the loading are the same as those calculated in Example 2.19. In terms of the angular coordinates θ and ψ in Figs. 2.40(c) and (d), these are

$$M_P(\theta) = (18.75 + 19.95 \sin (\theta) - 42.47 \sin^2 (\theta)) \text{ MNm}$$

$$M_P(\psi) = (18.75 - 19.95 \sin (\psi)) \text{ MNm} \quad .$$

The axial force N_P in the tie is zero. The bending moments m_1 in the arch due to the unit self-equilibrating release forces are of the same form but of opposite sign to those used for calculating d_2 in Example 3.9. Thus

$$m_1(\theta) = -(53.21 \cos (\theta) - 18.20)$$

$$m_1(\psi) = -(53.21 \cos (\psi) - 18.20) \quad .$$

The axial force n_1 in the tie is 1.0.

As in Example 3.9, the integrations in (4.28) and (4.29) are taken round the curve of the arch, using $dx = r d\theta$ or $r d\psi$ as appropriate. $r = 53.21$ m and the integration limits are 0 and 1.222 rads. Thus

$$u_{P1} = \frac{1}{EI} \underset{\substack{\uparrow \\ \text{arch}}}{} \int_0^{1.222} \Bigl(-(53.21 \cos (\theta) - 18.20)(18.75 + 19.95 \sin (\theta) -$$

$$- 42.47 \sin^2 (\theta)) \times 10^6 \times 53.21 \Bigr) d\theta +$$

$$+ \frac{1}{EI} \underset{\substack{\uparrow \\ \text{arch}}}{} \int_0^{1.222} \Bigl(-(53.21 \cos (\psi) - 18.20)(18.75 - 19.95 \sin (\psi)) \times$$

$$\times 10^6 \times 53.21 \Bigr) d\psi = \frac{-2.519 \times 10^{10} - 1.547 \times 10^{10}}{30.0 \times 10^9 \times 1.34} = -1.01 \text{ m}$$

and

$$f_{11} = \frac{1}{EA} \underset{\substack{\uparrow \\ \text{tie}}}{} \int_0^{100.0} 1.0^2 \, dx + \frac{2}{EI} \underset{\substack{\uparrow \\ \text{arch}}}{} \int_0^{1.222} (53.21 \cos (\theta) - 18.2)^2 \times 53.21 \, d\theta$$

$$= \frac{100.0}{200.0 \times 10^9 \times 3.926 \times 10^{-3}} + \frac{2 \times 40.93 \times 10^3}{30.0 \times 10^9 \times 1.34}$$

$$= 2.164 \times 10^{-6} \text{ m/N} \quad .$$

Therefore

$$X_1 = \frac{-u_{P1}}{f_{11}} = 467.4 \text{ kN} \; .$$

The corresponding bending moments in the arch are calculated from the above analytical expressions, using

$$M = M_P + m_1 X_1 \tag{4.38}$$

and are shown in Fig. 4.43. Compared with the solution for the arch with one foot free to move horizontally, shown in Fig. 2.41(c), it is clear that the structural action of the arch has altered, and that restraining the feet has considerably improved the bending moment distribution by reducing the large sagging moments. The structure has developed its so-called **arching action**.

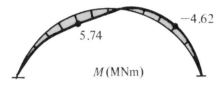

Fig. 4.43

4.8 SPACE FRAMEWORKS

As a final example of the flexibility method, we illustrate its use in the analysis of a space framework. We shall consider an encastered framework derived from the cranked cantilever analysed in Examples 2.20 and 3.10.

Example 4.14

The encastered framework shown in Fig. 4.44(a) has the same section properties as the cantilever in Examples 2.20 and 3.10. Thus it is of steel and of uniform section with $I_z = 30.0 \times 10^6 \text{ mm}^4$, $I_y = 15.0 \times 10^6 \text{ mm}^4$, and $J = 20.0 \times 10^6 \text{ mm}^4$. The joint and member notation and local coordinate systems are the same as in Fig. 2.43(a).

A primary structure can be produced by making a complete cut anywhere in the framework, and in this case, in order to be able to use previous results, we shall cut the framework at support 1, thereby producing six external releases. The corresponding release forces are the three direct support reactions X_1, X_2, X_3 in the global coordinate directions, and the three couple reactions X_4, X_5 and X_6 about axes parallel to these directions, shown in Fig. 4.44(b). As in the

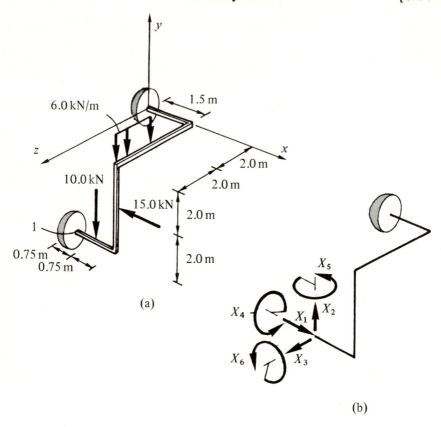

Fig. 4.44

previous analyses of the cranked cantilever we shall consider only bending and torsion in solving the problem. Thus the equations for $\mathbf{u}_P$ and $\mathbf{f}$ take the form

$$\mathbf{u}_P = \oint \left(\frac{\mathbf{m}_y M_{Py}}{EI_y} + \frac{\mathbf{m}_z M_{Pz}}{EI_z} + \frac{\mathbf{t}\, T_P}{GJ} \right) \, \mathrm{d}x \qquad (4.39)$$

$$f_{ij} = \oint \left(\frac{m_{yi} m_{yj}}{EI_y} + \frac{m_{zi} m_{zj}}{EI_z} + \frac{t_i t_j}{GJ} \right) \, \mathrm{d}x \qquad (4.40)$$

and

$$\mathbf{f} = \oint \left(\frac{\mathbf{m}_y \mathbf{m}_y^{\mathrm{T}}}{EI_y} + \frac{\mathbf{m}_z \mathbf{m}_z^{\mathrm{T}}}{EI_z} + \frac{\mathbf{t}\mathbf{t}^{\mathrm{T}}}{GJ} \right) \, \mathrm{d}x \qquad (4.41)$$

where M_{Py}, M_{Pz} and T_P are the stress resultants due to the loading on the primary structure. These have been calculated in Example 2.20. $\mathbf{m}_y$, $\mathbf{m}_z$ and $\mathbf{t}$ are the matrices of stress resultants corresponding to unit external forces or couples acting in the six release directions in turn. Three examples of these stress resultants, m_{y1}, m_{z3} and t_6 are shown in Fig. 4.45(a), (b) and (c).

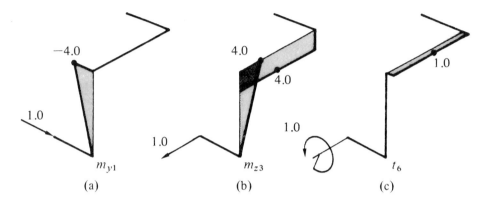

Fig. 4.45

It is clear that the evaluation of $\mathbf{u}_P$ and $\mathbf{f}$ is extremely long and laborious and the details will not be presented here. However, the results are as follows:

$$\mathbf{u}_P = \begin{bmatrix} -0.513 \\ -0.266 \\ -0.470 \\ 0.097 \\ -0.070 \\ -0.068 \end{bmatrix} \begin{matrix} m \\ m \\ m \\ rad \\ rad \\ rad \end{matrix} \, ,$$

$$\mathbf{f} = \begin{bmatrix} 67.82 & -20.35 & 5.50 & 0 & 4.67 & 14.07 \\ -20.35 & 28.38 & 20.93 & -5.23 & 0 & -6.28 \\ 5.50 & 20.93 & 39.42 & -7.90 & 6.65 & 0 \\ 0 & -5.23 & -7.90 & 3.28 & 0 & 0 \\ 4.67 & 0 & 6.65 & 0 & 4.93 & 0 \\ 14.07 & -6.28 & 0 & 0 & 0 & 4.43 \end{bmatrix} \times 10^{-6} \begin{matrix} m/N \\ rad/Nm \\ etc. \end{matrix} .$$

Thus

$$\mathbf{X} = -\mathbf{f}^{-1}\mathbf{u}_P = \begin{bmatrix} 12.76 \\ 20.09 \\ -1.32 \\ -0.63 \\ 3.90 \\ 3.25 \end{bmatrix} \begin{array}{l} \text{kN} \\ \text{kN} \\ \text{kN} \\ \text{kNm} \\ \text{kNm} \\ \text{kNm} \end{array}$$

The final stress resultants can then again be assembled from equations of the form

$$M_y = M_{Py} + \mathbf{m}_y^T\mathbf{X}, \quad M_z = M_{Pz} + \mathbf{m}_z^T\mathbf{X}, \quad T = T_P + \mathbf{t}^T\mathbf{X} \quad .(4.42\text{a,b,c})$$

Examples 4.13 and 4.14 demonstrate the strengths and weaknesses of the flexibility method. Thus whereas Example 4.13 shows how useful the method can be for structures with few redundancies, Example 4.14 emphasises the need for a computer-oriented method for solving most practical frameworks. Such a method will be discussed in the next two chapters.

4.9 MATRIX FORMULATION OF FLEXIBILITY METHOD

We have already used matrix notation in the preceding sections for frameworks where a restricted number of stress resultants have been considered. If in the general case, we wish to consider all the stress resultants, then as in Section 3.14, an elegant formulation of the flexibility method can be obtained by introducing the matrices $\mathbf{S}, \mathbf{s}, \boldsymbol{\phi}, \mathbf{n}, \mathbf{r}$ and $\mathbf{d}_s$, defined in that section. Thus

$$\mathbf{u}_P = \oint (\mathbf{s}\boldsymbol{\phi}\mathbf{S}_P + \mathbf{n}\alpha\Delta\theta)\mathrm{d}x - \mathbf{r}\mathbf{d}_s \qquad (4.43)$$

$$\mathbf{f} = \oint (\mathbf{s}\boldsymbol{\phi}\mathbf{s}^T)\mathrm{d}x \quad . \qquad (4.44)$$

The solution is then given by (4.23) as

$$X = -\mathbf{f}^{-1}\mathbf{u}_P \qquad (4.23)*$$

and the final stress resultants by

$$\mathbf{S} = \mathbf{S}_P + \mathbf{s}^T\mathbf{X} \quad . \qquad (4.45)$$

In the above we have included the effects of support settlements $\mathbf{d}_s$. $\mathbf{r}$ refers to the matrix of reactions at the displaced supports due to unit self-equilibrating forces across each release in turn. They are calculated by considering equilibrium at the supports in the same way as in Example 3.11.

4.10 DISPLACEMENTS OF STATICALLY INDETERMINATE FRAMEWORKS

We have devoted the previous chapter to a discussion of the virtual work method for finding the displacements of statically determinate frameworks. Having in this chapter solved statically indeterminate frameworks by the flexibility method, it now remains to consider how the displacements of these structures are found.

Firstly, we note that in the derivation of the virtual work equation for a framework, no reference is made as to whether or not the framework is statically determinate. Thus the general formula for the displacements (3.74) given in Section 3.14 still applies, and

$$\mathbf{d} = \oint (\mathbf{s}^* \boldsymbol{\phi} \mathbf{S} + \mathbf{n}^* \alpha \Delta\theta) dx - \mathbf{r}^* \mathbf{d_s} \ . \tag{4.46}$$

$\mathbf{s}^*$, $\mathbf{n}^*$ and $\mathbf{r}^*$ are *any* set of stress resultants and reactions which are in equilibrium with unit forces corresponding to the required displacements $\mathbf{d}$. In this section, the asterisk has been introduced to distinguish these stress resultants from those due to the unit release forces in the primary structure. $\mathbf{S}$ are the stress resultants due to the loading, calculated by the flexibility method. The use of the word 'any' in the description of $\mathbf{s}^*$, etc., requires explanation. In the case of a statically determinate framework, there is only one set of stress resultants in equilibrium with the unit forces. However, in the case of a statically indeterminate framework, there are usually many possibilities. Equilibrium force systems can be found by isolating any convenient statically determinate **substructure** from the framework and finding the stress resultants in this substructure using the methods of Chapter 2. These stress resultants must by definition, be in equilibrium with the unit forces. It is important to note that it is not necessary to find the *actual* stress resultants that would occur in the complete framework due to the unit forces, and we therefore avoid having to reanalyse a statically indeterminate structure. The following examples illustrate the method.

Example 4.15

Consider the pin-jointed framework solved in Example 4.6. For convenience, we reproduce the problem in Fig. 4.46 with the calculated stress resultants N as shown. Suppose say we wish to calculate the vertical deflection d_1 of joint 3. Three possible substructures are shown in Fig. 4.47(a), (b), and (c), together with the corresponding equilibrium force systems $n_1^*(i)$, $(i = 1, 2, 3)$. d_1 is given by the usual simplified equation for pin-jointed frameworks (3.16) as

$$d_1 = \sum_m \{n_1^* \Delta\}_m \ . \tag{3.16}*$$

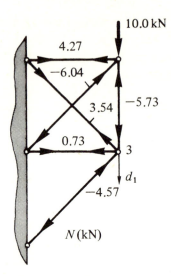

Fig. 4.46

It is calculated for the three alternative force systems in Table 4.13. We thus confirm, that despite the fact that we have used three apparently totally unrelated equilibrium force systems, the value for d_1 (= 0.197 mm) comes out to be the same in each case. A physical explanation for this is simply that the displacements of the three substructures in Fig. 4.47 have to be compatible.

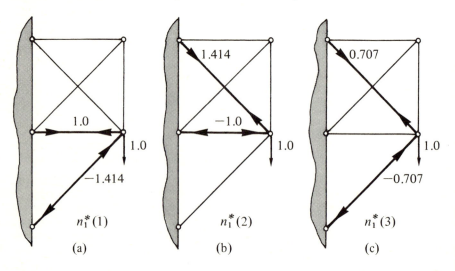

Fig. 4.47

Table 4.13

Calculations for Example 4.15

Member		a	b	c	d	e	f	Σ m
l		3.0	3.0	4.24	3.0	4.24	4.24	
A	10^{-3}	0.5	0.5	0.75	0.75	0.5	0.75	
N	10^3	4.27	−5.73	−6.04	0.73	3.54	−4.57	
Δ	10^{-3}	0.128	−0.172	−0.171	0.015	0.150	−0.130	
$n_1(1)$		0	0	0	1.00	0	−1.414	
$n_1\Delta$	10^{-3}	0	0	0	0.015	0	0.183	$0.197 (= d_1)$
$n_1(2)$		0	0	0	−1.00	1.414	0	
$n_1\Delta$	10^{-3}	0	0	0	−0.015	0.212	0	$0.197 (= d_1)$
$n_1(3)$		0	0	0	0	0.707	−0.707	
$n_1\Delta$	10^{-3}	0	0	0	0	0.106	0.091	$0.197 (= d_1)$

Example 4.16

As an example of a building framework, suppose that we wish to find the vertical deflection d_1 of the crown of the fixed-footed portal frame solved in Example 4.10. The relevant equation for the plane framework in bending is

$$d_1 = \oint \left(\frac{m_1^* M}{EI} \right) dx \ . \tag{3.41}*$$

In this case, we can take $m_1^* (i)$, $(i = 1, 2)$, to be the bending moments generated by unit vertical forces acting on the two alternative cantilever substructures shown in Figs. 4.48(a) and (b). M is given in Fig. 4.26. The calculations for d_1 are carried out in Table 4.14, and again the solutions agree.

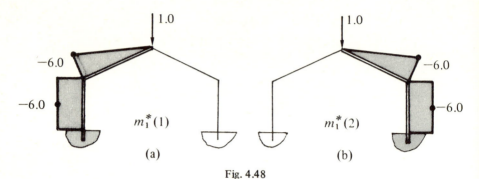

Fig. 4.48

Table 4.14
Calculations for Example 4.16

member	a	b	c	d	Σ m
l	4.0	6.5	6.5	4.0	
M $\quad 10^3$	-19.35 $9.74 +$ 4.0	-19.35 $+8.16$ 15.0	-22.02 $8.16 +$ 12.0	-22.02 23.07	
$m_1(1)$	-6.0	-6.0			
$\int m_1 M dx \; 10^3$	51.2	52.2	0	0	103.4 $= d_1(EI)$
$m_1(2)$			-6.0	-6.0	
$\int m_2 M dx \; 10^3$			116.1	-12.7	103.4 $= d_1(EI)$

The substructure method for obtaining simple equilibrium force systems can be extended to frameworks of any complexity. Thus if we wished to find the horizontal displacement of point A of the building frame in Fig. 4.49(a), we should need only to consider the bending moments in the column shown in Fig. 4.49(b).

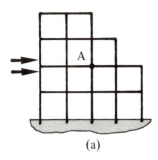

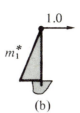

(a)   (b)

Fig. 4.49

4.11 PROBLEMS

Young's moduli are as follows:
E(steel) = 200.00 GN/m^2, E(concrete) = 25.0 GN/m^2, E(aluminium alloy) = 70.0 GN/m^2.

4.1   In the suspended pin-jointed framework shown in Fig. 4.50, all the members are of the same material and have the same cross-sectional area. Calculate the force in member a due to the loading shown in the figure.

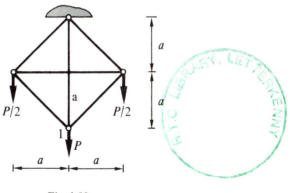

Fig. 4.50

4.2   In the steel pin-jointed truss shown in Fig. 4.51, the cross-sectional areas of the horizontal and vertical members are the same and equal to 1.0 × 10^{-3} m^2. The cross-sectional areas of the diagonal members are equal to 1.5 × 10^{-3} m^2. Obtain the forces in the framework due to the 20.0 kN and 40.0 kN forces acting as shown in the figure.

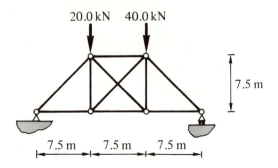

Fig. 4.51

4.3   In the steel pin-jointed bridge truss shown in Fig. 4.52, all the members have the same cross-sectional area equal to 0.75×10^{-3} m^2. Obtain the flexibility matrix for release forces in members a and b. Thence determine the forces in the framework due to the loading shown in the figure.

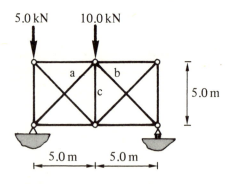

Fig. 4.52

4.4   The truss in Fig. 4.52 is constructed so that the bottom chord is shaded by a road deck and sidewalks. Calculate the additional forces in the framework if the remaining members undergo a temperature rise of 20 °C.

4.5   It is decided to reduce the compressive forces in members a and b of the truss in Fig. 4.52 by constructing the vertical member c, Δ_C too long. Find Δ_C if the forces are to be reduced by 5.0 kN.

4.6   In the truss shown in Fig. 4.53, all the members are of the same material. The cross-sectional areas of members a and b are equal and 1.5 times the cross-sectional areas of all the other members. Calculate the forces in the framework due to the 20.0 kN force acting as shown in the figure.

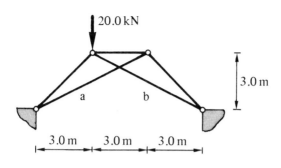

Fig. 4.53

4.7   In the guyed tripod shown in Fig. 4.54, member a is a 7.5 mm diameter steel wire and members b, c and d are aluminium alloy tubes, 30.0 mm in external diameter and 2.0 mm in wall thickness. A horizontal force of 5.0 kN is applied at the top of the structure in the x-direction as shown. Calculate the force in the steel wire.

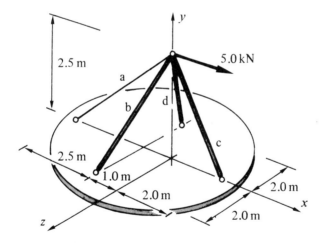

Fig. 4.54

4.8   In the space framework in Fig. 4.55 the members are all of the same cross-sectional area. The four supports are at the corners of a square whose sides are parallel to the coordinate directions and the member joining the two free joints is parallel to the z axis. Calculate the forces in the framework due to the 10.0 kN vertical force acting as shown in the figure.

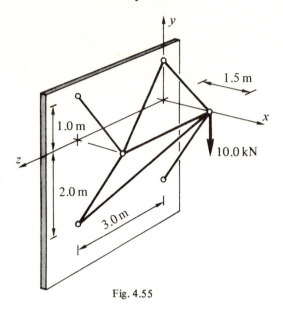

Fig. 4.55

4.9 A steel cantilever beam is restrained by the two steel ties shown in Fig. 4.56. The relevant sectional properties of the members are as follows:

(i)   second moment of area of the beam about a horizontal axis:
      30×10^6 mm^4,

(ii) cross-sectional area of the beam: 4.0×10^3 mm^2,

(iii) cross sectional areas of the ties: 500.0 mm^2.

The cantilever is loaded by the single concentrated force of 10.0 kN midway between the points of connection of the two ties. Calculate the forces in the ties. Hence draw the bending moment diagram for the cantilever.

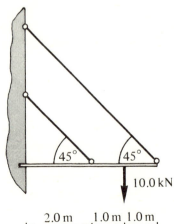

Fig. 4.56

4.10 A concrete beam is reinforced by the steel pin-jointed structure shown in Fig. 4.57. All the inclined members of the support structure are at 30° to the horizontal. They all have the same cross-sectional areas equal to 750.0 mm². The second moment of area of the beam section about a horizontal axis is 500.0×10^6 mm⁴. Neglecting axial effects in the beam, calculate the force in member a due to the distributed force of 15.0 kN/m acting over half the beam, as shown in the figure.

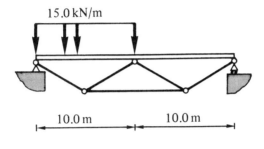

15.0 kN/m

10.0 m   10.0 m

Fig. 4.57

4.11 A rectangular beam of depth d is subjected to a linear temperature differential $\Delta\theta$, between its upper and lower surfaces (the upper being at the higher temperature). Show that this temperature distribution should be accounted for in the virtual work equation for a beam (3.25) by including the following additional term for the internal virtual work done by the moment M^*:

$$W_I^* = -\int_0^l \left(M^* \frac{\alpha\Delta\theta}{d} \right) dx \ .$$

The propped cantilever in Fig. 4.58 is subjected to just such a temperature differential. Show that an upward force P is induced at the prop, given by

$$P = \frac{3\alpha\Delta\theta}{2d} \left(\frac{EI}{l} \right) \ .$$

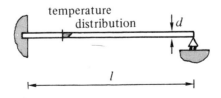

temperature
distribution d

l

Fig. 4.58

In Problems 4.12 to 4.20, axial effects are assumed to be negligible.

4.12 The encastered beam in Fig. 4.59 is subjected to the concentrated force *P*. Determine the couples exerted by the supports at the ends of the beam.

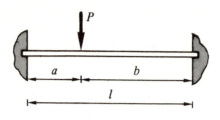

Fig. 4.59

4.13 In the concrete portal frame bridge shown in Fig. 4.60, the relevant second moment of area of the beam is uniform and twice that of the inclined columns. The beam is restrained vertically at its ends by roller supports (which can exert reactions up or down). Calculate the flexibility matrix for the two reactions at these restrained ends, and thence determine the reactions and the bending moments in the beam at joints 1 and 2.

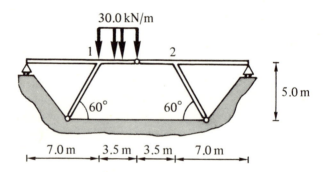

Fig. 4.60

4.14 A long concrete box culvert is symmetrically loaded by the earth pressures shown in Fig. 4.61. Determine the bending moments (per unit axial length) at the corners of the culvert. (Note that because of the symmetry of the problem, the shear force at the centre of the top of the culvert is zero.)

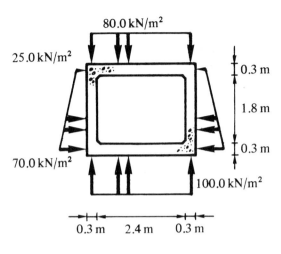

Fig. 4.61

4.15 In the **Vierendeel** girder shown in Fig. 4.62, all the members have the same material and cross-sectional properties E and I. What is the degree of redundancy? Determine the flexibility matrix for a primary structure generated by making cuts at joints 1 and 2.

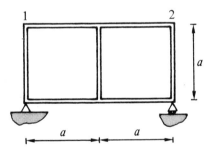

Fig. 4.62

4.16 Give formulae for the degrees of redundancy of the following structures:
   (i)   A rigid-jointed plane building frame containing a total of I beam members, each spanning between two columns.
   (ii) A Vierendeel girder as shown in Fig. 4.62, but with the number of bays increased from 2 to I.
   (iii) A rigid-jointed space frame in the form of a rectangular grid such as that shown in Fig. 4.63, the frame containing I beam members.

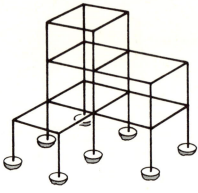

Fig. 4.63

4.17 The propped cantilever shown in Fig. 4.64 is composed of a ¼-circle section of 2.0 m radius, rigidly connected to a beam 2.0 m in length. The **bending stiffness** (EI) of the curved section is half that of the beam. Calculate the horizontal and vertical reactions at support 1. Thence plot the bending moment diagram for the cantilever.

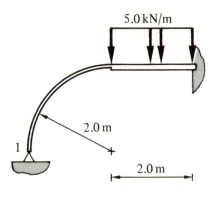

Fig. 4.64

4.18 The semicircular ring beam shown in Fig. 4.65 lies in a horizontal plane. It is of uniform material and cross-section. It is subjected to a uniformly distributed vertical force of p/unit length over its entire length as shown. It is fully encastered at one end and the other end is restrained by the axially rigid pin-ended members shown in the figure, these members being parallel to the x and y coordinate directions. Obtain expressions in terms of E, G, I and J for the reactions on the beam at this end. If the beam were fully encastered at this end, which of the reactions would be non-zero?

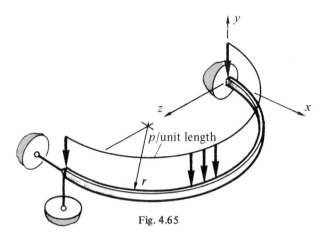

Fig. 4.65

4.19 Calculate the following displacements:
   (i)   The vertical displacement of joint 1 in the suspended structure described in Problem 4.1 (Fig. 4.50).
   (ii) The horizontal displacement of the 20.0 kN force on the truss described in Problem 4.6 (Fig. 4.53).
   (iii) The vertical displacement of the centre of the beam described in Problem 4.10 (Fig. 4.57).

4.20 Wind loading on the multistorey building frame in Fig. 4.66 induces the bending moments in the left-hand column shown in the figure. The bending stiffness (EI) of the column is 15.0 MNm^2. Calculate the horizontal displacement of point A. If the bending stiffness of the central column is twice that of the left-hand column, what bending moments are induced by the wind loading in the central column?

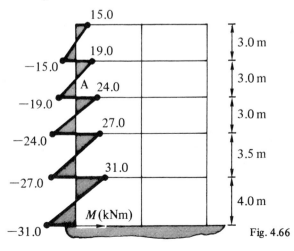

Fig. 4.66

4.21 In the steel pin-jointed structure shown in Fig. 4.67, members a, b and e have cross-sectional areas equal to 500.0 mm^2 and members c, d and f have cross-sectional areas equal to 750.0 mm^2. Under the loading, support 1 moves 1.0 mm horizontally to the right and support 2 moves 1.5 mm downwards. Calculate the forces in members e and f due to these support settlements.

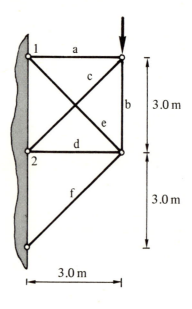

Fig. 4.67

4.22 The right-hand support of the encastered beam in Fig. 4.59 moves d_s downwards. Neglecting axial effects, determine the additional couples exerted by the supports at the ends of the beam due to this settlement.

REFERENCE

[4.1] Robinson, J. (1966), *Structural Matrix Analysis for the Engineer*, Wiley, New York.

CHAPTER 5

Stiffness Method I: Basic Structural Theory

5.1 INTRODUCTION

We have seen in the previous chapter that the flexibility method for analysing a statically indeterminate framework treats the release forces $\mathbf{X}$ as the unknowns. The **stiffness method** can basically be regarded as the opposite of this in that it treats particular displacements of the framework as the unknowns. For most frameworks, these displacement unknowns, called the **degrees of freedom**, greatly exceed the force unknowns, and it is not usually possible to solve interesting practical cases without a computer. However, the method is much more easily programmed than the flexibility method, and it is therefore much more important for general structural analysis.

The basis of the stiffness method can be very simply described. Thus suppose a framework is subjected to N forces P_i, and the corresponding displacements are d_i, as in Fig. 5.1(a) (with $N = 3$). Previously in Chapter 3, Section 3.13, we related $\mathbf{P}$ and $\mathbf{d}$ by the flexibility matrix $\mathbf{f}$ as follows:

$$\mathbf{d} = \mathbf{f}\mathbf{P} . \tag{5.1}$$

However, we can also define a **stiffness matrix** $\mathbf{k}$, equal to $\mathbf{f}^{-1}$, which gives the inverse relation

$$\mathbf{P} = \mathbf{k}\mathbf{d} . \tag{5.2}$$

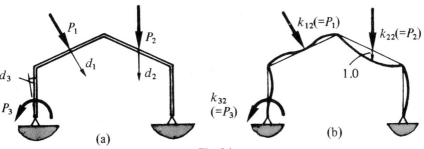

Fig. 5.1

The stiffness method makes use of the fact that it is possible to calculate k for statically indeterminate frameworks by easily programmed techniques. Thus when k is known, the displacements of a framework subject to a set of forces P, can be obtained as

$$d = k^{-1}P \ . \tag{5.3}$$

These displacements are regarded as the solution, since the internal forces and thus the complete structural response, are then calculable.

The object of this chapter is to introduce the stiffness method by illustrative examples of simple pin-jointed frameworks and building frames. In the case of the building frames, we shall employ a version of the stiffness method which is equivalent to a well-known classical method of structural analysis based on slope-deflection equations. The problem of programming the stiffness method will be discussed in detail in Chapter 6.

We shall begin this introduction to the stiffness method by considering a precise definition of the stiffness matrix k.

5.2 STIFFNESS OF A FRAMEWORK

5.2.1 Definition of Stiffness
Let us again consider the framework in Fig. 5.1(a). First it is helpful to recall from Section 3.13, that the flexibility matrix f is composed of flexibility co-efficients f_{ij}, being displacements caused by unit forces. They are defined as

$$f_{ij} = d_i \ [P_j = 1.0, P_k = 0, \ k \neq j] \quad . \tag{5.4}$$

The stiffness matrix k is composed of **stiffness coefficients** k_{ij} which have the opposite meaning. Thus suppose we arrange the three forces P so that $d_2 = 1.0$ and $d_1 = d_3 = 0$ as in Fig. 5.1(b). This displacement system can be written concisely as $[d_2 = 1.0, d_k = 0, k \neq 2]$. The particular set of forces needed to cause these displacements is unique and k_{22} is defined as being equal to P_2, k_{12} equal to P_1 and k_{32} equal to P_3. The stiffness coefficients are therefore the forces required to cause unit displacements, and their general definition is as follows:

$$k_{ij} = P_i \ [d_j = 1.0, d_k = 0, k \neq j] \quad . \tag{5.5}$$

An important difference between the flexibility and stiffness coefficients should be noted. Whereas the values of the flexibility coefficients are constant for a particular framework, the values of the stiffness coefficients depend on the number of degrees of freedom considered in the problem. Thus suppose a fourth force-displacement pair P_4, d_4 were added to those previously considered, as in Fig. 5.2(a). d_4 would then also have to be constrained to be zero in determining k_{12}, k_{22}, k_{32} and k_{42}, as in Fig. 5.2(b). P_1, P_2 and P_3 would clearly need to be changed, thus altering the corresponding stiffness coefficients.

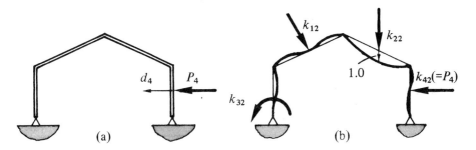

Fig. 5.2

From the definition of stiffness coefficients, we can find the forces required to cause a particular set of displacements $\mathbf{d}$ of the framework. Thus P_1 say, is given by the sum of the products of stiffness coefficients and displacements as

$$P_1 = k_{11}d_1 + k_{12}d_2 + \ldots + k_{1N}d_N \ . \tag{5.6}$$

Similar expressions can be written for P_2 to P_N, and collected they form the matrix equation

$$\mathbf{P} = \mathbf{k}\,\mathbf{d} \ . \tag{5.7}$$

(5.7) can be regarded as expressing the equilibrium between the framework and its loading, and it is called the **global equilibrium equation**.

5.2.2 The Equation of Virtual Work

We have seen that the equation of virtual work is indispensable in determining structural flexibilities. It does not, however, have any useful application in determining stiffnesses. It is instructive to see why this is so. Consider again the framework in Fig. 5.1(a), in which, for simplicity, we shall assume only the bending moments are important. The virtual work equation then becomes

$$\mathbf{P}^* \mathbf{d} = \oint \left(\frac{M^* M}{EI} \right) \, \mathrm{d}x \ . \tag{5.8}$$

In determining k_{ij}, we need to calculate the external forces required to cause a particular set of displacements. Thus let $\mathbf{P}^*$ be the external forces causing the displacements $d_j^* = 1.0$, $d_k^* = 0$, $k \neq j$; and m_j^* be the internal bending moments in equilibrium with $\mathbf{P}^*$. P_i^* is then the required stiffness coefficient k_{ij}. We can isolate P_i^* on the left-hand side of (5.8) by combining it with *any* set of compatible displacements for which $d_i = 1.0$, $d_k = 0$, $k \neq i$ and the corresponding internal bending moments are m_i. We then have

$$P_i^* \times 1.0 = k_{ij} = \oint \left(\frac{m_j^* m_i}{EI} \right) dx \ . \tag{5.9}$$

This equation, however, is useless for calculating the stiffness coefficients, for m_j^* are no less difficult to determine than the forces **P*** themselves. Nevertheless, it does give us some useful information, namely that the stiffness matrix is symmetrical, for if m_i are chosen to be the bending moments corresponding to the *actual* set of compatible displacements for which $d_i = 1.0, d_k = 0, k \neq i$, the right-hand side of (5.9) is identical to that which would be produced by the same method, for determining k_{ji}. This leads us to a complementary version of the *reciprocal theorem* discussed in Chapter 3, Section 3.13. Thus the force P_i among the set of forces required to cause a unit displacement d_j, is equal to the force P_j among the set of forces required to cause a unit displacement d_i.

5.2.3 Degrees of Freedom and Kinematic Indeterminacy
We have noted that the number of the degrees of freedom chosen to describe the behaviour of a framework determines the particular values of the stiffness coefficients. More significantly, this number is also closely associated with the ease of calculating the stiffness matrix. For this latter purpose, it is important to specify enough degrees of freedom to completely define the deformation of the framework. This is possible using a finite number of degrees of freedom providing it is known how the framework deforms between the points whose displacements are thus determined. We shall illustrate this concept with a simple example.

Example 5.1
Consider again the pin-jointed framework discussed in Example 4.1 and reproduced for convenience in Fig. 5.3. Let us take as the degrees of freedom, the four displacements **d** of joints 1 and 2 in the global coordinate directions as shown in Fig. 5.4. If **d** are known, the positions of joints 1 and 2 of the framework are determined. We can then deduce the complete deformed shape of the structure, by assuming that the members remain straight and undergo uniform axial strain to fit the displaced positions of the joints. This seems an obvious intuitive assumption, but it should also be noted that it agrees with the engineering theory of beams, for we know that the members are subject to uniform axial forces N and therefore that their strains are uniform and given by

$$\epsilon_{xx} = \left(\frac{N}{EA} + \alpha \Delta \theta \right) \ . \tag{5.10}$$

The minimum number of degrees of freedom needed to describe the deformation of a framework, is called its **kinematic indeterminacy**. Thus in the case of the above pin-jointed framework, the kinematic indeterminacy is equal to four. If fewer degrees of freedom are used, then calculating the stiffness matrix becomes a difficult analytical problem. We shall see however, that in the case of general frameworks it is quite possible and sometimes advantageous to work with a greater number of degrees of freedom.

We now go on to consider the solution of pin-jointed frameworks by the stiffness method.

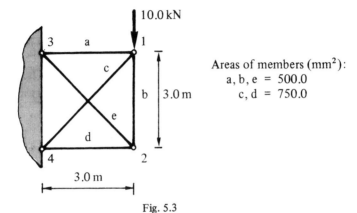

Areas of members (mm^2):
a, b, e = 500.0
c, d = 750.0

Fig. 5.3

5.3 THE DIRECT SOLUTION OF A PIN-JOINTED FRAMEWORK

We shall illustrate the stiffness method by solving the pin-jointed framework discussed in Example 5.1. For simplicity in this first presentation, the stiffness matrix will be determined by a direct method involving consideration of the deformation of the framework as a whole and its corresponding equilibrium. Later we shall see that a more convenient calculation for computer programming will involve considering the stiffnesses of the individual members.

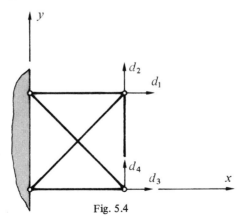

Fig. 5.4

Example 5.2

The pin-jointed framework loaded by the 10.0 kN force is composed of steel members with the sectional areas shown in Fig. 5.3. The stiffness of the framework can be calculated by giving each of the four degrees of freedom in Fig. 5.4, a unit value in turn, each time holding the other degrees of freedom zero. From the deformation of the framework, it is then possible to determine the extensions of the members Δ and their axial forces N. Equilibrium of the joints then enables us to calculate the external forces **P** on the framework. Thus consider the framework with $d_1 = 1.0$ and $d_k = 0$, $k \neq 1$, shown in Fig. 5.5. (A short digression at this point is perhaps necessary. The actual displacements of the framework under the loading are very small, of the order of 1.0 mm. Thus in order to find its stiffness, the unit displacment should be of this order of magnitude. However, in the SI system of units, the unit displacement is 1.0 m. This presents no difficulty, provided we neglect any non-linear effects due to changes in the geometry of the structure, which would occur if we actually tried to impose a displacement of 1.0 m at joint 1. Again in Fig. 5.5, the unit displacement has been depicted as being relatively large compared with the structure, simply for clarity.) If we then consider an enlarged view of joint 1 moving through a *small* unit displacement as in Fig. 5.6, it is apparent that only members a and c are extended to accommodate this displacement and

$$\Delta_a = 1.0, \quad \Delta_c = 1.0 \cos(45°) = 0.707 \quad .$$

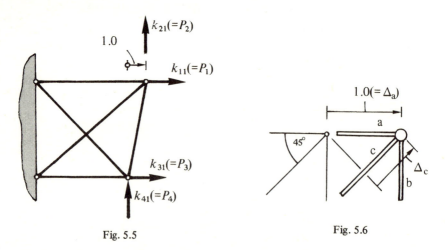

Fig. 5.5   Fig. 5.6

The member forces come from inverting equation (3.10) for the elastic extensions given in Chapter 3, Section 3.3. Thus

$$N = \frac{EA\Delta}{l} . \tag{5.11}$$

Whence inserting the particular values for the member properties of a and c we obtain

$$N_a = 33.33 \text{ MN/m}, \quad N_c = 25.0 \text{ MN/m}† .$$

The axial forces in all the other members are zero. We then calculate the applied external forces by considering equilibrium at joints 1 and 2 in turn. The free body diagram for joint 1 is shown in Fig. 5.7. Thus resolving horizontally and vertically we have

$$\Sigma \rightarrow \quad -33.33 - 25.0 \cos(45°) + P_1 = 0$$

$$\Sigma \uparrow \quad -25.0 \cos(45°) + P_2 = 0 .$$

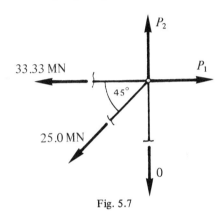

Fig. 5.7

Therefore

$$k_{11} = P_1 = 51.01 \text{ MN/m}$$

$$k_{21} = P_2 = 17.68 \text{ MN/m} .$$

At joint 2, members b, d, and e contain zero axial forces, so

$$k_{31} = P_3 = 0, \quad k_{41} = P_4 = 0 .$$

The stiffness coefficients for the other degrees of freedom can be found in a similar manner, and the calculations are presented in Table 5.1. Whence

$$k = \begin{bmatrix} 51.01 & 17.68 & 0 & 0 \\ 17.68 & 51.01 & 0 & -33.33 \\ 0 & 0 & 61.79 & -11.79 \\ 0 & -33.33 & -11.79 & 45.12 \end{bmatrix} \text{ MN/m} .$$

† The extensions Δ corresponding to unit displacements are considered to be dimensionless, (i.e. m/m), whence N in (5.11) has the dimensions N/m, being a force/unit displacement.

Table 5.1
Direct calculation of stiffness coefficients

Unit Displacement	Deformed framework	Non-zero member extensions	Non-zero member forces(MN/m)	Stiffness coefficients (MN/m)
d_1		$\Delta_a = 1.0$ $\Delta_c = 0.707$	$N_a = 33.33$ $N_c = 25.00$	$k_{11} = 51.01$ $k_{21} = 17.68$ $k_{31} = 0$ $k_{41} = 0$
d_2		$\Delta_b = 1.0$ $\Delta_c = 0.707$	$N_b = 33.33$ $N_c = 25.00$	$k_{12} = 17.68$ $k_{22} = 51.01$ $k_{32} = 0$ $k_{42} = -33.33$
d_3		$\Delta_d = 1.0$ $\Delta_e = 0.707$	$N_d = 50.0$ $N_e = 16.67$	$k_{13} = 0$ $k_{23} = 0$ $k_{33} = 61.79$ $k_{43} = -11.79$
d_4		$\Delta_b = -1.0$ $\Delta_e = -0.707$	$N_b = -33.33$ $N_e = -16.67$	$k_{14} = 0$ $k_{24} = -33.33$ $k_{34} = -11.79$ $k_{44} = 45.12$

The force matrix **P** contains the single non-zero term P_2, corresponding to the 10.0 kN force at joint 1. Since this force acts vertically downwards, P_2 is negative and equal to -10.0 kN. Thus

$$\mathbf{P} = \begin{bmatrix} 0 \\ -10.0 \\ 0 \\ 0 \end{bmatrix} \text{ kN } .$$

As discussed in Section 5.1, the solution in terms of the degrees of freedom **d** is then given as

$$\mathbf{d} = \mathbf{k}^{-1}\mathbf{P} \tag{5.3*}$$

and is found by standard numerical routines for solving simultaneous equations. Thus

$$\mathbf{d} = \begin{bmatrix} 0.183 \\ -0.527 \\ -0.078 \\ -0.410 \end{bmatrix} \text{ mm } .$$

The corresponding member forces are found by referring back to Table 5.1, which contains the member forces due to *unit* displacements. It is then apparent that

$$N_a = 33.33\, d_1 \times 10^6 = 6.09 \text{ kN}$$

$$N_b = (33.33\, d_2 - 33.33\, d_4) \times 10^6 = -3.91 \text{ kN}$$

$$N_c = (25.0\, d_1 + 25.0\, d_2) \times 10^6 = -8.61 \text{ kN}$$

$$N_d = 50.0\, d_3 \times 10^6 = -3.91 \text{ kN}$$

$$N_e = (16.67\, d_3 - 16.67\, d_4) \times 10^6 = 5.53 \text{ kN } .$$

These results agree with those produced in Example 4.2, by the flexibility method.

Comparing the two methods, the flexibility method employs one unknown release force for this problem, the stiffness method employs four unknown degrees of freedom. Thus even for this simple case, the stiffness method requires computing facilities. The flexibility method, however, involves the more sophisticated analysis, since solving the statically determinate primary structure is a more complex problem than finding the stiffnesses. The procedure presented above for finding stiffnesses by considering the overall structural deformation is simple in concept but is structure-oriented. For programming purposes it is better to consider the stiffnesses of the individual members, and assemble them to produce the overall stiffness matrix. This approach is considered in the next section.

5.4 SOLUTION OF A PIN-JOINTED FRAMEWORK USING MEMBER STIFFNESSES†

5.4.1 Introduction
In this section we shall consider the calculation of the overall stiffness matrix k of a pin-jointed framework by assembling the stiffnesses of the individual members. For this we need to define the stiffness matrix of each member, and a further property of the member called the transformation matrix. The latter defines the compatibility of the member within the framework. We shall then show that simple matrix operations lead to the automatic assembly of k.

5.4.2 Stiffness Matrix of a Pin-ended Member
We first recall that a member in a framework is described in terms of a member coordinate system, chosen in a particular way with respect to the joint numbering. Thus the x_m coordinate runs from the end with the lower joint number to the end with the higher. The y_m and z_m coordinates are at right angles to this, and coincide with the principal axes of the cross-section. In the case of a member in a pin-jointed framework, however, we are only concerned with the axial coordinate. Thus for the members in the framework of Example 5.2 for example, the axial coordinates are oriented as in Fig. 5.8.

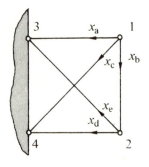

Fig. 5.8

---

† From this section onwards it is important to make a clear distinction between member and global parameters. Thus the former will be denoted throughout by the subscript m.

We next consider the member in the pin-jointed framework as an isolated entity. It too has stiffness properties which are defined by the **member stiffness matrix** k_m. For clarity, we can distinguish this matrix from the overall structural stiffness matrix k by calling the latter, the **global stiffness matrix**. k_m is obtained with respect to the **member degrees of freedom** d_m which are the set of displacements necessary to completely define the deformation of the member. Thus for member m in Fig. 5.9(a), only two degrees of freedom are needed, being the axial displacements d_{m1} and d_{m2} at ends 1 and 2 respectively, where end 1 coincides with the origin of the x_m coordinate. The transverse displacements at the ends are not considered, because they only effect the *position* of the member, not its deformation. The corresponding **member forces** P_{m1} and P_{m2} act as shown. Like the calculation for k in Section 5.3, k_m can be found by making d_{m1} and d_{m2} unity in turn, and calculating the corresponding member forces. Thus suppose $d_{m1} = 1.0$ and $d_{m2} = 0$, as in Fig. 5.9(b). We then have $\Delta_m = -1.0$ and from (5.11) obtain

$$N_m = - \left(\frac{EA}{l} \right)_m \quad .$$

(a)

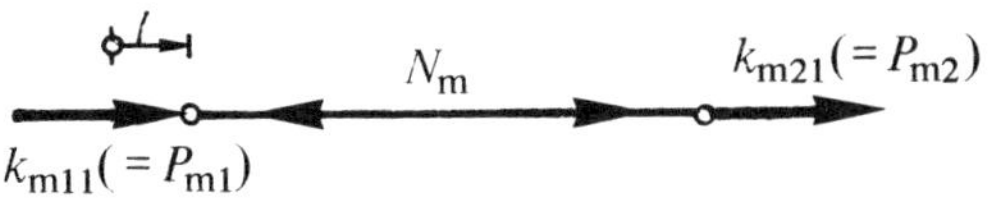

(b)

Fig. 5.9

Whence considering equilibrium at the two ends gives

$$k_{m11} = P_{m1} = \left(\frac{EA}{l} \right)_m , \quad k_{m21} = P_{m2} = -\left(\frac{EA}{l} \right)_m \quad .$$

A similar calculation with d_{m2} equal to unity gives

$$k_{m12} = -\left(\frac{EA}{l} \right)_m \qquad k_{m22} = \left(\frac{EA}{l} \right)_m \quad .$$

Thus the member stiffness matrix takes the form

$$\mathbf{k}_m = \begin{bmatrix} 1.0 & -1.0 \\ -1.0 & 1.0 \end{bmatrix} \left\{ \frac{EA}{l} \right\}_m \qquad (5.12)$$

Having obtained $\mathbf{k}_m$, we can then derive the **member equilibrium equation** in the same manner as the global equilibrium equation (5.7). Thus the member forces required to cause a particular set of member displacements, $\mathbf{d}_m$, are given by

$$\mathbf{P}_m = \mathbf{k}_m \mathbf{d}_m \qquad (5.13)$$

5.4.3 Member Transformation Matrix

For each member in a framework, there is a unique relationship between the member degrees of freedom, and the degrees of freedom used to describe the deformation of the complete framework. For clarity, the latter will now be called the **global degrees of freedom**, and the corresponding forces the **global forces**. We can illustrate this relationship, by considering member e say, in the framework in Fig. 5.3. With the chosen joint numbering, end 1 of member e is attached to joint 2, and end 2 is attached to the rigid support at joint 3. Suppose joint 2 moves to a new position corresponding to particular values of the global degrees of freedom d_3 and d_4 as in Fig. 5.10. Then by simple consideration of the geometry of this figure, we see that

$$d_{e1} = -d_3 \cos(45°) + d_4 \sin(45°) = -0.707\, d_3 + 0.707\, d_4$$

d_{e1} is unaffected by changes in d_1 and d_2. Also, d_{e2} at the end attached to joint 3 is zero and hence is independent of $\mathbf{d}$. The relationship between $\mathbf{d}_e$ and $\mathbf{d}$ for the member can thus be expressed in the following form

$$\mathbf{d}_e = \begin{bmatrix} 0 & 0 & -0.707 & 0.707 \\ 0 & 0 & 0 & 0 \end{bmatrix} \mathbf{d}$$

and this is written as

$$\mathbf{d}_e = \mathbf{t}_e \mathbf{d} \qquad (5.14)$$

$\mathbf{t}_e$ in (5.14) is called the **member transformation matrix** for member e. It should be noted that it has the same number of rows as there are member degrees of freedom, and the same number of columns as there are global degrees of freedom. The transformation matrices for the other members in the framework in Fig. 5.3 can be found in the same way by considering the geometry of the deformation. They are given by

$$t_a = \begin{bmatrix} -1.0 & 0 & 0 & 0 \\ 0 & 0 & 0 & 0 \end{bmatrix}, \quad t_b = \begin{bmatrix} 0 & -1.0 & 0 & 0 \\ 0 & 0 & 0 & -1.0 \end{bmatrix}$$

$$t_c = \begin{bmatrix} -0.707 & -0.707 & 0 & 0 \\ 0 & 0 & 0 & 0 \end{bmatrix}, \quad t_d = \begin{bmatrix} 0 & 0 & -1.0 & 0 \\ 0 & 0 & 0 & 0 \end{bmatrix}.$$

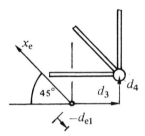

Fig. 5.10

The member transformation matrices can be regarded as expressing compatibility between the member degrees of freedom and the global degrees of freedom by the equation

$$d_m = t_m d .\tag{5.15}$$

They also, however, have the important role of expressing equilibrium between the member forces and the global forces. To see this, we consider again the end of member e at joint 2. Parts of the total global forces P_3 and P_4 are taken up in maintaining equilibrium with the member force P_{e1}. Let us call these $P_3^{(e)}$ and $P_4^{(e)}$ say. Then the relationship between P_{e1} and $P_3^{(e)}, P_4^{(e)}$ can be most clearly demonstrated by considering the mechanics of joint 2 in the expanded view of Fig. 5.11. Thus if P_{e1} acts on the member as in Fig. 5.11(a), an equal and opposite force acts on the axle of the pinned joint as in Fig. 5.11(b). As noted in Chapter 1, Section 1.4, the global forces act on the structure through the axle, and equilibrium therefore has to be maintained between the three forces on the axle treated as a free body as in the figure. We thus have

$$\Sigma \rightarrow \quad P_{e1} \cos(45°) + P_3^{(e)} = 0$$

$$\Sigma \uparrow \quad P_{e1} \sin(45°) - P_4^{(e)} = 0 .$$

Therefore

$$P_3^{(e)} = -0.707\,P_{e1}$$

$$P_4^{(e)} = 0.707\,P_{e1} \quad.$$

P_1 and P_2 do not enter into the equilibrium of member e. Also P_{e2} is in equilibrium with unknown support reactions which are not considered. Thus the relationship between those parts $\mathbf{P}^{(e)}$ of the global forces needed to keep member e in equilibrium and the member forces $\mathbf{P}_e$, is given by

$$\mathbf{P}^{(e)} = \begin{bmatrix} 0 & 0 \\ 0 & 0 \\ -0.707 & 0 \\ 0.707 & 0 \end{bmatrix} \mathbf{P}_e = \mathbf{t}_e^T\,\mathbf{P}_e \quad. \tag{5.16}$$

It is thus apparent that the transpose of the member transformation matrix expresses the equilibrium between the global and member forces. The total global forces necessary to maintain the complete structure in equilibrium are found by summing (5.16) for all the members. Thus

$$\mathbf{P} = \sum_m (\mathbf{t}_m^T \mathbf{P}_m) \quad. \tag{5.17}$$

Equation (5.17) is an important result in the stiffness method of analysis, and applies for all types of element in any structure. The above derivation has referred specifically to pin-jointed frameworks. A more general proof can be quickly derived from the virtual work equation, and this is given in the first appendix to this chapter, Appendix A5.1.

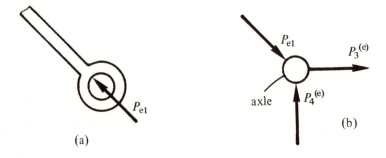

(a)

(b)

Fig. 5.11

5.4.4 Assembly of the Global Stiffness Matrix

Having obtained expressions for the member stiffness matrices and their transformation matrices, a simple series of matrix operations can be performed to assemble the global stiffness matrix. The underlying theory will now be presented. In this derivation of the relationship between k, k_m and t_m it is assumed that the reader has some facility in matrix manipulation. If in doubt, he should write out the expressions in full for say two of the members in the framework of Example 5.2.

Suppose the global degrees of freedom d of a framework take a particular set of values. The corresponding degrees of freedom d_m of member m are then given by

$$d_m = t_m d \tag{5.15}*$$

and the member forces by (5.13) and (5.15) as

$$P_m = k_m d_m = k_m t_m d . \tag{5.18}$$

Thus from (5.17) and (5.18) the global forces are

$$P = \sum_m (t_m^T P_m) = \sum_m (t_m^T k_m t_m) d . \tag{5.19}$$

The right-hand side of (5.19) is a summation carried out for all the members in the framework. However, in this summation d is a common factor, and it can be factorised out to give

$$P = \sum_m (t_m^T k_m t_m)) d . \tag{5.20}$$

We can now compare this with the global equilibrium equation (5.7)

$$P = k d \tag{5.7}*$$

and see that

$$k = \sum_m (t_m^T k_m t_m) . \tag{5.21}$$

(5.21) is then the desired relationship between k, k_m and t_m. It is called the **assembly equation** and it underlies all stiffness methods of structural analysis. Thus it should be noted that the above derivation is completely general, applying to all types of element in any structure.

We can see how (5.21) works for the pin-jointed framework of Example 5.2, by working out the contribution of an individual member to the global stiffness matrix by hand. Thus the contribution of member e is found as follows:

$$\mathbf{k_e} = \begin{bmatrix} 1.0 & -1.0 \\ -1.0 & 1.0 \end{bmatrix} \left\{ \frac{EA}{l} \right\}_e = \begin{bmatrix} 1.0 & -1.0 \\ -1.0 & 1.0 \end{bmatrix} \times 23.57 \text{ MN/m}$$

$$\mathbf{t_e} = \begin{bmatrix} 0 & 0 & -0.707 & 0.707 \\ 0 & 0 & 0 & 0 \end{bmatrix} .$$

Therefore

$$\mathbf{t_e^T k_e t_e} = \begin{bmatrix} 0 & 0 \\ 0 & 0 \\ -0.707 & 0 \\ 0.707 & 0 \end{bmatrix} \begin{bmatrix} 1.0 & -1.0 \\ -1.0 & 1.0 \end{bmatrix} \begin{bmatrix} 0 & 0 & -0.707 & 0.707 \\ 0 & 0 & 0 & 0 \end{bmatrix} \times 23.57$$

$$= \begin{bmatrix} 0 & 0 & 0 & 0 \\ 0 & 0 & 0 & 0 \\ 0 & 0 & 11.79 & -11.79 \\ 0 & 0 & -11.79 & 11.79 \end{bmatrix} \text{MN/m} .$$

Referring back to the complete global stiffness matrix calculated in Example 5.2, p. 269, it is then apparent that k_{34} for example, depends exclusively on member e. The full calculation and assembly of the global stiffness matrix for the five members is given in Table 5.2.

The above calculation, though somewhat laborious by hand, is easy to program for a computer. The basis of this calculation is the assembly equation (5.21), and its usefulness relies on the fact that $\mathbf{k_m}$ and $\mathbf{t_m}$ can be found from simple structural data. The details of the programming problem will be discussed in the next chapter.

Table 5.2

Assembly of global stiffness matrix

Member	$k_m\,10^6$	t_m	$(t_m^T k_m t_m)\,10^6$	$k\,10^6$
a	$33.33\begin{bmatrix}1.0 & -1.0\\ -1.0 & 1.0\end{bmatrix}$	$\begin{bmatrix}-1.0 & 0 & 0 & 0\\ 0 & 0 & 0 & 0\end{bmatrix}$	$\begin{bmatrix}33.33 & 0 & 0 & 0\\ 0 & 0 & 0 & 0\\ 0 & 0 & 0 & 0\\ 0 & 0 & 0 & 0\end{bmatrix}$	$\begin{bmatrix}33.33 & 0 & 0 & 0\\ 0 & 0 & 0 & 0\\ 0 & 0 & 0 & 0\\ 0 & 0 & 0 & 0\end{bmatrix}$
b	$33.33\begin{bmatrix}1.0 & -1.0\\ -1.0 & 1.0\end{bmatrix}$	$\begin{bmatrix}0 & -1.0 & 0 & 0\\ 0 & 0 & 0 & -1.0\end{bmatrix}$	$\begin{bmatrix}0 & 0 & 0 & 0\\ 0 & 33.33 & 0 & -33.33\\ 0 & 0 & 0 & 0\\ 0 & -33.33 & 0 & 33.33\end{bmatrix}$	$\begin{bmatrix}33.33 & 0 & 0 & 0\\ 0 & 33.33 & 0 & -33.33\\ 0 & 0 & 0 & 0\\ 0 & -33.33 & 0 & 33.33\end{bmatrix}$
c	$35.36\begin{bmatrix}1.0 & -1.0\\ -1.0 & 1.0\end{bmatrix}$	$\begin{bmatrix}-0.707 & -0.707 & 0 & 0\\ 0 & 0 & 0 & 0\end{bmatrix}$	$\begin{bmatrix}17.68 & 17.68 & 0 & 0\\ 17.68 & 17.68 & 0 & 0\\ 0 & 0 & 0 & 0\\ 0 & 0 & 0 & 0\end{bmatrix}$	$\begin{bmatrix}51.01 & 17.68 & 0 & 0\\ 17.68 & 51.01 & 0 & -33.33\\ 0 & 0 & 0 & 0\\ 0 & -33.33 & 0 & 33.33\end{bmatrix}$
d	$50.0\begin{bmatrix}1.0 & -1.0\\ -1.0 & 1.0\end{bmatrix}$	$\begin{bmatrix}0 & 0 & -1.0 & 0\\ 0 & 0 & 0 & 0\end{bmatrix}$	$\begin{bmatrix}0 & 0 & 0 & 0\\ 0 & 0 & 0 & 0\\ 0 & 0 & 50.0 & 0\\ 0 & 0 & 0 & 0\end{bmatrix}$	$\begin{bmatrix}51.01 & 17.68 & 0 & 0\\ 17.68 & 51.01 & 0 & -33.33\\ 0 & 0 & 50.0 & 0\\ 0 & -33.33 & 0 & 33.33\end{bmatrix}$
e	$23.57\begin{bmatrix}1.0 & -1.0\\ -1.0 & 1.0\end{bmatrix}$	$\begin{bmatrix}0 & 0 & -0.707 & 0.707\\ 0 & 0 & 0 & 0\end{bmatrix}$	$\begin{bmatrix}0 & 0 & 0 & 0\\ 0 & 0 & 0 & 0\\ 0 & 0 & 11.79 & -11.79\\ 0 & 0 & -11.79 & 11.79\end{bmatrix}$	$\begin{bmatrix}51.01 & 17.68 & 0 & 0\\ 17.68 & 51.01 & 0 & -33.33\\ 0 & 0 & 61.79 & -11.79\\ 0 & -33.33 & -11.79 & 45.12\end{bmatrix}$

5.4.5 Member Forces

Having found k using the assembly procedure discussed above, the solution of the structure in terms of the global degrees of freedom is again given by

$$d = k^{-1} P .$$ (5.3)*

It then remains to calculate the internal forces in the member. In fact, the member forces P_m have already been given explicitly in terms of d in (5.18) as

$$P_m = k_m t_m d$$ (5.18)*

and simply by considering equilibrium at the ends of the member in Fig. 5.9 we have

$$N_m = -P_{m1} = +P_{m2} .$$ (5.22)

Thus the internal forces can again be calculated by simple matrix multiplication, and considering for example member e in the framework of Example 5.2, we have

$$P_e = 23.57 \times 10^6 \begin{bmatrix} 1.0 & -1.0 \\ -1.0 & 1.0 \end{bmatrix} \begin{bmatrix} 0 & 0 & -0.707 & 0.707 \\ 0 & 0 & 0 & 0 \end{bmatrix} \begin{bmatrix} 0.183 \\ -0.527 \\ -0.078 \\ -0.410 \end{bmatrix} \times 10^{-3}$$

$$= \begin{bmatrix} -5.53 \\ 5.53 \end{bmatrix} kN .$$

Thus

$$N_e = 5.53 \text{ kN} .$$

Similar calculations yield the other member forces N_a, N_b, N_c and N_d.

5.5 SOLUTION OF A CONTINUOUS BEAM

5.5.1 Introduction

The simplest problem for demonstrating the application of the stiffness method to frameworks containing bending moments is the problem of the continuous beam. An example already solved by the flexibility method is the continuous bridge of Example 4.11, reproduced in Fig. 5.12(a). The members are of concrete for which $E = 25.0 \times 10^9$ N/m^2, and all have the same uniform section with $I = 0.4$ m^4. Again we choose a limited number of global degrees of freedom to describe the deformation of the structure, and in this case, if we neglect axial displacements, the global degrees of freedom **d** are the five clockwise rotations about the z axis of the beam at the supports, as in Fig. 5.12(b). The corresponding global forces are the couples **P**. The global stiffness matrix **k** is then composed of the couples required to cause unit rotations at each of the supports in turn. The members of the structure are the individual spans a, b, c and d of the beam, and the member coordinates are oriented as shown in Fig. 5.12(c). The first step in the stiffness solution is then to determine the member stiffness matrices.

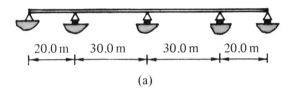

(a)

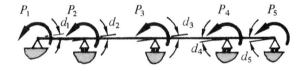

(b)

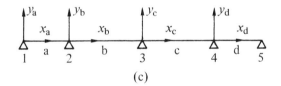

(c)

Fig. 5.12

5.5.2 Stiffness Matrix of a Member in Bending

Consider an isolated member m as in Fig. 5.13. The deformation of m is completely described by the two end rotations d_{m1}, d_{m2} as shown, and the corresponding member forces are the end couples P_{m1} and P_{m2}. A relationship between the member forces and displacements can be found by the flexibility method, by solving the problem as a simply supported beam subject to end couples. The details of this calculation are presented in the first part of the second appendix to this chapter, Appendix A5.2. There it is shown that

$$P_{m1} = \left\{\frac{EI}{l}\right\}_m (4.0\, d_{m1} + 2.0\, d_{m2}) \tag{5.23a}$$

$$P_{m2} = \left\{\frac{EI}{l}\right\}_m (2.0\, d_{m1} + 4.0\, d_{m2}) \ . \tag{5.23b}$$

These equations are a restricted version of equations that are well known in classical structural analysis as the **slope-deflection equations**. The combination of the member parameters $\left\{\dfrac{EI}{l}\right\}_m$ will be used very frequently in the subsequent text, and for conciseness we introduce the notation

$$\psi_m = \left\{\frac{EI}{l}\right\}_m \ . \tag{5.24}$$

If we consider d_{m1} and d_{m2} unity in turn in (5.23), we then obtain the member stiffness coefficients as follows

$$k_{m11} = 4.0\, \psi_m, \quad k_{m21} = 2.0\, \psi_m$$
$$k_{m12} = 2.0\, \psi_m, \quad k_{m22} = 4.0\, \psi_m \ . \tag{5.25 a,b,c,d}$$

Thus

$$k_m = \begin{bmatrix} 4.0 & 2.0 \\ 2.0 & 4.0 \end{bmatrix} \psi_m \ . \tag{5.26}$$

and (5.23) becomes the member equilibrium equation (5.13):

$$P_m = k_m\, d_m \ . \tag{5.13}^*$$

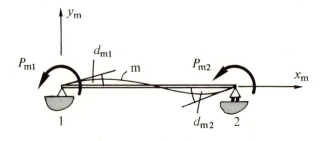

Fig. 5.13

5.5.3 Continuous Beam Subject to Concentrated Couples

The first example we shall consider is that of the continuous beam subject to concentrated couples at the joints. This is an academic problem, but gives a simple illustration of the stiffness method applied to beams.

Example 5.3

The continuous beam shown in Fig. 5.12 is loaded by concentrated couples as shown in Fig. 5.14. We require the rotations of the beam at the supports, and the final bending moment and shear force diagrams.

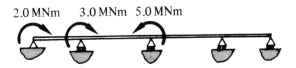

2.0 MNm   3.0 MNm   5.0 MNm

Fig. 5.14

First we assemble the global stiffness matrix from the member stiffness matrices and transformation matrices using (5.21). In this particular problem, the member transformation matrices can be obtained by inspection since by comparing Figs. 5.12 and 5.13, it is apparent that the member displacements and global displacements are related as follows:

$$\mathbf{d_a} = \begin{bmatrix} d_1 \\ d_2 \end{bmatrix} \quad , \quad \mathbf{d_b} = \begin{bmatrix} d_2 \\ d_3 \end{bmatrix} \quad , \quad \mathbf{d_c} = \begin{bmatrix} d_3 \\ d_4 \end{bmatrix} \quad , \quad \mathbf{d_d} = \begin{bmatrix} d_4 \\ d_5 \end{bmatrix} \quad . \tag{5.27}$$

Thus the member transformation matrices are given by

$$\mathbf{t_a} = \begin{bmatrix} 1.0 & 0 & 0 & 0 & 0 \\ 0 & 1.0 & 0 & 0 & 0 \end{bmatrix} , \quad \mathbf{t_b} = \begin{bmatrix} 0 & 1.0 & 0 & 0 & 0 \\ 0 & 0 & 1.0 & 0 & 0 \end{bmatrix}$$

$$\mathbf{t_c} = \begin{bmatrix} 0 & 0 & 1.0 & 0 & 0 \\ 0 & 0 & 0 & 1.0 & 0 \end{bmatrix} , \quad \mathbf{t_d} = \begin{bmatrix} 0 & 0 & 0 & 1.0 & 0 \\ 0 & 0 & 0 & 0 & 1.0 \end{bmatrix} \quad .$$

From (5.21) the contribution of member b for example to the global stiffness matrix is then

$$
t_b^T k_b t_b = \begin{bmatrix} 0 & 0 \\ 1.0 & 0 \\ 0 & 1.0 \\ 0 & 0 \\ 0 & 0 \end{bmatrix} \begin{bmatrix} 4.0 & 2.0 \\ 2.0 & 4.0 \end{bmatrix} \begin{bmatrix} 0 & 1.0 & 0 & 0 & 0 \\ 0 & 0 & 1.0 & 0 & 0 \end{bmatrix} \psi_b
$$

$$
= \begin{bmatrix} 0 & 0 & 0 & 0 & 0 \\ 0 & 4.0\psi_b & 2.0\psi_b & 0 & 0 \\ 0 & 2.0\psi_b & 4.0\psi_b & 0 & 0 \\ 0 & 0 & 0 & 0 & 0 \\ 0 & 0 & 0 & 0 & 0 \end{bmatrix}.
$$

Thus carrying out this calculation for all the members and summing the results, leads to the total global matrix **k** in the following form:

$$
k = \begin{bmatrix} 4.0\psi_a & 2.0\psi_a & 0 & 0 & 0 \\ 2.0\psi_a & 4.0(\psi_a + \psi_b) & 2.0\psi_b & 0 & 0 \\ 0 & 2.0\psi_b & 4.0(\psi_b + \psi_c) & 2.0\psi_c & 0 \\ 0 & 0 & 2.0\psi_c & 4.0(\psi_c + \psi_d) & 2.0\psi_d \\ 0 & 0 & 0 & 2.0\psi_d & 4.0\psi_d \end{bmatrix} \quad (5.28)
$$

The above calculation illustrates the use of (5.21). However, it is worth noting that in this simple case, the global stiffness matrix could have been calculated by inspection. Thus in the second column say, the stiffness coefficients k_{12}, k_{22}, k_{32}, k_{42} and k_{52} are the global forces causing a unit rotation at joint 2, with $d_k = 0$, $k \neq 2$, as in Fig. 5.15. In this case, only members a and b are deformed and

$$
d_{a1} = 0, \qquad d_{a2} = 1.0
$$
$$
d_{b1} = 1.0, \qquad d_{b2} = 0 .
$$

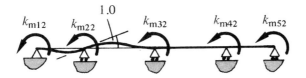

Fig. 5.15

The global forces are therefore in equilibrium with the appropriate member forces as follows

$$k_{12} = k_{a12} = 2.0\psi_a$$
$$k_{22} = (k_{a22} + k_{b11}) = 4.0(\psi_a + \psi_b)$$
$$k_{32} = k_{b21} = 2.0\psi_b$$

giving the non-zero terms in the second column of **k**. The particular values of k_{ij} for the present problem can be found by inserting the member parameters for the bridge into (5.28). Thus

$$\mathbf{k} = \begin{bmatrix} 2.0 & 1.0 & 0 & 0 & 0 \\ 1.0 & 3.33 & 0.67 & 0 & 0 \\ 0 & 0.67 & 2.67 & 0.67 & 0 \\ 0 & 0 & 0.67 & 3.33 & 1.0 \\ 0 & 0 & 0 & 1.0 & 2.0 \end{bmatrix} \times 10^9 \ \text{Nm/rad} \quad .$$

The loading takes the form of concentrated couples which directly correspond to the global degrees of freedom. The global force matrix is therefore given by

$$\mathbf{P} = \begin{bmatrix} -2.0 \\ -3.0 \\ 5.0 \\ 0 \\ 0 \end{bmatrix} \text{MNm} \quad .$$

Whence using

$$\mathbf{d} = \mathbf{k}^{-1} \mathbf{P} \qquad (5.3)^*$$

we obtain the solution in the form of the rotations of the continuous beam at the joints. Thus

$$
\mathbf{d} = \begin{bmatrix} -0.37 \\ -1.25 \\ 2.33 \\ -0.55 \\ 0.27 \end{bmatrix} \times 10^{-3} \text{ rad} \quad .
$$

It then remains to find the bending moment and shear force diagrams. As before, the member forces are given by (5.18) as

$$
\mathbf{P}_m = \mathbf{k}_m \mathbf{t}_m \mathbf{d} \tag{5.18}*
$$

where $\mathbf{t}_m \mathbf{d}$ are the member displacements. However, recalling that the member displacements in this case are very simply related to $\mathbf{d}$ by (5.27) we can calculate $\mathbf{P}_b$ say, as follows

$$
\mathbf{P}_b = 0.333 \times 10^9 \begin{bmatrix} 4.0 & 2.0 \\ 2.0 & 4.0 \end{bmatrix} \begin{bmatrix} -1.25 \\ 2.33 \end{bmatrix} \times 10^{-3} = \begin{bmatrix} -0.12 \\ 2.27 \end{bmatrix} \text{ MNm} \quad .
$$

The member forces for the other three members are calculated similarly and are given by

$$
\mathbf{P}_a = \begin{bmatrix} -2.0 \\ -2.88 \end{bmatrix} \text{ MNm}, \quad \mathbf{P}_c = \begin{bmatrix} 2.74 \\ 0.82 \end{bmatrix} \text{ MNm}, \quad \mathbf{P}_d = \begin{bmatrix} -0.82 \\ 0 \end{bmatrix} \text{ MNm} \quad .
$$

The bending moment diagram for the continuous beam is made up of the corresponding bending moment diagrams for the members. In determining the latter, each member can be treated as an individual simply supported beam subject to the above known end couples. This type of problem has already been analysed in Chapter 2, as Example 2.16. There it is shown that the bending moment diagram is a straight line between the moments at the ends due to the couples, and that a positive couple on the left-hand end produces a negative bending moment, and on the right-hand end, a positive bending moment. Thus the complete bending moment diagram corresponding to the member forces above, is shown in Fig. 5.16(a). Note that the step changes in the diagram at the joints, are caused by the present rather unusual loading in the form of concentrated couples.

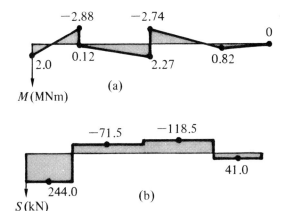

Fig. 5.16

In order to obtain the shear force diagram, we first have to calculate the reactions on the members. These can be called $\mathbf{R}_m$, acting as in Fig. 5.17. Thus resolving vertically and taking moments about end 1 of the member leads to

$$\mathbf{R}_m = \frac{(P_{m1} + P_{m2})}{l} \begin{bmatrix} 1.0 \\ -1.0 \end{bmatrix} \quad . \tag{5.29}$$

The reactions on the individual members in this present example are therefore

$$\mathbf{R}_a = \begin{bmatrix} -244.0 \\ 244.0 \end{bmatrix} \text{ kN}, \quad \mathbf{R}_b = \begin{bmatrix} 71.5 \\ -71.5 \end{bmatrix} \text{ kN},$$

$$\mathbf{R}_c = \begin{bmatrix} 118.5 \\ -118.5 \end{bmatrix} \text{ kN}, \quad \mathbf{R}_d = \begin{bmatrix} -41.0 \\ 41.0 \end{bmatrix} \text{ kN} \quad .$$

By considering the equilibrium of sub-sections cut from the members, it can be shown that the shear force is constant in each member and equal to $-R_{m1}$. Again, the shear force diagram for the beam is made up of the shear force diagrams of the members, and this is shown in Fig. 5.16(b).

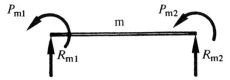

Fig. 5.17

5.5.4 Continuous Beam Subject to Intermediate Loading

In Example 5.3, we solved a continuous beam subject only to forces that corresponded to the global degrees of freedom. It was therefore a simple matter to determine the global force matrix. However, it is usual for a beam to be loaded by forces that do *not* correspond to the degrees of freedom; forces such as, for example, the distributed forces and concentrated force on the bridge in Example 4.11, and shown in Fig. 5.18. We shall call these forces **intermediate forces**, and in order to distinguish them from the global forces **P** corresponding to the degrees of freedom, we shall denote them by **Q** or **q**, the former referring to concentrated forces, the latter to distributed forces.

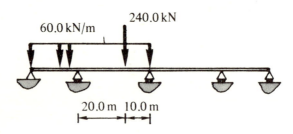

Fig. 5.18

The first objective of a stiffness analysis involving intermediate forces, is to obtain the **equivalent global forces**. These are defined as the global forces, which when applied alone to the structure, cause the same changes in the global degrees of freedom as those caused by the intermediate forces. For clarity we shall call these equivalent forces P_Q. Thus if P_Q can be found, then by definition the global degrees of freedom **d** are given by

$$d = k^{-1} P_Q . \tag{5.30}$$

The method of obtaining P_Q is based on the concept of **global fixing forces**. These are defined as the global forces, which when applied to the structure loaded by the intermediate forces, reduce the global degrees of freedom to zero. These we shall call P_F. In the case of the continuous beam, they take the form of the fixing couples shown in Fig. 5.19(a). If we then consider the effects of the intermedicate forces **Q** and of P_F on the structure in the combinations shown in Table 5.3 and invoke superposition, we can show that the forces $-P_F$ cause the same changes in the global degrees of freedom as **Q**. Thus

$$P_Q = -P_F . \tag{5.31}$$

The reason for introducing the fixing forces P_F, is that they can be easily calculated . Thus they are in equilibrium with the **member fixing forces** P_{Fm}, which are the forces required to reduce the member degrees of freedom to zero.

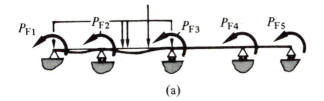

(a)

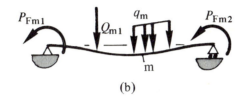

(b)

Fig. 5.19

Table 5.3
Loading combinations to demonstrate that $P_Q = -P_F$

Loading	Corresponding displacements
Q	d
$Q + P_F$	0
$\therefore\ P_F$	$-d$
$\therefore\ -P_F$	d

For a member of the continuous beam, shown in Fig. 5.19(b), these are the **fixed-end couples** needed to restrain the end rotations. It will be shown below that P_{Fm} can be calculated for any type of intermediate loading on the member. Thus if P_{Fm} are known, the equilibrium equation (5.17) between the global and member forces gives

$$P_F = \sum_m (t_m^T P_{Fm}) \ . \tag{5.32}$$

In the case of the continuous beam, for example, (5.32) is the formal expression of the simple equilibrium equations at each joint, which take the form

$$P_{F1} = P_{Fa1}$$

$$P_{F2} = P_{Fa2} + P_{Fb1}$$

$$P_{F3} = P_{Fb2} + P_{Fc1} \qquad (5.33)$$

$$P_{F4} = P_{Fc2} + P_{Fd1}$$

$$P_{F5} = P_{Fd2} \ .$$

The fixed-end couples on a member can be obtained by carrying out a flexibility analysis for the member treated as a simply supported beam, subject to end couples *and* to intermediate loading. This calculation is presented in the second part of Appendix A5.2. There it is shown that the member equilibrium equations of (5.13) are extended to

$$\mathbf{P_m} = \mathbf{k_m}\,\mathbf{d_m} + \mathbf{P_{Fm}} \qquad (5.34)$$

and $\mathbf{P_{Fm}}$ are calculable in terms of the bending moments M_Q due to the intermediate forces acting alone on the member. Thus

$$\mathbf{P_{Fm}} = -\mathbf{k_m}\,\mathbf{d_{Qm}} \qquad (5.35)$$

where

$$\mathbf{d_{Qm}} = \left\{\frac{1}{EI}\right\}_m \left[\begin{array}{c} -\int_0^l \left(1 - \frac{x}{l}\right) M_Q dx \\[2ex] \int_0^l \left(\frac{x}{l}\right) M_Q dx \end{array} \right]_m \qquad (5.36)$$

(For conciseness in Chapters 5 and 6, we shall extend the suffix notation of (3.7), so that m used as a subscript to a matrix as in (5.36), means that all the parameters inside the matrix take the values associated with member m.)

The fixed-end couples can be worked out in advance of the stiffness solution, for any particular type of intermediate loading on the member. Two important cases, the concentrated force Q_m, and the distributed force q_m per unit length over the whole member, will be considered here. Thus for the concentrated force shown in Fig. 5.20(a), we have

$$M_Q = Q_m \left\{\frac{bx}{l}\right\}_m \ [x < a] \ \text{ or } \ Q_m \left\{\frac{a(l-x)}{l}\right\}_m \ [x > a]$$

as in the figure. Therefore

$$\mathbf{d_{Qm}} = Q_m \left\{\frac{ab}{6EIl}\right\}_m \left[\begin{array}{c} -(l+b) \\[1ex] (l+a) \end{array} \right]_m$$

and

$$\mathbf{P}_{Fm} = Q_m \left\{ \frac{ab}{l^2} \right\}_m \begin{bmatrix} b \\ -a \end{bmatrix}_m .$$ (5.37)

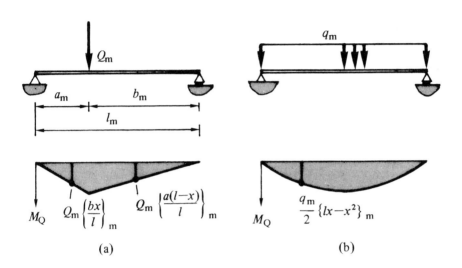

(a)   (b)

Fig. 5.20

For the distributed force shown in Fig. 5.20(b), we have

$$M_Q = \frac{q_m}{2} \{lx - x^2\}_m$$

as in the figure. Therefore

$$d_{Qm} = q_m \left\{ \frac{l^3}{24EI} \right\}_m \begin{bmatrix} -1.0 \\ 1.0 \end{bmatrix}$$

and

$$\mathbf{P}_{Fm} = \frac{q_m l_m^2}{12} \begin{bmatrix} 1.0 \\ -1.0 \end{bmatrix} .$$ (5.38)

Fixed-end couples for other types of loading are given in Table 5.4.

Table 5.4
Fixed-end couples

Load type	P_{Fm1}	P_{Fm2}
	$\dfrac{qa^2}{6l}(3l - 2a)$	$\dfrac{-qa^2}{6l}(3l - 2a)$
	$\dfrac{q}{12l^2}[e^3(4l - 3e) - \\ -c^3(4l - 3c)]$	$\dfrac{-q}{12l^2}[d^3(4l - 3d) - \\ -a^3(4l - 3a)]$
	$\dfrac{qal}{12}r(3r^2 - 8r + 6)$ $(r = a/l)$	$-\dfrac{qal}{12}r^2(4 - 3r)$
	$\dfrac{ql^2}{30}$	$\dfrac{-ql^2}{20}$
	$\dfrac{5}{96}ql^2$	$-\dfrac{5}{96}ql^2$
	$\dfrac{Qb}{l^2}(3a - l)$	$\dfrac{Qa}{l^2}(3b - l)$

Example 5.4
The process of assembling the equivalent global force matrix and solving the
continuous beam, will be illustrated for the bridge-beam with the same properties
as that in Example 5.3, but loaded as in Fig. 5.18.

We first evaluate the member fixing forces for the continuous beam using
(5.37) and (5.38) as appropriate. Thus

$$\mathbf{P}_{Fa} = \frac{60.0 \times 10^3 \times 20.0^2}{12} \begin{bmatrix} 1.0 \\ -1.0 \end{bmatrix} = \begin{bmatrix} 2.0 \\ -2.0 \end{bmatrix} \text{ MNm}$$

$$\mathbf{P}_{Fb} = \frac{60.0 \times 10^3 \times 30.0^2}{12} \begin{bmatrix} 1.0 \\ -1.0 \end{bmatrix} +$$

$$+ \frac{240.0 \times 10^3 \times 20.0 \times 10.0}{30.0^2} \begin{bmatrix} 10.0 \\ -20.0 \end{bmatrix} = \begin{bmatrix} 5.03 \\ -5.57 \end{bmatrix} \text{ MNm}$$

$$\mathbf{P}_{Fc} = \mathbf{P}_{Fd} = 0 \ .$$

Whence using the equilibrium equations (5.33), we assemble the global fixing
force matrix as follows:

$$\mathbf{P}_F = \begin{bmatrix} 2.0 \\ -2.0 + 5.03 \\ -5.57 + 0 \\ 0 + 0 \\ 0 \end{bmatrix} = \begin{bmatrix} 2.0 \\ 3.03 \\ -5.57 \\ 0 \\ 0 \end{bmatrix} \text{ MNm} \ .$$

The equivalent global force matrix $\mathbf{P}_Q$ is therefore given by

$$\mathbf{P}_Q = -\mathbf{P}_F = \begin{bmatrix} -2.0 \\ -3.03 \\ 5.57 \\ 0 \\ 0 \end{bmatrix} \text{ MNm} \ .$$

The global stiffness matrix of the beam is the same as that for the beam in Example 5.3. The solution of the present problem, in terms of the global displacements calculated by (5.30) then comes out to be

$$
\mathbf{d} = \begin{bmatrix} -0.34 \\ -1.32 \\ 2.57 \\ -0.61 \\ 0.30 \end{bmatrix} \times 10^{-3} \text{ rad} .
$$

It then remains to determine the bending moment and shear force diagrams for the continuous beam, this time taking the intermediate forces into account. The member forces are now given by (5.34) as

$$
\mathbf{P_m} = \mathbf{k_m} \, \mathbf{d_m} + \mathbf{P_{Fm}} \tag{5.34}*
$$

and using (5.17), they are expressed in terms of the known global displacements as

$$
\mathbf{P_m} = \mathbf{k_m} \, \mathbf{t_m} \, \mathbf{d} + \mathbf{P_{Fm}} . \tag{5.39}
$$

However, again in this simple case, the member displacements are given in terms of $\mathbf{d}$ by (5.27). We can therefore calculate $\mathbf{P_b}$ say as follows

$$
\mathbf{P_b} = 0.333 \times 10^9 \begin{bmatrix} 4.0 & 2.0 \\ 2.0 & 4.0 \end{bmatrix} \begin{bmatrix} -1.32 \\ 2.57 \end{bmatrix} \times 10^{-3} +
$$

$$
+ \begin{bmatrix} 5.03 \\ -5.57 \end{bmatrix} \times 10^6 = \begin{bmatrix} 4.98 \\ -3.02 \end{bmatrix} \text{ MNm} .
$$

The member forces on the other three members are calculated similarly and are given by

$$
\mathbf{P_a} = \begin{bmatrix} 0 \\ -4.98 \end{bmatrix} \text{ MNm}, \quad \mathbf{P_c} = \begin{bmatrix} 3.02 \\ 0.91 \end{bmatrix} \text{ MNm}, \quad \mathbf{P_d} = \begin{bmatrix} 0.91 \\ 0 \end{bmatrix} \text{ MNm} .
$$

The bending moment diagram for the continuous beam is again made up of the bending moment diagrams for the members, where each member is considered as a simply supported beam subject to the above known end couples and to the intermediate forces. In Example 2.16, it is shown that the bending moment diagram for such a case is found by superimposing the diagram for the bending moments M_P† due to the end couples alone, and the diagram for the bending moments M_Q due to the intermediate forces. M_P and M_Q are shown in Fig. 5.21, together with the final bending moment diagram M, the latter agreeing with that produced by the flexibility method for the same structure in Example 4.11. Note that, unlike the bending moment diagram produced by the concentrated couples in Example 5.3, this diagram is continuous over the supports.

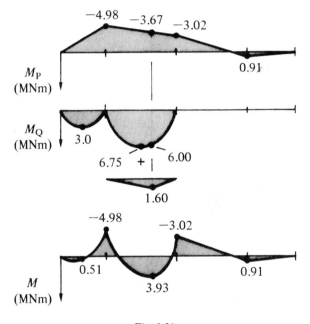

Fig. 5.21

In order to calculate the shear force diagram, we have to obtain the reactions on the members, $\mathbf{R}_m$, due both to the end couples and to the intermediate forces. The latter can be calculated for any particular forces $\mathbf{Q}$ by resolving and taking moments for the members. Calling these $\mathbf{R}_{Qm}$, we then modify (5.29) to give

$$\mathbf{R}_m = \left(\frac{P_{m1} + P_{m2}}{l}\right)\begin{bmatrix} 1.0 \\ -1.0 \end{bmatrix} + \mathbf{R}_{Qm} \,. \tag{5.40}$$

---

† Note that this use of M_P is distinct from that employed in Chapter 4 (for the moments due to the loading on the primary structure).

Thus the reactions on the four members due to the intermediate forces alone come out to be

$$\mathbf{R_{Qa}} = q_a \frac{l}{2} \begin{bmatrix} 1.0 \\ 1.0 \end{bmatrix} = \begin{bmatrix} 600.0 \\ 600.0 \end{bmatrix} \text{kN}$$

$$\mathbf{R_{Qb}} = q_b \frac{l}{2} \begin{bmatrix} 1.0 \\ 1.0 \end{bmatrix} + \frac{Q_b}{l} \begin{bmatrix} b \\ a \end{bmatrix} = \begin{bmatrix} 900.0 \\ 900.0 \end{bmatrix} + \begin{bmatrix} 80.0 \\ 160.0 \end{bmatrix} = \begin{bmatrix} 980.0 \\ 1060.0 \end{bmatrix} \text{kN}$$

$$\mathbf{R_{Qc}} = \mathbf{R_{Qd}} = \mathbf{0} \ ,$$

and the total member reactions are

$$\mathbf{R_a} = \begin{bmatrix} 350.8 \\ 849.1 \end{bmatrix} \text{kN}, \ \mathbf{R_b} = \begin{bmatrix} 1045.4 \\ 994.6 \end{bmatrix} \text{kN},$$

$$\mathbf{R_c} = \begin{bmatrix} 131.0 \\ -131.0 \end{bmatrix} \text{kN}, \ \mathbf{R_d} = \begin{bmatrix} -45.3 \\ 45.3 \end{bmatrix} \text{kN} \ .$$

The member shear force diagrams and thence the total diagram for the continuous beam are conveniently constructed by the rule of thumb discussed in Chapter 2, Section 2.5.2. The final diagram is shown in Fig. 5.22. Note that the total step change at each joint is equal to the total reaction provided by the support.

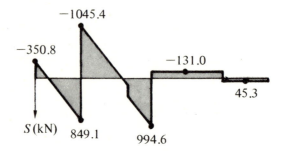

Fig. 5.22

5.6 RIGID-JOINTED PLANE FRAMEWORKS

5.6.1 Introduction

The stiffness method when applied to rigid-jointed plane frameworks, can be carried out at different levels of complexity, depending on the type of framework. Thus a classical method, called the **slope-deflection equation method**, analyses building frames for which axial forces and deflections can be ignored. Such frameworks can be solved with relatively few degrees of freedom. The method forms a useful introduction to the more general stiffness method for plane frameworks, and it will be explained below — together with an associated relaxation method called the **method of moment distribution**.

5.6.2 Building Frames with Sidesway Prevented

Example 5.5

Consider the framework in Fig. 5.23(a) with member and joint notation as in Fig. 5.23(b). The members are of steel, and all have the same cross-section with $I = 30.0 \times 10^6$ mm^4. We shall consider only bending actions in the framework, and ignore joint displacements due to the axial deformation of the members. Further, the presence of the prop at joint 3 restrains sidesway of the framework due to column bending. Thus joints 1, 2 and 3 can be assumed to remain stationary. The deformation of the framework is therefore completely defined by the three joint rotations d_1, d_2 and d_3. These then are the global degrees of freedom, and the three couples P_1, P_2 and P_3 are the global forces.

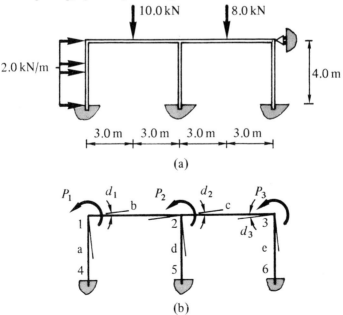

(a)

(b)

Fig. 5.23

The solution method is very similar to that for the continuous beam. Thus the global stiffness matrix is given by

$$k = \sum_m (t_m^T k_m t_m) \qquad (5.21)*$$

where k_m is the member stiffness matrix given by

$$k_m = \begin{bmatrix} 4.0 & 2.0 \\ 2.0 & 4.0 \end{bmatrix} \psi_m \qquad (5.26)*$$

and

$$\psi_m = \left\{ \frac{EI}{l} \right\}_m \cdot \qquad (5.24)*$$

Again, in the present example, the member transformation matrices are easily found by inspection and are given by

$$t_a = \begin{bmatrix} 1.0 & 0 & 0 \\ 0 & 0 & 0 \end{bmatrix}, \; t_b = \begin{bmatrix} 1.0 & 0 & 0 \\ 0 & 1.0 & 0 \end{bmatrix}, \; t_c = \begin{bmatrix} 0 & 1.0 & 0 \\ 0 & 0 & 1.0 \end{bmatrix}$$

$$t_d = \begin{bmatrix} 0 & 1.0 & 0 \\ 0 & 0 & 0 \end{bmatrix}, \; t_e = \begin{bmatrix} 0 & 0 & 1.0 \\ 0 & 0 & 0 \end{bmatrix} \cdot$$

Thus

$$k = \begin{bmatrix} 4.0(\psi_a + \psi_b) & 2.0\psi_b & \\ 2.0\psi_b & 4.0(\psi_b + \psi_c + \psi_d) & 2.0\psi_c \\ & 2.0\psi_c & 4.0(\psi_c + \psi_e) \end{bmatrix}$$

and inserting the particular member parameters for the structure, we obtain

$$k = \begin{bmatrix} 10.0 & 2.0 & 0 \\ 2.0 & 14.0 & 2.0 \\ 0 & 2.0 & 10.0 \end{bmatrix} \; \text{MNm/rad} \; \cdot$$

The equivalent global forces $\mathbf{P_Q}$ are equal to minus the fixing forces $\mathbf{P_F}$, which again are assembled from the member fixing forces using

$$\mathbf{P_F} = \sum_m (\mathbf{t_m^T P_{Fm}}) \quad . \tag{5.32}*$$

In the present example this gives

$$P_{F1} = P_{Fa1} + P_{Fb1}$$

$$P_{F2} = P_{Fb2} + P_{Fc1} + P_{Fd1}$$

$$P_{F3} = P_{Fc2} + P_{Fe1} \quad .$$

The member fixing forces, given by (5.37) and (5.38) are

$$\mathbf{P_{Fa}} = \frac{q_a l_a^2}{12} \begin{bmatrix} 1.0 \\ -1.0 \end{bmatrix} = \frac{-2.0 \times 10^3 \times 4.0^2}{12} \begin{bmatrix} 1.0 \\ -1.0 \end{bmatrix} = \begin{bmatrix} -2.67 \\ 2.67 \end{bmatrix} \text{ kNm}$$

$$\mathbf{P_{Fb}} = Q_b \left(\frac{ab}{l^2}\right)_b \begin{bmatrix} b \\ -a \end{bmatrix}_b = \frac{10.0 \times 10^3 \times 3.0^2}{6.0^2} \begin{bmatrix} 3.0 \\ -3.0 \end{bmatrix} = \begin{bmatrix} 7.5 \\ -7.5 \end{bmatrix} \text{ kNm}$$

$$\mathbf{P_{Fc}} = Q_c \left(\frac{ab}{l^2}\right)_c \begin{bmatrix} b \\ -a \end{bmatrix}_c = \frac{8.0 \times 10^3 \times 3.0^2}{6.0^2} \begin{bmatrix} 3.0 \\ -3.0 \end{bmatrix} = \begin{bmatrix} 6.0 \\ -6.0 \end{bmatrix} \text{ kNm}$$

$$\mathbf{P_{Fd}} = \mathbf{P_{Fl}} = 0 \quad .$$

(The reader should note that a minus sign is assigned to the uniformly distributed force in member a because the force is directed in the *opposite* direction to that assumed for q_m in the derivation of (5.38).) The global fixing force matrix is therefore given by

$$\mathbf{P_F} = \begin{bmatrix} 4.83 \\ -1.50 \\ -6.00 \end{bmatrix} \text{ kNm} \quad .$$

The solution in terms of the three joint rotations is given by

$$d = k^{-1}P_Q = -k^{-1}P_F = \begin{bmatrix} -0.503 \\ 0.096 \\ 0.581 \end{bmatrix} \times 10^{-3} \text{ rad} .$$

Finally, the member forces given by (5.39) as

$$P_m = k_m t_m d + P_{Fm} \qquad (5.39)*$$

are calculated as follows

$$P_a = 1.5 \times 10^6 \begin{bmatrix} 4.0 & 2.0 \\ 2.0 & 4.0 \end{bmatrix} \begin{bmatrix} 1.0 & 0 & 0 \\ 0 & 0 & 0 \end{bmatrix} \begin{bmatrix} -0.503 \\ 0.096 \\ 0.581 \end{bmatrix} \times 10^{-3} +$$

$$+ \begin{bmatrix} -2.67 \\ 2.67 \end{bmatrix} \times 10^3 = \begin{bmatrix} -5.68 \\ 1.16 \end{bmatrix} \text{kNm},$$

$$P_b = \begin{bmatrix} 5.68 \\ -8.12 \end{bmatrix} \text{kNm}, \quad P_c = \begin{bmatrix} 7.55 \\ -3.48 \end{bmatrix} \text{kNm},$$

$$P_d = \begin{bmatrix} 0.58 \\ 0.29 \end{bmatrix} \text{kNm}, \quad P_e = \begin{bmatrix} 3.48 \\ 1.74 \end{bmatrix} \text{kNm} .$$

The corresponding bending moment diagram for the framework is then constructed using the method discussed in Example 5.4. It is shown in Fig. 5.24.

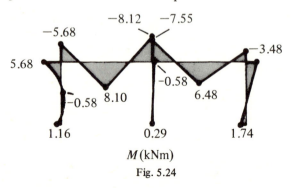

M (kNm)

Fig. 5.24

The method followed in Example 5.5 can be applied to building frames with any number of bays or storeys, provided sidesway is prevented at every floor level. However, it should also be noted that the method is applicable to frameworks which are free to sway, provided the structure and loading are symmetrical about a vertical centre line, for in this case, the structure has no tendency to sway even when unpropped.

5.6.3 Moment Distribution

The method of **moment distribution** is essentially a way of analysing frameworks of the type considered in Example 5.5, but avoiding the direct solution of the simultaneous equilibrium equations. It is extremely useful therefore, for hand analysis. The method is mathematically equivalent to starting from an **initial state** where the joints of the framework are fixed against rotation, and releasing each joint in turn until the structure reaches a **final state** with the joints free.

A central concept in the method is the way a framework responds to an external couple applied to a joint. This can be examined for a simple case by considering the framework in Fig. 5.25(a), which has a single free joint, and whose deformation is defined by the single rotational degree of freedom d_1. We wish to determine how the corresponding couple P_1 is distributed into the framework. The global stiffness matrix is composed of the single term k_{11}, and this is assembled from the member stiffnesses in the usual way. Formally we have

$$k = \sum_m (t_m^T k_m t_m) \tag{5.21}*$$

where

$$t_a = \begin{bmatrix} 1.0 \\ 0 \end{bmatrix}, \ t_b = \begin{bmatrix} 1.0 \\ 0 \end{bmatrix}, \ t_c = \begin{bmatrix} 1.0 \\ 0 \end{bmatrix}$$

and

$$k_m = \begin{bmatrix} 4.0 & 2.0 \\ 2.0 & 4.0 \end{bmatrix} \psi_m \ . \tag{5.26}*$$

Therefore

$$k_{11} = 4.0(\psi_a + \psi_b + \psi_c) \ .$$

The solution of the framework in terms of d_1 is then given by

$$d_1 = \frac{P_1}{k_{11}} = \frac{P_1}{4.0(\psi_a + \psi_b + \psi_c)}$$

and the member forces such as $\mathbf{P}_a$ are obtained as

$$\mathbf{P}_a = \mathbf{k}_a t_a d_1 = \begin{bmatrix} 1.0 \\ 0.5 \end{bmatrix} \frac{4.0\psi_a P_1}{4.0(\psi_a + \psi_b + \psi_c)} \quad . \tag{5.41}$$

The function $4.0\psi_a/(4.0(\psi_a + \psi_b + \psi_c))$ is called the **distribution factor** of member a at joint 1. Suppose we call this δ_a. Since $4.0\psi_m = k_{m\,11}$, it is then apparent that δ_a is the stiffness of member a at joint 1 divided by the sum of the stiffnesses of all the members at the joint. Using a similar notation for the distribution factors for members b and c, the forces in all three members then take the form

$$\mathbf{P}_a = \begin{bmatrix} \delta_a \\ \delta_a/2 \end{bmatrix} P_1, \ \mathbf{P}_b = \begin{bmatrix} \delta_b \\ \delta_b/2 \end{bmatrix} P_1, \ \mathbf{P}_c = \begin{bmatrix} \delta_c \\ \delta_c/2 \end{bmatrix} P_1 \quad . \tag{5.42}$$

These forces are the couples shown in Fig. 5.25(b). We then see that the external couple P_1 is distributed into the members at joint 1 in the ratio of their distribution factors and that couples equal to half the member couples at 1 are induced at the remote fixed ends of the members at joints 2, 3 and 4. These secondary couples are called the **carry-over couples**.

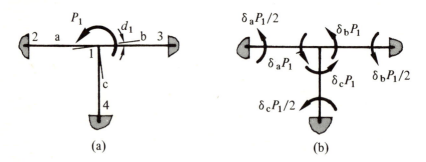

(a)   (b)

Fig. 5.25

Example 5.6
We shall illustrate the moment distribution method by reworking Example 5.5. We first calculate the distribution factors for the three joints that are free to rotate as in Table 5.5. The moment distribution is then carried out entirely in terms of the member couples. It is undertaken in a tabular format as in Table 5.6(a), each column representing a member force. For clarity in relating the distribution to the actual framework, the columns are identified by the member and the

global joint to which the particular end of the member is attached. Thus column a4 for example, refers to the member couple P_{a2}, because 'a' is oriented with end 1 at joint 1, and end 2 at joint 4. For ease of reference the distribution factors for the members are then inserted into the appropriate columns at the free joints. These are taken directly from Table 5.5. The moment distribution starts from the initial state when all the joints are fixed against rotation. The corresponding member couples are the fixed-end couples calculated from the intermediate loading using again (5.37) and (5.38). These have already been obtained in Example 5.5, and they are inserted in the first row of the distribution. The initial state is produced by the global fixing forces, which are equal to the amount by which the member forces at each joint are out of balance. Thus at joint 1 for example the out-of-balance couple is $(-2.67 + 7.50) = 4.83$ kNm. The first step in the distribution is to release joint 1, while keeping joints 2 and 3 fixed. Note that at this stage the framework then has a single free joint, and behaves like the simple framework considered above.

Table 5.5
Distribution Factors

Joint	Member	ψ_m 10^6	$\Sigma\psi_m$ 10^6	δ_m
1	a	1.5	2.5	0.60
	b	1.0		0.40
2	b	1.0		0.286
	c	1.0	3.5	0.286
	d	1.5		0.429
3	c	1.0	2.5	0.40
	e	1.5		0.60

The release of joint 1 is achieved by adding an external couple of -4.83 kNm to the joint, and inducing corresponding couples in members a and b in the ratio of the distribution factors. These are entered in row 2 of the table. A line is then drawn under the couples at joint 1, and indicates that all the couples above this line are in balance. The addition of -4.83 kNm to joint 1 also induces carry-over couples at the remote ends of members a and b, at joints 4 and 2. These are entered in row 3 of the table. The next step in the distribution is to release joint 2, keeping joints 1 and 3 fixed. The out-of-balance couple at joint 2 is $(-7.5 + 6.0 - 0.97) = -2.47$ kNm and joint 2 is therefore released by adding an external couple of 2.47 kNm to the joint. The corresponding member couples are entered in row 4. Carry-over couples are then induced at the remote ends of

Table 5.6(a)
Moment distribution (moments in kNm)

Row no. ↓	a4	a1	b1	b2	d2	c2	c3	e3	d5	e6
		0.60	0.40	0.286	0.429	0.286	0.40	0.60		
1	2.67	−2.67	7.5	−7.5	0	6.0	−6.0	0	0	0
2		−2.90	−1.93							
3	−1.45			−0.97						
4				0.71	1.06	0.71				
5			0.36				0.36		0.53	
6							2.26	3.38		
7						1.13				1.69
8				−0.32	−0.48	−0.32				
9			−0.16				−0.16		−0.24	
10		−0.12	−0.08				0.06	0.10		
11	−0.06			−0.04		0.03				0.05
12	1.16	−5.69	5.69	−8.12	0.58	7.55	−3.48	3.48	0.29	1.74

Table 5.6(b)
Moment distribution (condensed format)

Row no. ↓	a4	a1	b1	b2	d2	c2	c3	e3	d5	e6
1	2.67	−2.67	7.5	−7.5	0	6.0	−6.0	0	0	0
2		−2.90	−1.93	0.43	0.64	0.43	2.40	3.60		
3	−1.45		0.22	−0.97		1.20	0.22		0.32	1.80
4		−0.13	−0.09	−0.07	−0.10	−0.07	−0.09	−0.13		
5	−0.07		−0.04	−0.05		−0.05	−0.04		−0.05	−0.07
6		0.02	0.02	0.03	0.04	0.03	0.02	0.02		
7	0.01		0.01	0.01		0.01	0.01		0.02	0.01
8	1.16	−5.68	5.69	−8.12	0.58	7.55	−3.48	3.49	0.29	1.74

members b, c and d and these are entered in row 5. Note that at this stage joint 1 is again out of balance by 0.36 kNm. Joint 3 is then released as in row 6, with carry-over couples going to joints 2 and 6 as in row 7. We next consider joints 1 and 2 again, which are out of balance as a result of the carry-over couples from the previous releases. Joint 2 has the bigger out-of-balance couple of 1.13 kNm and this is released next in row 8, sending carry-over couples to joints 1 and 3. The calculation then proceeds by releasing joints 1 and 3. It is apparent that the moment distribution method is a rapidly converging **iterative** method that will not yield an exact solution, since any given release will always produce carry-over couples. The distribution is therefore terminated when the carry-over couples or out-of-balance couples are of the same order of magnitude as the required accuracy for the solution. In this particular problem the distribution is terminated at row 11, when the out-of-balance couple at joint 2 is -0.01 kNm. The final member forces are determined by summing all the couples produced by the various operations of the distribution. These are shown in row 12. They agree to the chosen order of accuracy with the couples produced by the direct solution of the equilibrium equations in Example 5.5.

The above version of the moment distribution method follows a physical process of fixing and releasing joints in turn that can be carried out in a laboratory. This process, however, does give rise to an extended layout in the distribution table, and it is possible to produce a more concise presentation by releasing all the out-of-balance couples simultaneously in one row, and then inserting the carry-over couples simultaneously in the next row. This process is shown in Table 5.6(b) where the out-of-balance couples in row 1 are released together in row 2. The carry-over couples are inserted in row 3, and the corresponding out-of-balance couples released in row 4. This process, which cannot be modelled physically, does, however, converge to the same result as that produced by the previous moment distribution, as shown in row 8 of the table.

5.6.4 Frameworks Subject to Sidesway
Example 5.5 demonstrates the slope-deflection equation method in a restricted form applicable to frameworks propped against sidesway. The method can, however, be extended to deal with structures where sidesway is allowed. Thus consider the framework in Fig. 5.23, but with the lateral prop at joint 3 removed. An extra degree of freedom d_4, is then needed to define the deformation of the framework, this being the horizontal displacement of the beam members b and c as in Fig. 5.26. The corresponding global force is the equivalent horizontal force P_4 on the beam. The column members a, d and e are subject to transverse displacements at their ends, and the member stiffness matrices therefore have to be recalculated with the transverse displacements as extra degrees of freedom. The corresponding member equilibrium equations are then equivalent to the full slope-deflection equations of the classical method of analysis.

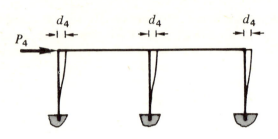

Fig. 5.26

Consider a typically deformed member subject both to end rotations and transverse end displacements as in Fig. 5.27. The enlarged member stiffness matrix corresponding to the four degrees of freedom can be determined directly from the stiffness matrix already calculated for the two rotational degrees of freedom d_2 and d_4. Thus we begin by subjecting the member to unit end rotations in turn. Unknown coefficients in the stiffness matrix can then be found simply by taking moments about axes through the ends of the member.

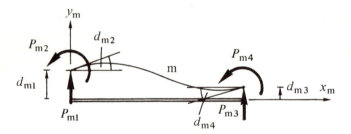

Fig. 5.27

Thus in Fig. 5.28(a), where $d_{m2} = 1.0$, $d_{mk} = 0$, $k \neq 2$, the end couples are $4.0\{EI/l\}_m$ and $2.0\{EI/l\}_m$ and act as shown. These are now the stiffness coefficients k_{m22} and k_{m42}. The transverse forces at the ends of the member are by definition the stiffness coefficients k_{m12} and k_{m32}. Thus taking moments about each end in turn, we have

$$k_{m12} = \left\{\frac{6.0EI}{l^2}\right\}_m , \quad k_{m32} = \left\{\frac{-6.0EI}{l^2}\right\}_m .$$

Similarly in Fig. 5.28(b), where $d_{m4} = 1.0$, $d_{mk} = 0$, $k \neq 4$, the stiffness coefficients k_{m14} and k_{m34} are given by

$$k_{m14} = \left\{\frac{6.0EI}{l^2}\right\}_m , \quad k_{m34} = \left\{\frac{-6.0EI}{l^2}\right\}_m .$$

We next constrain the member to deform with $d_{m1} = 1.0$, $d_{mk} = 0$, $k \neq 1$, as in Fig. 5.28(c). In this case, the end couples are k_{m21} and k_{m41}, and assuming that the member stiffness matrix is symmetrical, they are already known from the above results. Thus

$$k_{m21} = k_{m41} = \left\{ \frac{6.0EI}{l^2} \right\}_m .$$

These couples are shown explicitly in Fig. 5.28(c). Whence, taking moments again, we obtain

$$k_{m11} = \left\{ \frac{12.0EI}{l^3} \right\}_m , \quad k_{m31} = \left\{ \frac{-12.0EI}{l^3} \right\}_m .$$

Similarly, in Fig. 5.28(d), where $d_{m3} = 1.0$, $d_{mk} = 0$, $k \neq 3$ the end couples are

$$k_{m23} = k_{m43} = \left\{ \frac{-6.0EI}{l^2} \right\}_m$$

and therefore

$$k_{m13} = \left\{ \frac{-12.0EI}{l^3} \right\}_m , \quad k_{m33} = \left\{ \frac{12.0EI}{l^3} \right\}_m .$$

Collecting all the above results together, the member stiffness matrix is given by

$$\mathbf{k}_m = \begin{bmatrix} 12.0EI/l^3 & 6.0EI/l^2 & -12.0EI/l^3 & 6.0EI/l^2 \\ 6.0EI/l^2 & 4.0EI/l & -6.0EI/l^2 & 2.0EI/l \\ -12.0EI/l^3 & -6.0EI/l^2 & 12.0EI/l^3 & -6.0EI/l^2 \\ 6.0EI/l^2 & 2.0EI/l & -6.0EI/l^2 & 4.0EI/l \end{bmatrix}_m \qquad (5.43)$$

or using the notation of (5.24)

$$\mathbf{k}_m = \begin{bmatrix} 12.0/l^2 & 6.0/l & -12.0/l^2 & 6.0/l \\ 6.0/l & 4.0 & -6.0/l & 2.0 \\ -12.0/l^2 & -6.0/l & 12.0/l^2 & -6.0/l \\ 6.0/l & 2.0 & -6.0/l & 4.0 \end{bmatrix}_m \psi_m . \qquad (5.44)$$

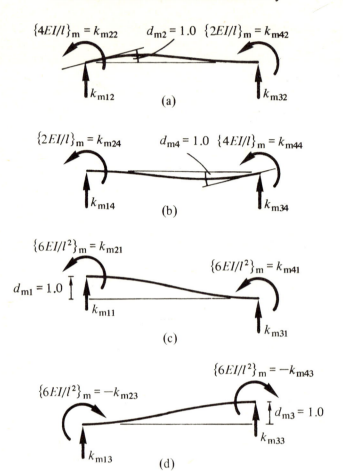

Fig. 5.28

In the presence of intermediate loading, the member equilibrium equations are again given by (5.34) as

$$\mathbf{P}_m = \mathbf{k}_m \mathbf{d}_m + \mathbf{P}_{Fm} \qquad (5.34)^*$$

where $\mathbf{P}_{Fm}$ are the four fixing forces needed to keep all the member degrees of freedom zero as in Fig. 5.29. The fixed-end couples P_{Fm2} and P_{Fm4} are the same as before and are given by (5.37) and (5.38) or Table 5.4, for particular intermediate forces. The transverse forces P_{Fm1} and P_{Fm3} are calculated by considering the rotational equilibrium of the member about the two ends. Thus for a member loaded by a concentrated force Q_m as in Fig. 5.20(a)

$$P_{Fm1} = Q_m \left\{ \frac{ab}{l^3}(b-a) + \frac{b}{l} \right\}_m \quad , \quad P_{Fm3} = Q_m \left\{ \frac{ab}{l^3}(a-b) + \frac{a}{l} \right\}_m .$$

After rearranging and collecting terms, the complete fixing force matrix then comes out to be

$$\mathbf{P}_{Fm} = \frac{Q_m}{l_m^2} \begin{bmatrix} \frac{b^2}{l}(l+2a) \\ ab^2 \\ \frac{a^2}{l}(l+2b) \\ -a^2 b \end{bmatrix}_m \tag{5.45}$$

For the member loaded by a uniformly distributed force q_m as in Fig. 5.20(b)

$$P_{Fm1} = P_{Fm3} = q_m l_m / 2$$

and

$$\mathbf{P}_{Fm} = q_m l_m \begin{bmatrix} 1/2 \\ l/12 \\ 1/2 \\ -l/12 \end{bmatrix}_m \tag{5.46}$$

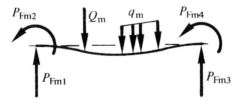

Fig. 5.29

The global stiffness matrix is calculated from the usual assembly equation

$$\mathbf{k} = \sum_m (\mathbf{t}_m^T \mathbf{k}_m \mathbf{t}_m) \tag{5.21}*$$

where for the columns such as a, d and e in Fig. 5.26, the transformation matrices t_m now contain four rows. Finally the equivalent global force matrix P_Q is again equal to $-P_F$ where P_F is given by

$$P_F = \sum_m (t_m^T P_{Fm}) \ . \tag{5.32}*$$

Example 5.7
We shall consider again the framework in Fig. 5.23, but this time with the prop at joint 3 removed. The degrees of freedom are d_1, d_2 and d_3 in Fig. 5.23 and d_4 in Fig. 5.26. Members a, d, and e are described by four member degrees of freedom, oriented with respect to the usual member coordinate systems. The member degrees of freedom for 'a' say, are then as shown in Fig. 5.30. Members b and c are described by two rotational degrees of freedom as before.

Fig. 5.30

The member transformation matrices are found by inspection to be as follows:

$$t_a = \begin{bmatrix} 0 & 0 & 0 & 1.0 \\ 1.0 & 0 & 0 & 0 \\ 0 & 0 & 0 & 0 \\ 0 & 0 & 0 & 0 \end{bmatrix}, \quad t_b = \begin{bmatrix} 1.0 & 0 & 0 & 0 \\ 0 & 1.0 & 0 & 0 \end{bmatrix},$$

$$t_c = \begin{bmatrix} 0 & 1.0 & 0 & 0 \\ 0 & 0 & 1.0 & 0 \end{bmatrix},$$

$$t_d = \begin{bmatrix} 0 & 0 & 0 & 1.0 \\ 0 & 1.0 & 0 & 0 \\ 0 & 0 & 0 & 0 \\ 0 & 0 & 0 & 0 \end{bmatrix}, \; t_e = \begin{bmatrix} 0 & 0 & 0 & 1.0 \\ 0 & 0 & 1.0 & 0 \\ 0 & 0 & 0 & 0 \\ 0 & 0 & 0 & 0 \end{bmatrix}.$$

Thus the contribution of member a for example, to the global stiffness matrix is calculated as

$$t_a^T k_a t_a = \begin{bmatrix} 0 & 1.0 & 0 & 0 \\ 0 & 0 & 0 & 0 \\ 0 & 0 & 0 & 0 \\ 1.0 & 0 & 0 & 0 \end{bmatrix} \begin{bmatrix} 12.0/l_a^2 & 6.0/l_a & -12.0/l_a^2 & 6.0/l_a \\ 6.0/l_a & 4.0 & -6.0/l_a & 2.0 \\ -12.0/l_a^2 & -6.0/l_a & 12.0/l_a^2 & -6.0/l_a \\ 6.0/l_a & 2.0 & -6.0/l_a & 4.0 \end{bmatrix} \times$$

$$\times \begin{bmatrix} 0 & 0 & 0 & 1.0 \\ 1.0 & 0 & 0 & 0 \\ 0 & 0 & 0 & 0 \\ 0 & 0 & 0 & 0 \end{bmatrix} \psi_a = \begin{bmatrix} 4.0 & 0 & 0 & 6.0/l_a \\ 0 & 0 & 0 & 0 \\ 0 & 0 & 0 & 0 \\ 6.0/l_a & 0 & 0 & 12.0/l_a^2 \end{bmatrix} \psi_a.$$

Whence, summing the contributions of the five members produces the assembled global stiffness matrix **k** in the following form:

$$k = \begin{bmatrix} 4.0(\psi_a + \psi_b) & 2.0\psi_b & 0 & 6.0\psi_a/l_a \\ 2.0\psi_b & 4.0(\psi_b + \psi_c + \psi_d) & 2.0\psi_c & 6.0\psi_d/l_d \\ 0 & 2.0\psi_c & 4.0(\psi_c + \psi_e) & 6.0\psi_e/l_e \\ 6.0\psi_a/l_a & 6.0\psi_d/l_d & 6.0\psi_e/l_e & 12.0\left(\dfrac{\psi_a}{l_a^2} + \dfrac{\psi_d}{l_d^2} + \dfrac{\psi_e}{l_e^2}\right) \end{bmatrix}$$

$$= \begin{bmatrix} 10.0 & 2.0 & 0 & 2.25 \\ 2.0 & 14.0 & 2.0 & 2.25 \\ 0 & 2.0 & 10.0 & 2.25 \\ 2.25 & 2.25 & 2.25 & 3.375 \end{bmatrix} \begin{array}{l} \text{MNm/rad,} \\ \text{MN/m} \\ \text{etc.} \end{array}$$

(Note that when the degrees of freedom include both linear *and* rotational displacements, the dimensions of the stiffness coefficients are mixed.)

The member fixing forces for the beam members b and c are the same as those calculated in Example 5.5. For member a we now produce the particular values of the four terms in the column matrix (5.46) using $q = -2.0$ kN/m. Thus

$$
\mathbf{P}_{Fa} = \begin{bmatrix} -4.0 \\ -2.67 \\ -4.0 \\ 2.67 \end{bmatrix} \begin{array}{l} \text{kN} \\ \text{kNm} \\ \text{kN} \\ \text{kNm} \end{array}
$$

For the unloaded columns d and e

$$
\mathbf{P}_{Fd} = \mathbf{P}_{Fe} = \mathbf{0} \quad .
$$

The contribution of member a to the global fixing force matrix is given by

$$
\mathbf{t}_a^T \mathbf{P}_{Fa} = \begin{bmatrix} 0 & 1.0 & 0 & 0 & -4.0 \\ 0 & 0 & 0 & 0 & -2.67 \\ 0 & 0 & 0 & 0 & -4.0 \\ 1.0 & 0 & 0 & 0 & 2.67 \end{bmatrix} = \begin{bmatrix} -2.67 \\ 0 \\ 0 \\ -4.0 \end{bmatrix} \begin{array}{l} \text{kNm} \\ \\ \\ \text{kN} \end{array} \quad .
$$

The contributions of members b and c are

$$
\mathbf{t}_b^T \mathbf{P}_{Fb} + \mathbf{t}_c^T \mathbf{P}_{Fc} = \begin{bmatrix} 1.0 & 0 \\ 0 & 1.0 \\ 0 & 0 \\ 0 & 0 \end{bmatrix} \begin{bmatrix} 7.5 \\ -7.5 \end{bmatrix} + \begin{bmatrix} 0 & 0 \\ 1.0 & 0 \\ 0 & 1.0 \\ 0 & 0 \end{bmatrix} \begin{bmatrix} 6.0 \\ -6.0 \end{bmatrix}
$$

$$
= \begin{bmatrix} 7.5 \\ -1.5 \\ -6.0 \\ 0 \end{bmatrix} \begin{array}{l} \text{kNm} \\ \text{kNm} \\ \text{kNm} \\ \end{array} \quad .
$$

Whence summing, we obtain the following global fixing force matrix

$$
\mathbf{P}_F = \begin{bmatrix} 4.83 \\ -1.5 \\ -6.0 \\ -4.0 \end{bmatrix} \begin{matrix} \text{kNm} \\ \text{kNm} \\ \text{kNm} \\ \text{kN} \end{matrix}
$$

The solution in terms of the three joint rotations and the sidesway is then

$$
\mathbf{d} = \mathbf{k}^{-1}\mathbf{P}_Q = -\mathbf{k}^{-1}\mathbf{P}_F = \begin{bmatrix} -0.834 \\ -0.070 \\ 0.249 \\ 1.622 \end{bmatrix} \times 10^{-3} \begin{matrix} \text{rad} \\ \text{rad} \\ \text{rad} \\ \text{m} \end{matrix}
$$

The member forces given by

$$
\mathbf{P}_m = \mathbf{k}_m \mathbf{t}_m \mathbf{d} + \mathbf{P}_{Fm} \tag{5.39}*
$$

are calculated as follows

$$
\mathbf{P}_a = 1.5 \times 10^6 \begin{bmatrix} 0.75 & 1.5 & -0.75 & 1.5 \\ 1.5 & 4.0 & -1.5 & 2.0 \\ -0.75 & -1.5 & 0.75 & -1.5 \\ 1.5 & 2.0 & -1.5 & 4.0 \end{bmatrix} \begin{bmatrix} 0 & 0 & 0 & 1.0 \\ 1.0 & 0 & 0 & 0 \\ 0 & 0 & 0 & 0 \\ 0 & 0 & 0 & 0 \end{bmatrix} \times
$$

$$
\times \begin{bmatrix} -0.834 \\ -0.070 \\ 0.249 \\ 1.622 \end{bmatrix} \times 10^{-3} + \begin{bmatrix} -4.0 \\ -2.67 \\ -4.0 \\ 2.67 \end{bmatrix} \times 10^3 = \begin{bmatrix} -4.05 \\ -4.02 \\ -3.95 \\ 3.82 \end{bmatrix} \begin{matrix} \text{kN} \\ \text{kNm} \\ \text{kN} \\ \text{kNm} \end{matrix}
$$

$$\mathbf{P_b} = 1.0 \times 10^6 \begin{bmatrix} 4.0 & 2.0 \\ 2.0 & 4.0 \end{bmatrix} \begin{bmatrix} 1.0 & 0 & 0 & 0 \\ 0 & 1.0 & 0 & 0 \end{bmatrix} \begin{bmatrix} -0.834 \\ -0.070 \\ 0.249 \\ 1.622 \end{bmatrix} \times 10^{-3} +$$

$$+ \begin{bmatrix} 7.5 \\ -7.5 \end{bmatrix} \times 10^3 = \begin{bmatrix} 4.02 \\ -9.45 \end{bmatrix} \begin{matrix} \text{kNm} \\ \text{kNm} \end{matrix} \quad .$$

$$\mathbf{P_c} = \begin{bmatrix} 6.22 \\ -5.14 \end{bmatrix} \begin{matrix} \text{kNm} \\ \text{kNm} \end{matrix} \,, \quad \mathbf{P_d} = \begin{bmatrix} 1.67 \\ 3.23 \\ -1.67 \\ 3.44 \end{bmatrix} \begin{matrix} \text{kN} \\ \text{kNm} \\ \text{kN} \\ \text{kNm} \end{matrix} \,, \quad \mathbf{P_e} = \begin{bmatrix} 2.40 \\ 5.14 \\ -2.40 \\ 4.40 \end{bmatrix} \begin{matrix} \text{kN} \\ \text{kNm} \\ \text{kN} \\ \text{kNm} \end{matrix} \quad .$$

The corresponding bending moment diagram for the framework is shown in Fig.
5.31. Comparing this with the diagram for the propped framework in Fig. 5.24,
it is apparent that sidesway has caused a significant redistribution of the moments.

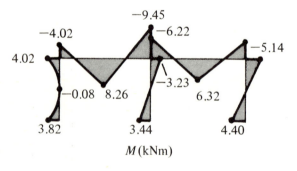

Fig. 5.31

The slope-deflection equation method with sidesway, can be carried out
for building frames with any number of bays or storeys, provided a separate
sidesway degree of freedom is assumed at every floor level. However, as the
height of frameworks increase, the method becomes inaccurate. This is because,
since axial deformation is ignored, it is incapable of representing the overall
cantilever-type bending action under lateral loading shown in Fig. 5.32. This

then is a further limitation of the classical slope-deflection equation method, even when it is applied to building frames. The method is also complex to apply to frameworks with inclined members such as the portal frame analysed in Example 4.10. Thus for the framework shown in Fig. 5.33(a), in addition to the three rotational degrees of freedom at joints 1, 2 and 3, *two* sidesway degrees of freedom d_4 and d_5 are needed to completely define the sidesway deformation of the structure. These are shown in Figs. 5.33(b) and (c). The derivation of the member transformation matrices for this case is a complicated, structure-oriented problem. However, all these difficulties are resolved by including axial deformations in the analysis. The resulting, completely general, plane framework analysis will now be discussed.

Fig. 5.32

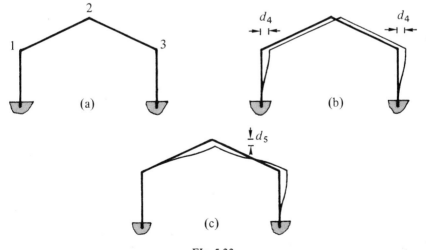

Fig. 5.33

5.6.5 Analysis of Rigid-jointed Plane Frameworks Including Axial Effects

The displacements of the joints of a rigid-jointed plane framework and the overall structural deformation, are completely defined by three degrees of freedom at each joint, these being the linear displacements in the global x and y directions, and the clockwise rotation about the z axis. The corresponding global forces are the three equivalent forces acting at each joint, in the form of two direct forces and a couple. Thus the unpropped framework of Example 5.7 could be described by nine global degrees of freedom as in Fig. 5.34.

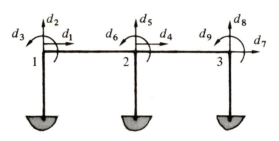

Fig. 5.34

The deformation of a member in the framework is similarly completely defined by six member degrees of freedom, these being the linear displacements in the x_m and y_m directions and the clockwise rotation about the z_m axis, at each end, as in Fig. 5.35. The member stiffness matrix is found as usual by making each degree of freedom unity in turn. It is then easy to show that the stiffness coefficients for the axial deformation are the same as those found for the pin-ended member in (5.12); and for the bending deformation, the bending member in (5.43). Thus the combined member stiffness matrix is as follows

$$k_m = \begin{bmatrix} EA/l & 0 & 0 & -EA/l & 0 & 0 \\ 0 & 12.0EI/l^3 & 6.0EI/l^2 & 0 & -12.0EI/l^3 & 6.0EI/l^2 \\ 0 & 6.0EI/l^2 & 4.0EI/l & 0 & -6.0EI/l^2 & 2.0EI/l \\ -EA/l & 0 & 0 & EA/l & 0 & 0 \\ 0 & -12.0EI/l^3 & -6.0EI/l^2 & 0 & 12.0EI/l^3 & -6.0EI/l^2 \\ 0 & 6.0EI/l^2 & 2.0EI/l & 0 & -6.0EI/l^2 & 4.0EI/l \end{bmatrix}_m$$

$$(5.47)$$

Note that k_{m12}, k_{m13}, etc. are zero since there is no interaction between axial and bending effects in the member if the displacements are small.

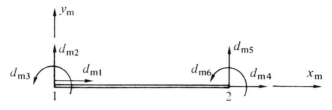

Fig. 5.35

The assembly of the global stiffness matrix from the member stiffness matrices is carried out using the usual assembly equation (5.21)

$$k = \sum_m (t_m^T k_m t_m) \qquad (5.21)^*$$

where in this case the member transformation matrices have six rows.

Fig. 5.36

The equivalent global force matrix is calculated from the member fixing forces, which again are the forces required to reduce the member degrees of freedom to zero in the presence of intermediate loading. When now considering axial deformation, two extra member fixing forces, P_{Fm1} and P_{Fm4} are required to restrain the axial degrees of freedom to zero, as shown in Fig. 5.36. These fixing forces are in equilibrium with the axial components of the intermediate loading, and they can be calculated using the flexibility method. Thus suppose first that the member is subject to a concentrated axial force Q_m at a distance a_m from end 1 as shown in Fig. 5.37(a). The end displacements d_{m1} and d_{m4} are then zero under the action of Q_m, P_{Fm1} and P_{Fm4}. The flexibility equation for the displacements of the single axially loaded member can be derived from (3.53) by selecting axial terms. Thus

$$d_{mi} = \int_0^{l_m} \left\{ \frac{n_i N}{EA} \right\}_m dx_m \qquad (5.48)$$

Therefore

$$d_{m1} = 0 = \int_0^{l_m} \left\{ \frac{n_1 N}{EA} \right\}_m dx_m$$

where N is the axial stress resultant at any point in the member due to the loading and n_1 is the stress resultant due to a unit force at end 1. By making a cut in the member at a distance x_m from end 1 and taking axial equilibrium, we find that

$$N = -P_{Fm1} \; [x \leqslant a] \quad \text{or} \quad -(P_{Fm1} + Q_m) \; [x > a]$$

and

$$n_1 = -1.0 \quad .$$

Thus

$$d_{m1} = 0 = (P_{Fm1} a_m + (P_{Fm1} + Q_m)b_m) \; \left\{ \frac{1}{EA} \right\}_m$$

and therefore

$$P_{Fm1} = -Q_m \{b/l\}_m \quad .$$

Either by carrying out a similar calculation for d_{m4} using the following expressions for N and the stress resultant n_4 due to a unit force at end 2,

$$N = (P_{Fm4} + Q_m) \; [x \leqslant a] \quad \text{or} \quad P_{Fm4} \; [x > a]$$

$$n_4 = 1.0$$

or simply by resolving for the whole member in the x_m direction, we obtain

$$P_{Fm4} = -Q_m \{a/l\}_m \quad .$$

Thus the axial fixing force matrix for a concentrated axial force Q_m is given by

$$\begin{bmatrix} P_{Fm1} \\ P_{Fm4} \end{bmatrix} = \frac{-Q_m}{l_m} \begin{bmatrix} b \\ a \end{bmatrix}_m \quad . \tag{5.49}$$

A similar calculation can be carried out for a uniformly distributed axial force q_m per unit length over the whole member as shown in Fig. 5.37(b). In this case

$$N = -(P_{Fm1} + q_m x_m)$$

and

$$n_1 = -1.0 \quad .$$

Thus

$$d_{m1} = 0 = (P_{Fm1} + q_m \, l/2) \; \left\{ \frac{l}{EA} \right\}_m$$

and therefore

$$P_{Fm1} = -q_m l_m / 2 .$$

Whence resolving for the whole member in the x direction, we obtain

$$P_{Fm4} = -q_m l_m / 2$$

and the axial fixing force matrix for the uniformly distributed force q_m is given by

$$\begin{bmatrix} P_{Fm1} \\ P_{Fm4} \end{bmatrix} = \frac{-q_m l_m}{2} \begin{bmatrix} 1.0 \\ 1.0 \end{bmatrix} . \tag{5.50}$$

The fixing forces for other axial force distributions, can easily be calculated in a similar manner.

The inclusion of axial degrees of freedom also allows us to account for two of the structural actions not yet considered in the stiffness method, namely temperature changes and construction errors. Thus if a particular member is subjected to a temperature rise of $\Delta\theta_m$, and it is unrestrained, the the extension of the member $\Delta_{\Theta m}$ is given by (3.11) as

$$\Delta_{\Theta m} = \{\alpha l \Delta\theta\}_m \quad . \tag{3.11}*$$

Compressive forces at each end of the member are then needed, to reduce the member degrees of freedom to zero. These are found simply by considering the forces required to produce $-\Delta_{\Theta m}$ in the member. Thus from (5.11)

$$P_{Fm1} = -P_{Fm4} = \left\{ \frac{EA \, \Delta_\Theta}{l} \right\}_m = \{EA\alpha\Delta\theta\}_m \quad . \tag{5.51}$$

Similarly if the member is Δ_{Cm} too long because of a construction error, then

$$P_{Fm1} = -P_{Fm4} = \left\{ \frac{EA \, \Delta_C}{l} \right\}_m \quad . \tag{5.52}$$

The equivalent global force matrix is again assembled from the member fixing force matrices using

$$P_Q = -P_F = - \sum_m (t_m^T P_{Fm}) \quad . \tag{5.31}*, (5.32)*$$

The stiffness solution of the framework then proceeds in the same way as before, but this time with many more degrees of freedom.

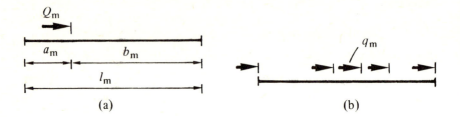

Fig. 5.37

Example 5.8

We shall summarise the solution of the framework of Example 5.7, taking axial deformation into account. For this purpose the members of the framework are all assumed to have the same cross-sectional area $A = 4.0 \times 10^3$ mm^2.

The global degrees of freedom are those shown in Fig. 5.34. Again, for this particular framework, the member transformation matrices can easily be found by inspection. Thus for member a with degrees of freedom oriented as shown in Fig. 5.38 we have

$$
t_a = \begin{bmatrix}
0 & -1.0 & 0 & & & \\
1.0 & 0 & 0 & & 0 & 0 \\
0 & 0 & 1.0 & & & \\
\hline
 & 0 & & & 0 & 0
\end{bmatrix} \longleftarrow (3 \times 3 \text{ null matrix})
$$

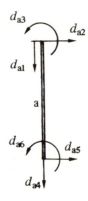

Fig. 5.38

The member stiffness matrix of 'a' is given by

$$k_a = \begin{bmatrix} 200.0 & 0 & 0 & -200.0 & 0 & 0 \\ 0 & 1.125 & 2.25 & 0 & -1.125 & 2.25 \\ 0 & 2.25 & 6.00 & 0 & -2.25 & 3.00 \\ -200.0 & 0 & 0 & 200.0 & 0 & 0 \\ 0 & -1.125 & -2.25 & 0 & 1.125 & -2.25 \\ 0 & 2.25 & 3.00 & 0 & -2.25 & 6.00 \end{bmatrix} \begin{array}{l} \\ \\ \text{MN/m} \\ \text{MN/rad} \\ \text{etc.} \\ \\ \end{array}$$

Thus the contribution of 'a' to the global stiffness matrix is given by

$$t_a^T k_a t_a = \begin{bmatrix} \begin{matrix} 1.125 & 0 & 2.25 \\ 0 & 200.0 & 0 \\ 2.25 & 0 & 6.0 \end{matrix} & \begin{matrix} \\ \mathbf{0} \\ \\ \end{matrix} & \begin{matrix} \\ \mathbf{0} \\ \\ \end{matrix} \\ \hline \mathbf{0} & \mathbf{0} & \mathbf{0} \\ \hline \mathbf{0} & \mathbf{0} & \mathbf{0} \end{bmatrix} \begin{array}{l} (3 \times 3 \text{ null matrix}) \\ \\ \text{MN/m} \\ \text{MN/rad} \\ \text{etc.} \\ \\ \end{array}$$

The global stiffness matrix assembled from all the members is then given by

$$k = \begin{bmatrix} 134.46 & 0 & 2.25 & -133.33 & 0 & 0 & 0 & 0 & 0 \\ & 200.33 & 1.0 & 0 & -0.33 & 1.0 & 0 & 0 & 0 \\ & & 10.0 & 0 & -1.0 & 2.0 & 0 & 0 & 0 \\ & & & 267.79 & 0 & 2.25 & -133.33 & 0 & 0 \\ & & & & 200.67 & 0 & 0 & -0.33 & 1.0 \\ & \text{Symmetrical} & & & & 14.0 & 0 & -1.0 & 2.0 \\ & & & & & & 134.46 & 0 & 2.25 \\ & & & & & & & 200.33 & 1.0 \\ & & & & & & & & 10.0 \end{bmatrix} \begin{array}{l} \\ \\ \\ \\ \text{MN/m} \\ \text{MNm/rad} \\ \text{etc.} \\ \\ \\ \end{array}$$

The member fixing forces $P_{\text{Fm2}}, P_{\text{Fm3}}, P_{\text{Fm5}}$ and P_{Fm6} due to the transverse intermediate loading have the same values as the components of $\mathbf{P}_{\text{Fm}}$ in (5.45) and (5.46). In the present example, where the axial intermediate loading is zero

$$\begin{bmatrix} P_{\text{Fm1}} \\ P_{\text{Fm4}} \end{bmatrix} = \mathbf{0} \ .$$

Thus the member fixing force matrix for 'a' comes out to be as follows

$$
\mathbf{P}_{Fa} = \begin{bmatrix} 0 \\ -4.0 \\ -2.67 \\ 0 \\ -4.0 \\ 2.67 \end{bmatrix} \begin{array}{l} kN \\ kN \\ kNm \\ kN \\ kN \\ kNm \end{array} \quad .
$$

and the contribution of 'a' to the global fixing force matrix is given by

(3 X 3 null matrix)

$$
\mathbf{t}_a^T \mathbf{P}_{Fa} = \left[\begin{array}{ccc|c} 0 & 1.0 & 0 & \\ -1.0 & 0 & 0 & 0 \\ 0 & 0 & 1.0 & \\ \hline & 0 & & 0 \\ \hline & 0 & & 0 \end{array} \right] \begin{bmatrix} 0 \\ -4.0 \\ -2.67 \\ \hline 0 \\ -4.0 \\ 2.67 \end{bmatrix} = \begin{bmatrix} -4.0 \\ 0 \\ -2.67 \\ \hline 0 \\ 0 \end{bmatrix} \begin{array}{l} kN \\ kN \\ kNm \\ \\ \end{array} \quad .
$$

(3 X 1 null matrix)

Carrying out this calculation for all the members and summing, leads to the global fixing force matrix given by

$$
\mathbf{P}_F = \begin{bmatrix} -4.0 \\ 5.0 \\ 4.83 \\ 0 \\ 9.0 \\ -1.5 \\ 0 \\ 4.0 \\ -6.0 \end{bmatrix} \begin{array}{l} kN \\ kN \\ kNm \\ kN \\ kN \\ kNm \\ kN \\ kN \\ kNm \end{array} \quad .
$$

This again is the formal derivation of the global fixing force matrix. In fact, the matrix can easily be assembled by inspection from the total member fixing forces at each joint, which are summed from a figure such as Fig. 5.39.

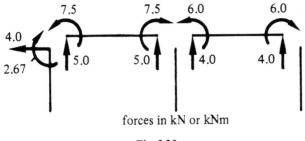

forces in kN or kNm

Fig. 5.39

The solution in terms of the global displacements is given as usual by

$$
d = k^{-1}P_Q = -k^{-1}P_F =
\begin{bmatrix}
1.648 \\
-0.020 \\
-0.843 \\
1.618 \\
-0.050 \\
-0.069 \\
1.600 \\
-0.019 \\
0.257
\end{bmatrix}
\times 10^{-3}
\begin{matrix}
m \\
m \\
rad \\
m \\
m \\
rad \\
m \\
m \\
rad
\end{matrix}
\quad .
$$

The member forces in 'a' say, are then given by

$$
P_a = k_a t_a d + P_{Fa} =
\begin{bmatrix}
4.10 \\
-0.04 \\
-1.35 \\
-4.10 \\
0.04 \\
1.18
\end{bmatrix}
+
\begin{bmatrix}
0 \\
-4.0 \\
-2.67 \\
0 \\
-4.0 \\
2.67
\end{bmatrix}
=
\begin{bmatrix}
4.10 \\
-4.04 \\
-4.02 \\
-4.10 \\
-3.96 \\
3.85
\end{bmatrix}
\begin{matrix}
kN \\
kN \\
kNm \\
kN \\
kN \\
kNm
\end{matrix}
\quad .
$$

A similar calculation gives the member forces in the other members. Thus

$$
\mathbf{P_{Fb}} = \begin{bmatrix} 4.04 \\ 4.10 \\ 4.02 \\ -4.04 \\ 5.90 \\ -9.43 \end{bmatrix} \begin{matrix} kN \\ kN \\ kNm \\ kN \\ kN \\ kNm \end{matrix}, \quad
\mathbf{P_{Fc}} = \begin{bmatrix} 2.38 \\ 4.18 \\ 6.21 \\ -2.38 \\ 3.82 \\ -5.14 \end{bmatrix} \begin{matrix} kN \\ kN \\ kNm \\ kN \\ kN \\ kNm \end{matrix},
$$

$$
\mathbf{P_{Fd}} = \begin{bmatrix} 10.08 \\ 1.67 \\ 3.23 \\ -10.08 \\ -1.67 \\ 3.43 \end{bmatrix} \begin{matrix} kN \\ kN \\ kNm \\ kN \\ kN \\ kNm \end{matrix}, \quad
\mathbf{P_{Fe}} = \begin{bmatrix} 3.82 \\ 2.38 \\ 5.14 \\ -3.82 \\ -2.38 \\ 4.37 \end{bmatrix} \begin{matrix} kN \\ kN \\ kNm \\ kN \\ kN \\ kNm \end{matrix}.
$$

Comparing the member couples with those produced in Example 5.7, it is apparent that including axial deformation has improved the solution by less than one per cent. However, the complete stiffness method is applicable to any rigid-jointed plane framework, of any height and including diagonal members. It therefore forms a suitable basis for a general plane framework computer program. The details of the programming problem will be discussed in Chapter 6.

5.6.6 Curved Members
Curved members can in theory be dealt with by the stiffness method by deriving the stiffness matrices for particular members in the same manner as in Appendix A5.2, the integrations for the member displacements then being carried out round the curved arcs as in Example 3.9. This is a complicated analysis and difficult to generalise. A simpler approximate method for including curved members in the standard stiffness analysis, is to divide them into a suitable number of sub-members which are then assumed to be straight. For example, the curved arch discussed in Example 3.9, can be approximated by 14 straight sub-members as in Fig. 5.40. The resulting structure is then solved in terms of a considerably increased number of members, joints and degrees of freedom.

Fig. 5.40

5.7 PLANE FRAMEWORKS CONTAINING HINGES

5.7.1 Introduction

Pin-jointed frameworks have been discussed in Sections 5.3 and 5.4 as a separate structural problem. In this section we shall consider plane frameworks which are basically rigid-jointed, but are also articulated by the inclusion of a limited number of hinges.

5.7.2 Articulated Supports

If a framework is hinged at its supports, then further degrees of freedom are needed to completely define its deformation. Thus a hinge at the foot of member a in the portal frame in Fig. 5.41(a) can be dealt with simply by including a rotational global degree of freedom at the support. Consider again the simplified slope-deflection equation analysis of Example 5.5; the extra degree of freedom is then d_4 shown in Fig. 5.41(b). The transformation matrices of the members each contain an extra column, and that for member a is given by

$$
\mathbf{t}_a = \begin{bmatrix} 1.0 & 0 & 0 & 0 \\ 0 & 0 & 0 & 1.0 \end{bmatrix} .
$$

The stiffness solution then proceeds as before, using an enlarged global stiffness matrix and equivalent global force matrix.

In programming the stiffness method, support restraints and settlements are included by a systematic numerical procedure. This will be discussed in Chapter 6.

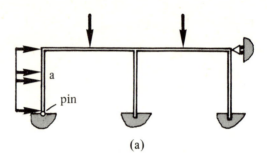

(a)

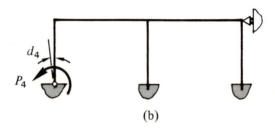

(b)

Fig. 5.41

5.7.3 Member with Both Ends Pinned

Suppose a rigid-jointed building frame is braced by a pin-ended member as in Fig. 5.42. Such a member can easily be included in the assembly equation (5.21)

$$k = \sum_m (t_m^T k_m t_m) \qquad (5.21)*$$

by defining its deformation by just the two axial degrees of freedom in Fig. 5.9, and its stiffness by the member stiffness matrix k_m

$$k_m = \begin{bmatrix} 1.0 & -1.0 \\ -1.0 & 1.0 \end{bmatrix} \left(\frac{EA}{l} \right)_m \qquad (5.12)*$$

The member degrees of freedom are independent of the global joint rotations at its ends, and the transformation matrix contains just two rows.

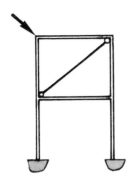

Fig. 5.42

Example 5.9

A simple illustration of a rigid framework containing a pin-ended member is the tied cantilever analysed by the flexibility method in Example 4.8 and reproduced in Fig. 5.43(a). Both members are of steel. The pin-ended tie has $A_a = 500.0$ mm^2. The cantilever has $A_b = 4.0 \times 10^3$ mm^2 and $I_b = 30.0 \times 10^6$ mm^4.

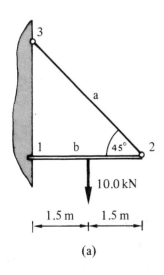

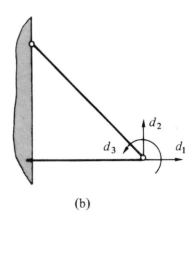

(b)

(a)

Fig. 5.43

The deformation of the framework is described by the three global degrees of freedom at joint 2, shown in Fig. 5.43(b). Members a and b then have two and six degrees of freedom respectively, oriented in the directions of the member coordinate systems. Thus

$$
t_a = \begin{bmatrix} -0.707 & 0.707 & 0 \\ 0 & 0 & 0 \end{bmatrix}, \quad t_b = \begin{bmatrix} 0 & 0 & 0 \\ 0 & 0 & 0 \\ 0 & 0 & 0 \\ 1.0 & 0 & 0 \\ 0 & 1.0 & 0 \\ 0 & 0 & 1.0 \end{bmatrix}.
$$

The member stiffness matrices from (5.12) and (5.47) are given by

$$
k_a = \begin{bmatrix} 1.0 & -1.0 \\ -1.0 & 1.0 \end{bmatrix} \times 23.57 \text{ MN/m}
$$

$$
k_b = \begin{bmatrix} 266.67 & 0 & 0 & -266.67 & 0 & 0 \\ 0 & 2.67 & 4.0 & 0 & -2.67 & 4.0 \\ 0 & 4.0 & 8.0 & 0 & -4.0 & 4.0 \\ -266.67 & 0 & 0 & 266.67 & 0 & 0 \\ 0 & -2.67 & -4.0 & 0 & 2.67 & -4.0 \\ 0 & 4.0 & 4.0 & 0 & -4.0 & 8.0 \end{bmatrix} \begin{matrix} \\ \\ \text{MN/m,} \\ \text{MNm/rad} \\ \text{etc.} \\ \end{matrix}
$$

Thus using (5.21)

$$
k = \begin{bmatrix} 278.45 & -11.78 & 0 \\ -11.78 & 14.45 & -4.0 \\ 0 & -4.0 & 8.0 \end{bmatrix} \begin{matrix} \text{MN/m} \\ \text{MNm/rad} \\ \text{etc.} \end{matrix}
$$

The fixing force matrix for member b is given by (5.45) and (5.49) as

$$
\mathbf{P}_{Fb} =
\begin{bmatrix}
0 \\
5.0 \\
3.75 \\
0 \\
5.0 \\
-3.75
\end{bmatrix}
\begin{matrix}
\text{kN} \\
\text{kN} \\
\text{kNm} \\
\text{kN} \\
\text{kN} \\
\text{kNm}
\end{matrix}
\quad .
$$

Thus from (5.32)

$$
\mathbf{P}_F =
\begin{bmatrix}
0 \\
5.0 \\
-3.75
\end{bmatrix}
\begin{matrix}
\text{kN} \\
\text{kN} \\
\text{kNm}
\end{matrix}
\quad .
$$

The solution is again obtained as follows

$$
\mathbf{d} = \mathbf{k}^{-1}\mathbf{P}_Q = -\mathbf{k}^{-1}\mathbf{P}_F =
\begin{bmatrix}
-0.011 \\
-0.261 \\
0.338
\end{bmatrix}
\times 10^{-3}
\begin{matrix}
\text{m} \\
\text{m} \\
\text{rad}
\end{matrix}
$$

and using (5.39) the member forces are given by

$$
\mathbf{P}_a =
\begin{bmatrix}
-4.17 \\
4.17
\end{bmatrix}
\text{kN} \ , \
\mathbf{P}_b =
\begin{bmatrix}
2.95 \\
7.05 \\
6.15 \\
-2.95 \\
2.95 \\
0
\end{bmatrix}
\begin{matrix}
\text{kN} \\
\text{kN} \\
\text{kNm} \\
\text{kN} \\
\text{kN} \\
\text{kNm}
\end{matrix}
$$

The force in the tie is then seen to agree with the redundant release force calculated in Example 4.8.

5.7.4 Member with One End Pinned and One End Rigidly Connected

Finally, we shall consider the more complex problem of the member with just one end pinned, as for example member c in the framework shown in Fig. 5.44. This member is asymmetric, and its stiffness matrix has to be calculated as a special case.

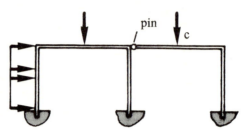

Fig. 5.44

Consider a member m with end 1 pinned as in Fig. 5.45. We shall define the the member deformation by the usual six degrees of freedom shown in Fig. 5.35. The member equilibrium equations are then given by

$$\mathbf{P_m} = \mathbf{k_m d_m} \tag{5.13}*$$

where $\mathbf{k_m}$ is the stiffness matrix given by (5.47). For the present member however, P_{m3} is zero throughout the calculation, thus the third equilibrium equation in (5.13), given by

$$P_{m3} = 0 = 6.0 \left\{\frac{EI}{l^2}\right\}_m d_{m2} + 4.0 \left\{\frac{EI}{l}\right\}_m d_{m3} - 6.0 \left\{\frac{EI}{l^2}\right\}_m d_{m5} + 2.0 \left\{\frac{EI}{l}\right\}_m d_{m6}$$

$$\tag{5.53}$$

demonstrates that d_{m3} is *not* now independent, but is related to d_{m2}, d_{m5} and d_{m6} as follows:

$$d_{m3} = \frac{-1.5}{l_m} d_{m2} + \frac{1.5}{l_m} d_{m5} - 0.5 d_{m6} \ . \tag{5.54}$$

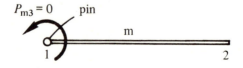

Fig. 5.45

Substituting this expression for d_{m3} into the other five equilibrium equations and collecting terms, then leads to

$$
\mathbf{P_m} =
\begin{bmatrix}
EA/l & 0 & 0 & -EA/l & 0 & 0 \\
0 & 3.0EI/l^3 & 0 & 0 & -3.0EI/l^3 & 3.0EI/l^2 \\
0 & 0 & 0 & 0 & 0 & 0 \\
-EA/l & 0 & 0 & EA/l & 0 & 0 \\
0 & -3.0EI/l^3 & 0 & 0 & 3.0EI/l^3 & -3.0EI/l^2 \\
0 & 3.0EI/l^2 & 0 & 0 & -3.0EI/l^2 & 3.0EI/l
\end{bmatrix}_m \mathbf{d_m} \quad (5.55)
$$

or

$$
\mathbf{P_m} = \mathbf{k_m^1 d_m} \tag{5.56}
$$

where $\mathbf{k_m^1}$ is the equivalent stiffness matrix for a member with end 1 pinned. This matrix can be regarded simply as a generator of the correct member forces corresponding to a particular set of member displacements $\mathbf{d_m}$ and is used in the usual assembly equation (5.21) to give the contribution of m to the global stiffness matrix. The displacement d_{m3} is in fact redundant, and can be eliminated. However, it simplifies the programming problem to retain d_{m3} in the calculation, since the member stiffness and transformation matrices are then the usual size, and the transformation matrix need not be modified.

Finally we shall merely quote the equivalent stiffness matrix $\mathbf{k_m^2}$ for a member with end 2 pinned, found in the same manner as above. Thus

$$
\mathbf{k_m^2} =
\begin{bmatrix}
EA/l & 0 & 0 & -EA/l & 0 & 0 \\
0 & 3.0EI/l^3 & 3.0EI/l^2 & 0 & -3.0EI/l^3 & 0 \\
0 & 3.0EI/l^2 & 3.0EI/l & 0 & -3.0EI/l^2 & 0 \\
-EA/l & 0 & 0 & EA/l & 0 & 0 \\
0 & -3.0EI/l^3 & -3.0EI/l^2 & 0 & 3.0EI/l^3 & 0 \\
0 & 0 & 0 & 0 & 0 & 0
\end{bmatrix}_m . \quad (5.57)
$$

5.8 SPACE FRAMEWORKS

The solution of space frameworks by the stiffness method is similar to the solution of plane frameworks, except that many more degrees of freedom are involved. Thus the displacements of the joints in a space framework, and the overall structural deformation are completely defined by six degrees of freedom at each joint, these being the linear displacements in the global x, y and z directions and the clockwise rotations about the three axes. The corresponding global forces are the six equivalent forces acting at each joint in the form of three direct forces and three couples. Thus the deformation of the space framework in Fig. 5.46(a) for example, would be defined by the 18 degrees of freedom shown in Fig. 5.46(b).

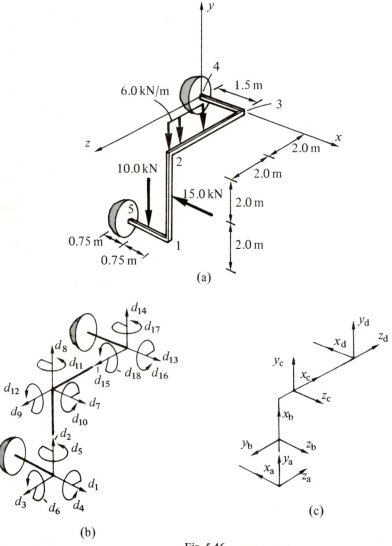

(a)

(b)

(c)

Fig. 5.46

The deformation of a member in a space framework is similarly completely defined by 12 member degrees of freedom, these being the linear displacements in the x_m, y_m and z_m directions and the clockwise rotations about the three axes at each end as in Fig. 5.47. Again the member stiffness matrix is found by making each degree of freedom unity in turn. The stiffness coefficients corresponding to bending about the y_m axis have the same form as the stiffness coefficients previously derived for bending about the z_m axis. Thus the only stiffness coefficients needing further consideration correspond to the twisting degrees of freedom d_{m4} and d_{m10}.

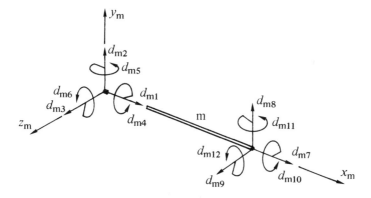

Fig. 5.47

Suppose end 1 of the member is given a unit twist about the x_m axis so that $d_{m4} = 1.0$, $d_{mk} = 0$, $k \neq 4$. The twisting couples at the two ends, P_{m4} and P_{m10} are the stiffness coefficients k_{m44} and $k_{m10,4}$. There are no other forces on the member. In calculating P_{m4} and P_{m10} we again use the flexibility method. Thus the displacement d_{mi} of a single member subject to twist can be derived from (3.53) by selecting twisting terms. Thus

$$d_{mi} = \int_0^{l_m} \left\{ \frac{t_i T}{GJ} \right\}_m dx_m \quad . \tag{5.58}$$

In the present case, by making a cut and taking moments for the subsection shown in Fig. 5.48 about the x_m axis, we have

$$T = -P_{m4} \quad .$$

The member twist d_{m4} is then obtained using the equilibrium force system t_4 corresponding to a unit twisting couple at end 1. Thus

$$t_4 = -1.0$$

and

$$d_{m4} = 1.0 = P_{m4} \left\{ \frac{l}{GJ} \right\}_m \quad .$$

Thus

$$P_{m4} = k_{m44} = \left\{ \frac{GJ}{l} \right\}_m$$

and from the rotational equilibrium of the member about the x_m axis

$$P_{m10} = k_{m10,4} = \left\{ \frac{-GJ}{l} \right\}_m \quad .$$

A similar calculation for $d_{m10} = 1.0, d_{mk} = 0, k \neq 10$, gives

$$k_{m4,10} = \left\{ \frac{-GJ}{l} \right\}_m \quad , \quad k_{m10,10} = \left\{ \frac{GJ}{l} \right\}_m \quad .$$

      Collecting all the stiffness coefficients together, and this time using subscripts y and z to indicate bending about the y_m and z_m axes respectively, the stiffness matrix for a space framework member comes out to be as shown in Table 5.7.

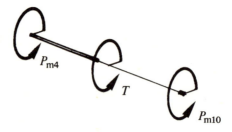

Fig. 5.48

      The fixing force matrix for a member in a space framework again has two components that need further consideration, namely the twisting fixing forces P_{Fm4} and P_{Fm10} at each end. These fixing forces are in equilibrium with twisting intermediate loading, which can take the form of a concentrated twisting couple Q_m at a particular point in the member as in Fig. 5.49(a), or a distributed twisting couple q_m. The latter is usually due to transverse distributed forces applied to the member in such a way that their line of action in any cross-section does not pass through the shear centre as in Fig. 5.49(b). The fixing couples can

Table 5.7

Member stiffness matrix for a member in a rigid-jointed space framework

EA/l	0	0	0	0	0	$-EA/l$	0	0	0	0	0
0	$12.0EI_z/l^3$	0	0	0	$6.0EI_z/l^2$	0	$-12.0EI_z/l^3$	0	0	0	$6.0EI_z/l^2$
0	0	$12.0EI_y/l^3$	0	$-6.0EI_y/l^2$	0	0	0	$-12.0EI_y/l^3$	0	$-6.0EI_y/l^2$	0
0	0	0	GJ/l	0	0	0	0	0	$-GJ/l$	0	0
0	0	$-6.0EI_y/l^2$	0	$4.0EI_y/l$	0	0	0	$6.0EI_y/l^2$	0	$2.0EI_y/l$	0
0	$6.0EI_z/l^2$	0	0	0	$4.0EI_z/l$	0	$-6.0EI_z/l^2$	0	0	0	$2.0EI_z/l$
$-EA/l$	0	0	0	0	0	EA/l	0	0	0	0	0
0	$-12.0EI_z/l^3$	0	0	0	$-6.0EI_z/l^2$	0	$12.0EI_z/l^3$	0	0	0	$-6.0EI_z/l^2$
0	0	$-12.0EI_y/l^3$	0	$6.0EI_y/l^2$	0	0	0	$12.0EI_y/l^3$	0	$6.0EI_y/l^2$	0
0	0	0	$-GJ/l$	0	0	0	0	0	GJ/l	0	0
0	0	$-6.0EI_y/l^2$	0	$2.0EI_y/l$	0	0	0	$6.0EI_y/l^2$	0	$4.0EI_y/l$	0
0	$6.0EI_z/l^2$	0	0	0	$2.0EI_z/l$	0	$-6.0EI_z/l^2$	0	0	0	$4.0EI_z/l$

again be calculated in the same way as the axial fixing forces in Section 5.6.5, but this time using the torsional equation (5.58). In fact the calculations for a concentrated couple Q_m at a distance a_m from end 1, and a uniformly distributed couple q_m per unit length over the whole member, lead to the same expressions for the fixing forces as before, namely

$$\begin{bmatrix} P_{Fm4} \\ P_{Fm10} \end{bmatrix} = \frac{-Q_m}{l_m} \begin{bmatrix} b \\ a \end{bmatrix}_m \quad , \tag{5.59}$$

$$\begin{bmatrix} P_{Fm4} \\ P_{Fm10} \end{bmatrix} = \frac{-q_m l_m}{2} \begin{bmatrix} 1.0 \\ 1.0 \end{bmatrix} \quad . \tag{5.60}$$

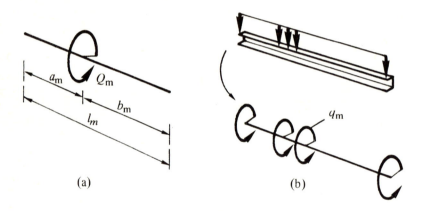

(a)   (b)

Fig. 5.49

Example 5.10
The sizes of the matrices make an explicit analysis of a space framework impracticable here. However, it is still instructive to review the steps of the analysis with reference to a particular example, such as the space framework in Fig. 5.46(a). Suppose the member coordinates are arranged as in Fig. 5.46(c), with the y_m coordinates coinciding with the minor principal axes of each member. The members are of steel and all have the same cross-section with $A = 4.0 \times 10^3$ mm^2, $I_y = 15.0 \times 10^6$ mm^4, $I_z = 30.0 \times 10^6$ mm^4, and $J = 20.0 \times 10^6$ mm^4.

The transformation matrix for member a is found by inspection to be

$$
t_a = \begin{bmatrix}
-1.0 & 0 & 0 & 0 & 0 & 0 & & & \\
0 & 1.0 & 0 & 0 & 0 & 0 & & & \\
0 & 0 & -1.0 & 0 & 0 & 0 & & \mathbf{0} & \mathbf{0} \\
0 & 0 & 0 & -1.0 & 0 & 0 & & & \\
0 & 0 & 0 & 0 & 1.0 & 0 & & & \\
0 & 0 & 0 & 0 & 0 & -1.0 & & & \\
\hline
& & & \mathbf{0} & & & & \mathbf{0} & \mathbf{0}
\end{bmatrix}
\quad \text{(6 } \times \text{ 6 null matrix)}
$$

The member stiffness matrix for 'a' is a twelve by twelve matrix of the form:

$$
k_a = \begin{bmatrix}
533.33 & 0 & 0 & \cdots & 0 \\
0 & 21.33 & 0 & \cdots & 16.0 \\
0 & 0 & 10.66 & \cdots & 0 \\
\vdots & \vdots & \vdots & \ddots & \vdots \\
0 & 16.0 & 0 & \cdots & 16.0
\end{bmatrix}
\quad
\begin{array}{l}
\text{MN/m} \\
\text{MNm/rad} \\
\text{etc.}
\end{array}
$$

Thus the contribution of member a to the global stiffness matrix is as follows

$$
t_a^T k_a t_a = \begin{bmatrix}
533.33 & 0 & 0 & 0 & 0 & 0 & & & \\
0 & 21.33 & 0 & 0 & 0 & -16.0 & & & \\
0 & 0 & 10.66 & 0 & 8.0 & 0 & & & \\
0 & 0 & 0 & 1.03 & 0 & 0 & & \mathbf{0} & \mathbf{0} \\
0 & 0 & 8.0 & 0 & 8.0 & 0 & & & \\
0 & -16.0 & 0 & 0 & 0 & 16.0 & & & \\
\hline
& & & \mathbf{0} & & & & \mathbf{0} & \mathbf{0} \\
\hline
& & & \mathbf{0} & & & & \mathbf{0} & \mathbf{0}
\end{bmatrix}
\quad
\begin{array}{l}
\text{(6 } \times \text{ 6 null} \\
\text{matrix)} \\
\\
\text{MN/m} \\
\text{MNm/rad} \\
\text{etc.}
\end{array}
$$

The complete global stiffness matrix is assembled from the contributions of all the members, each of which populate regions of the global stiffness matrix as indicated below.

The member fixing forces are calculated from (5.45) and (5.46) for transverse intermediate loading, taking account of the orientation of the forces with respect to the y_m and z_m axes, from (5.49) and (5.50) for axial intermediate loading, and from (5.59) and (5.60) for twisting intermediate loading. In the present case, the fixing force matrix for member a corresponds to the single concentrated transverse force of 10.0 kN shown in Fig. 5.46(a). Thus from (5.45)

$$
\begin{bmatrix} P_{Fa2} \\ P_{Fa6} \\ P_{Fa8} \\ P_{Fa12} \end{bmatrix} = \frac{10.0}{1.5^2} \begin{bmatrix} (0.75^2/1.5)(1.5+1.5) \\ 0.75^3 \\ (0.75^2/1.5)(1.5+1.5) \\ -0.75^3 \end{bmatrix} = \begin{bmatrix} 5.0 \\ 1.875 \\ 5.0 \\ -1.875 \end{bmatrix} \begin{matrix} \text{kN} \\ \text{kNm} \\ \text{kN} \\ \text{kNm} \end{matrix}
$$

Either by inspection or by using $P_F = \sum\limits_m (t_m^T P_{Fm})$ it can then be shown that the contribution of member a to the global fixing force matrix is

$$
t_a^T P_{Fa} = \begin{bmatrix} 0 \\ 5.0 \\ 0 \\ 0 \\ 0 \\ -1.875 \\ \hline 0 \\ \hline 0 \end{bmatrix} \begin{matrix} \\ \text{kN} \\ \\ \\ \\ \text{kNm} \\ \text{(6} \times \text{1 null matrix)} \\ \end{matrix}
$$

The complete global fixing force matrix is assembled from the contributions of all the members, each of which populate regions of $\mathbf{P}_F$ as shown below.

The solution of the space framework is found in terms of the displacements by the usual formulae

$$\mathbf{d} = \mathbf{k}^{-1}\mathbf{P}_Q = -\mathbf{k}^{-1}\mathbf{P}_F \ . \qquad\qquad (5.30)^* \ (5.31)^*$$

We then finally determine the 12 member forces in each member from

$$\mathbf{P}_m = \mathbf{k}_m \mathbf{t}_m \mathbf{d} + \mathbf{P}_{Fm} \ . \qquad\qquad\qquad (5.34)^*$$

The member bending moment and shear force diagrams are found by considering the bending and shear forces in the y_m and the z_m directions respectively. Thus for bending about the z_m axis, the relevant member forces are the end couples P_{m6} and P_{m12}, and the transverse end forces P_{m2} and P_{m8}. The diagrams are then plotted in the same manner as Example 5.4.

5.9 CONCLUSION

In this chapter, all the important structural aspects of the stiffness method for framework analysis have been described. It has been shown that the method usually employs many more degrees of freedom than the flexibility method. However, once $\mathbf{k}_m$, $\mathbf{t}_m$ and $\mathbf{P}_{Fm}$ have been determined for each member in the framework, the solution proceeds by a series of matrix operations which are very convenient for computer programming. Summarising we have

$$\mathbf{k} = \sum_m (\mathbf{t}_m^T \mathbf{k}_m \mathbf{t}_m) \qquad\qquad\qquad (5.61a)$$

$$\mathbf{P}_F = \sum_m (\mathbf{t}_m^T P_{Fm}) \qquad\qquad\qquad (5.61b)$$

$$\mathbf{P}_Q = -\mathbf{P}_F \qquad\qquad\qquad\qquad (5.61c)$$

and

$$\mathbf{P} + \mathbf{P_Q} = \mathbf{k}\,\mathbf{d} \tag{5.62}$$

(for the framework subject *both* to joint forces **P** and intermediate forces **Q**). Whence

$$\mathbf{d} = \mathbf{k}^{-1}(\mathbf{P} + \mathbf{P_Q}) \tag{5.63a}$$

and

$$\mathbf{P_m} = \mathbf{k_m}\,\mathbf{t_m}\,\mathbf{d} + \mathbf{P_{Fm}} . \tag{5.63b}$$

5.10 PROBLEMS

Many of the following problems are exercises in assembling the global stiffness matrix using Equation (5.21). All are plane frameworks, and the x and y coordinates are in the usual horizontal and vertical directions described in Chapter 1, Section 1.3. Young's moduli are as follows: $E(\text{steel}) = 200.0 \text{ GN/m}^2$, $E(\text{concrete}) = 25.0 \text{ GN/m}^2$, $E(\text{aluminium alloy}) = 70.0 \text{ GN/m}^2$.

In Problems 5.1 to 5.3 the global degrees of freedom are the two linear displacements in the x and y directions respectively of each free joint, assigned in the sequence of joint numbering shown in the figures.

5.1   A Ferris wheel in a fairground can be idealised as eight pin-ended spokes of length 10.0 m attached symmetrically to a rigid rim and meeting at a central hub of negligible dimensions, as shown in Fig. 5.50. The spokes are all of steel and have the same cross-sectional area equal to $5.00 \times 10^{-3} \text{ m}^2$. The loading on the wheel induces a total reactive force at the hub of 500.0 kN directed at 30° to the vertical as shown. Taking the hub to be the single free joint in the structure, calculate the global stiffness matrix, the displacements of the hub relative to the rim and the forces in the spokes.

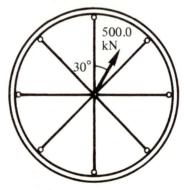

Fig. 5.50

5.2   In the steel pin-jointed frameworks shown in Figs. 5.51(a) and (b), all the members have the same cross-sectional area equal to 500.0 mm². The inclined members in Fig. 5.51(a) are at 60° to the horizontal. For each framework calculate the contribution of member a to the global stiffness matrix.

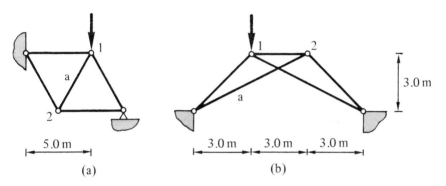

Fig. 5.51

5.3   In the suspended pin-jointed framework in Fig. 5.52, all the members have the same material and cross-sectional properties E and A. Calculate the global stiffness matrix in terms of *five* suitable global degrees of freedom. Why in the stiffness method, is it necessary to assume that joint 3 is restrained horizontally?

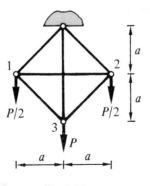

Fig. 5.52

In Problems 5.4 to 5.15, axial effects are assumed to be negligible. The global degrees of freedom are the rotations of the free joints about the z axis, again assigned in the sequence of joint numbering shown in the figures. Extra sway degrees of freedom are assigned as indicated in particular problems.

5.4   A rigid-jointed framework comprises a three-span continuous beam with two members cantilevering from the internal joints as shown in Fig. 5.53. The cantilever members are connected at their free ends by a wire which is tensioned up to a force P. Obtain the stiffness matrix for the continuous beam. Thence noting the symmetry of the problem, solve for the joint rotations and show that the bending moment in member a is constant and equal to $0.4Pa$.

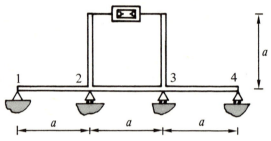

Fig. 5.53

5.5   A two-span continuous beam is encastered at the left-hand end and has the alternative support conditions at the right-hand end shown in Figs. 5.54(a), (b) and (c). It is of uniform material and cross-section. For each support condition, calculate the bending moment at joint 1.

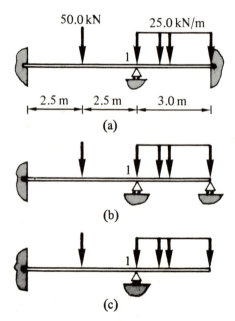

Fig. 5.54

5.6   The symmetrical three-span concrete bridge shown in Fig. 5.55 has a
       central span for which $I = 0.5 \text{ m}^4$ and two outer spans for which
       $I = 0.3 \text{ m}^4$. It is subjected to traffic loading comprising a uniformly distri-
       buted force of 50.0 kN/m over the complete left-hand span and part of the
       central span, and four 150.0 kN concentrated forces applied by the
       wheels of a heavy transporter. The centre of the transporter is 2.5 m from
       the right-hand end of the central span as shown. Calculate the global
       stiffness matrix of the bridge and the equivalent force matrix of the loading.
       Estimate the error in calculating the rotations at the supports if the two
       axle forces in each bogey of the transporter are approximated by a single
       300.0 kN force acting at the centre of the bogey.

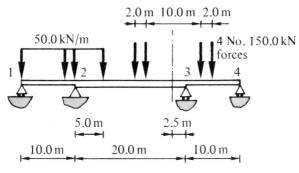

Fig. 5.55

5.7   The members of the framework in Fig. 5.56 are all of the same material.
       The second moments of area of the beams are equal and are three times
       those of the columns. Determine the bending moments at the encastered
       ends of the beams.

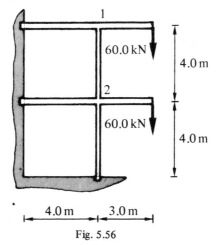

Fig. 5.56

5.8   The lean-to structure shown in Fig. 5.57, is composed of uniform-section steel members whose second moments of area are equal to 30.0×10^{-6} m^4. The beam is loaded by equally spaced forces of 2.5 kN, and the outer column by a distributed force of 2.0 kN/m. Calculate the rotations of the free joints and the bending moment in the members at joint 1.

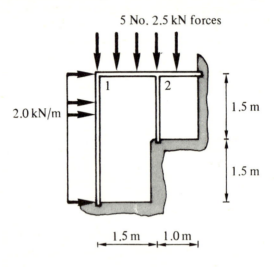

Fig. 5.57

5.9   Use the stiffness method to determine the bending moments at the corners of the box culvert described in Problem 4.14 (Fig. 4.61). (Note that because of the symmetry of the problem, the equilibrium equations contain only *two* independent unknowns.)

5.10 The composite portal frame in Fig. 5.58 is composed of concrete columns of 150.0 mm by 350.0 mm rectangular section, and a steel beam whose second moment of area about the major principal axis is 75.0×10^{-6} m^4. Assuming that the steel–concrete joints are rigid and the feet of the columns are encastered, use the slope-deflection equations to obtain the rotations of the joints and the sidesway of the beam. Thence determine the bending moments at the joints.

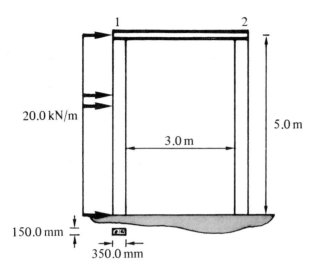

Fig. 5.58

5.11 In the two-storey single-bay portal frame shown in Fig. 5.59, the members
are all of steel and the second moments of area of the beams and columns
are 150.0×10^{-6} mm^4 and 50.0×10^{-6} mm^4 respectively. Taking d_5 and
d_6 to be the sway degrees of freedom of the upper and lower beams
respectively, calculate the global stiffness matrix and the equivalent global
force matrix.

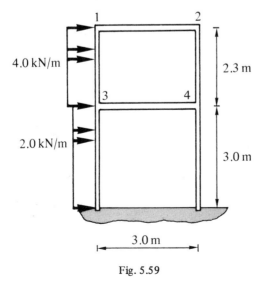

Fig. 5.59

5.12 In the framework in Fig. 5.60, the members all have the same material and cross-sectional properties with $EI = 10.0 \times 10^6$ Nm^2. The deformation is described by the rotations of joints 1 and 2 and by the horizontal movement of joint 1. (Note that neglecting axial deformation implies that joint 2 does *not* move horizontally.) Calculate the global stiffness matrix and the equivalent global force matrix.

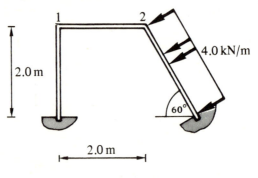

Fig. 5.60

5.13 Use the method of moment distribution to calculate the member forces in the lean-to structure described in Problem 5.8 (Fig. 5.57).

5.14 A beam in a rigid-jointed concrete building frame is to be analysed by isolating it and its neighbouring columns in the sub-structure shown in Fig. 5.61. The ends of the columns remote from the beam are assumed to be encastered. The second moment of area of the beam is twice that of the columns. Use the method of moment distribution to determine the bending moments in the beam caused by the uniformly distributed forces of 30.0 kN/m shown in the figure. (Note that it is not necessary to include the forces at the encastered ends of the columns in the moment distribution table.)

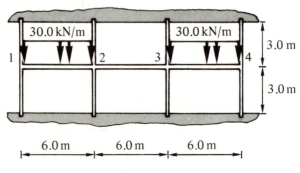

Fig. 5.61

5.15 In the concrete portal frame bridge shown in Fig. 5.62, the second moment
of area of the beam is twice that of the inclined columns. Use the method
of moment distribution to determine the bending moments in the beam at
joints 2 and 4. (Note that the pinned supports 1, 3, 5 and 6 can be treated
in the moment distribution table as joints free to rotate, with the distribution factor for the single member attached to them equal to 1.0.)

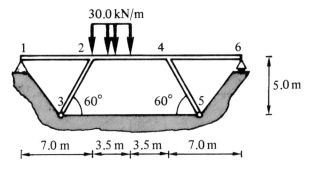

Fig. 5.62

In Problems 5.16 and 5.17 axial effects should be included. The global
degrees of freedom are the two linear displacements in the x and y directions
and the rotation about the z axis of each free joint, assigned in the sequence of
joint numbering shown in the figures.

5.16 In the composite portal frame described in Problem 5.10 (Fig. 5.58), the
cross-sectional area of the beam is 5.0×10^{-3} m^2. Calculate the global
stiffness matrix and the equivalent global force matrix.

5.17 In the framework described in Problem 5.12 (Fig. 5.60) the members all
have $EA = 5.0 \times 10^9$ N. Calculate the global stiffness matrix and the
equivalent global force matrix.

5.18 In the frameworks described in Problem 5.2 (Figs. 5.51(a) and (b)),
members a undergo a 20° rise in temperature relative to the remaining
members. Calculate the corresponding equivalent global force matrices for
the two frameworks.

5.19 An encastered roof beam is reinforced by two pin-ended ties at 45° to
the horizontal as shown in Fig. 5.63. The material and cross-sectional
properties of the ties are denoted by E_t, A_t, and of the beam by E_b, I_b.
Neglecting axial effects in the beam, obtain expressions for the vertical
displacement and rotation of the beam at the central supported point due
to the force P acting as shown in the figure.

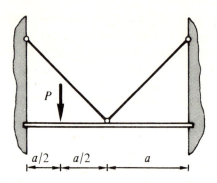

Fig. 5.63

5.20 The concrete beam in Fig. 5.64 is reinforced by a pin-jointed steel framework. The inclined members of the framework are at 30° to the horizontal.
The second moment of area of the beam is 15.0×10^{-3} m^4, and the
cross-sectional areas of the pin-ended members are 1.0×10^{-3} m^2. Show
that if the axial deformation of the beam is neglected, only six degrees of
freedom are needed to define the deformation of the structure. Determine
the corresponding global stiffness matrix and the equivalent global force
matrix.

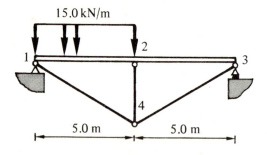

Fig. 5.64

APPENDIX A5.1 GENERAL DERIVATION OF THE EQUILIBRIUM
EQUATIONS BETWEEN MEMBER AND GLOBAL
FORCES

The virtual work equation for a framework (3.70), ignoring for the present
purpose support settlements and temperature changes, takes the form

$$\mathbf{P}^{*T}\mathbf{d} = \oint (\mathbf{S}^{*T}\boldsymbol{\phi}\mathbf{S})\mathrm{d}x = \sum_m \int_0^{l_m} \{\mathbf{S}^{*T}\boldsymbol{\phi}\,\mathbf{S}\}_m \,\mathrm{d}x_m \,. \qquad (A5.1.1)$$

The same equation can be applied to each member considered as a free body subject to the external member forces P_m^*. Thus

$$P_m^{*T} d_m = \int_0^{l_m} \{S^{*T} \phi\, S\}_m \, dx_m \ . \tag{A5.1.2}$$

Therefore

$$P^{*T} d = \sum_m (P_m^{*T} d_m) \ . \tag{A5.1.3}$$

But (5.15) relates d_m and d for each member by means of the member transformation matrix t_m as follows:

$$d_m = t_m d \ . \tag{5.15}*$$

Thus

$$P^{*T} d = \sum_m (P_m^{*T} t_m d) \ . \tag{A5.1.4}$$

In the summation over the members, d is a common factor, and it can therefore be factorised out. Thus

$$\sum_m (P_m^{*T} t_m d) = (\sum_m (P_m^{*T} t_m)) d \ . \tag{A5.1.5}$$

Whence comparing (A5.1.4) and (A5.1.5) we have

$$P^{*T} = \sum_m (P_m^{*T} t_m) \ . \tag{A5.1.6}$$

Finally taking the transpose of both sides of this equation, and dropping the star, leads to the following general equilibrium equation between member and global forces in a framework:

$$P = \sum_m (t_m^T P_m) \ . \tag{5.17}*$$

APPENDIX A5.2 PROPERTIES OF A MEMBER IN BENDING

A5.2.1 Member Stiffness Matrix
Consider member m in Fig. A5.1 whose deformation is defined by the end rotations d_{m1} and d_{m2}. The corresponding member forces are the couples P_{m1} and P_{m2}. In order to find the element stiffness matrix, we shall invert the member flexibility matrix which can be calculated using the virtual work equation

as described in Chapter 3. We thus need to find the displacements $\mathbf{d_m}$ due to the forces $\mathbf{P_m}$. Let M_P say, be the bending moments corresponding to $\mathbf{P_m}$. Then using (3.27) for the displacements of a beam in bending, we have

$$\mathbf{d_m} = \int_0^{l_m} \left\{ \frac{\mathbf{m} M_P}{EI} \right\}_m dx_m \qquad (A5.2.1)$$

where $\mathbf{m}$ is the matrix of the bending moments due to unit couples at the ends 1 and 2 in turn. The bending moment diagrams corresponding to M_P and m_1 and m_2 are shown in Fig. A5.2. Thus

$$M_P = \left\{ -P_1 \left(1 - \frac{x}{l} \right) + P_2 \frac{x}{l} \right\}_m \qquad (A5.2.2)$$

$$m_1 = - \left\{ 1 - \frac{x}{l} \right\}_m , \quad m_2 = \left\{ \frac{x}{l} \right\}_m . \qquad (A5.2.3a,b)$$

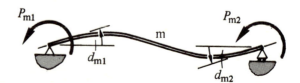

Fig. A5.1

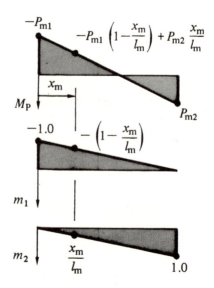

Fig. A5.2

Whence carrying out the integrations in (A5.2.1) leads to

$$d_{m1} = \left\{\frac{1}{EI}\right\}_m \int_0^{l_m} \left\{P_1 \left(1 - \frac{x}{l}\right)^2 - P_2 \frac{x}{l}\left(1 - \frac{x}{l}\right)\right\}_m dx_m$$

$$= \left\{\frac{l}{EI}\right\}_m \left(\frac{P_{m1}}{3} - \frac{P_{m2}}{6}\right) \tag{A5.2.4a}$$

$$d_{m2} = \left\{\frac{1}{EI}\right\}_m \int_0^{l_m} \left\{-P_1 \frac{x}{l}\left(1 - \frac{x}{l}\right) + P_2 \left(\frac{x}{l}\right)^2\right\}_m dx_m$$

$$= \left\{\frac{l}{EI}\right\}_m \left(\frac{-P_{m1}}{6} + \frac{P_{m2}}{3}\right) \quad . \tag{A5.2.4b}$$

Thus

$$\mathbf{d_m} = \mathbf{f_m P_m} \tag{A5.2.5}$$

where the **member flexibility matrix** $\mathbf{f_m}$ is given by

$$\mathbf{f_m} = \left\{\frac{l}{EI}\right\}_m \begin{bmatrix} 1/3 & -1/6 \\ -1/6 & 1/3 \end{bmatrix} \quad . \tag{A5.2.6}$$

Equations (A5.2.4) can be inverted by solving for P_{m1} and P_{m2} in terms of d_{m1} and d_{m2} giving

$$P_{m1} = \left\{\frac{EI}{l}\right\}_m (4.0 d_{m1} + 2.0 d_{m2}) \quad , \tag{A5.2.7a}$$

$$P_{m2} = \left\{\frac{EI}{l}\right\}_m (2.0 d_{m1} + 4.0 d_{m2}) \quad . \tag{A5.2.7b}$$

Thus

$$\mathbf{P_m} = \mathbf{k_m d_m} \tag{A5.2.8}$$

where the **member stiffness matrix** $\mathbf{k}_m$ is given by

$$\mathbf{k}_m = \mathbf{f}_m^{-1} = \left\{ \frac{EI}{l} \right\}_m \begin{bmatrix} 4.0 & 2.0 \\ 2.0 & 4.0 \end{bmatrix} . \qquad (A5.2.9)$$

A5.2.2 Fixed-end Couples

Consider the member in Fig. A5.1, but this time additionally subject to arbitrary intermediate forces Q_m, q_m as in Fig. A5.3. Suppose the bending moments due to Q_m, q_m acting alone on the member are M_Q. These can be determined in advance for any particular arrangement of forces, by solving the member as a simply supported beam. The member displacements due both to $\mathbf{P}_m$ and to Q_m, q_m are given by

$$\mathbf{d}_m = \int_0^{l_m} \left\{ \frac{\mathbf{m}(M_P + M_Q)}{EI} \right\}_m dx_m \qquad (A5.2.10)$$

where M_P and $\mathbf{m}$ are again given by (A5.2.2) and (A5.2.3). Thus using the results already calculated in (A5.2.4) we obtain

$$d_{m1} = \left\{ \frac{l}{EI} \right\}_m \left(\frac{P_{m1}}{3} - \frac{P_{m2}}{6} \right) - \left\{ \frac{1}{EI} \right\}_m \int_0^{l_m} \left\{ \left(1 - \frac{x}{l}\right) M_Q \right\}_m dx_m$$

$$(A5.2.11a)$$

$$d_{m2} = \left\{ \frac{1}{EI} \right\}_m \left(\frac{-P_{m1}}{6} + \frac{P_{m2}}{3} \right) + \left\{ \frac{1}{EI} \right\}_m \int_0^{l_m} \left\{ \left(\frac{x}{l}\right) M_Q \right\}_m dx_m .$$

$$(A5.2.11b)$$

The terms involving M_Q in (A5.2.11) are the end displacements of the member due to Q_m, q_m acting alone on the member. For conciseness we shall call these $\mathbf{d}_{Qm}$. Then

$$\mathbf{d}_m = \mathbf{f}_m \mathbf{P}_m + \mathbf{d}_{Qm} \qquad (A5.2.12)$$

where

$$\mathbf{d}_{Qm} = \left\{ \frac{1}{EI} \right\}_m \begin{bmatrix} -\int_0^l \left(1 - \frac{x}{l}\right) M_Q dx \\ \int_0^l \left(\frac{x}{l}\right) M_Q dx \end{bmatrix}_m . \qquad (A5.2.13)$$

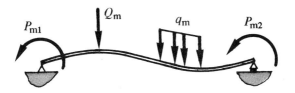

Fig. A5.3

Rearranging (A5.2.12) we have

$$f_m P_m = d_m - d_{Qm} \qquad (A5.2.14)$$

and therefore

$$P_m = f_m^{-1} (d_m - d_{Qm}) = k_m d_m - k_m d_{Qm} . \qquad (A5.2.15)$$

The term involving d_{Qm} in (A5.2.15) represents the end couples caused by the intermediate forces, when the member displacements d_m are zero. Calling these P_{Fm}, then P_{Fm} are the **fixed-end couples** required to maintain d_m zero in the presence of the intermediate loading. Thus

$$P_m = k_m d_m + P_{Fm} \qquad (A5.2.16)$$

where

$$P_{Fm} = -k_m d_{Qm} . \qquad (A5.2.17)$$

Stiffness Method II: Computation

6.1 INTRODUCTION

In Chapter 5 we discussed the physical aspects of the stiffness method of analysis. In this chapter we shall consider the geometrical and numerical aspects of the method, and its use as a basis for programming a computer to solve frameworks. The analysis of pin-jointed frameworks, rigid-jointed plane frameworks and space frameworks will be described in turn, and the details of the programming problem will be illustrated by two computer programs presented in appendices to the chapter.

6.2 PIN-JOINTED FRAMEWORKS

6.2.1 Introduction

In this section we shall consider the various steps necessary for programming the stiffness method to solve one of the simplest structural problems, the pin-jointed plane framework loaded by direct forces on its joints. The discussion will be illustrated by the square panel framework, solved in Example 5.1 and reproduced in Fig. 6.1(a). Again the members are of steel, and their cross-sectional areas in square millimetres are as shown in the figure.

The numerical analysis follows the procedure summarised in Chapter 5, Section 5.9. Thus we calculate t_m, and k_m for each member, assemble the global stiffness and force matrices, solve the equilibrium equations for the displacements, and calculate the member forces. The analysis is concluded by obtaining one further structural response of importance to the designer of foundations — the reactions at the supports.

6.2.2 Definition of the Framework

The first task is to define the framework in a numerical form suitable for presentation to the computer. A typical list of data, that for the pin-jointed framework in Fig. 6.1 is given in Table 6.1. (The table is a print-out of a FORTRAN data file.)

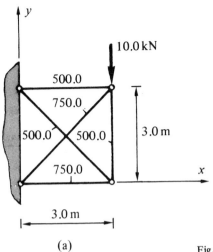

(a)

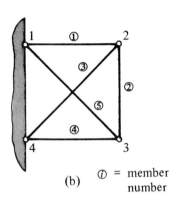

(b) ⓘ = member number

Fig. 6.1

Table 6.1
Data for a pin-jointed plane framework

Line number	Data			
1	4	5	4	1
2	1	0.0	3.0	
3	2	3.0	3.0	
4	3	3.0	0.0	
5	4	0.0	0.0	
6	1	1	2	
7	2	2	3	
8	3	2	4	
9	4	3	4	
10	5	1	3	
11	1	200.0E9	500.0E-6	
12	2	200.0E9	500.0E-6	
13	3	200.0E9	750.0E-6	
14	4	200.0E9	750.0E-6	
15	5	200.0E9	500.0E-6	
16	1	1	0.0	
17	1	2	0.0	
18	4	1	0.0	
19	4	2	0.0	
20	2	2	-10.0E3	

The first line of the data list in Table 6.1 specifies the **basic structural data**, comprising the numbers of joints, members, joint restraints and joint forces. These numbers allow blocks of information of the correct length to be read in by the computer, defining the joint coordinates, member properties, supports and loading.

The next part of the data list relates to the joints, and specifies their positions with respect to an arbitrarily situated global x, y coordinate system. All the joints, including the supported joints, are numbered by the engineer in an arbitrary order, using the sequence 1, 2, 3 , etc., as in Fig. 6.1(b). The position of each joint is then specified by a line in the data list containing a joint number, and the x and y coordinates of that joint relative to the global coordinates. Thus we generate lines 2 to 5 of Table 6.1.

We then define the members, specifying their position relative to the joints, and their physical properties. For this purpose, the members are identified by *numbering* them in some arbitrary order, again using the sequence 1, 2, 3 etc., as in Fig. 6.1(b). The positions of the members are then determined by their **connectivity**, being the joints to which the members are attached. Thus a list of data is provided, each line of which contains a member number, and the two global joint numbers of that member, as in lines 6 to 10 of Table 6.1. The physical properties of the members needed for calculating the member stiffness matrices given by (5.12) are E, A and l. However, since the joint positions and member connectivity are known, l can easily be calculated in the program. Thus, only the values of E and A are needed for each member, and these are specified by a data list as in lines 11 to 15 of Table 6.1.

We next specify the supports of the framework in terms of the **joint restraints**. A joint restraint is defined as a known fixed value assigned to one of the displacement components of a joint. Thus for a joint supported by a pinned support as in Fig. 6.2(a), the x and y components of the displacement are either zero, or equal to known support settlements. For a joint supported by a roller support, as in Figs. 6.2(b) or (c), either the x *or* y component of the displacement is zero. The direction of a joint restraint can be defined numerically by the integers 1 or 2 referring to the x or y directions respectively. The joint restraints are then specified by a list of data, each line of which contains a joint number, a direction number and the known joint displacement (positive if the displacement is in the positive coordinate direction), as in lines 16 to 19 of Table 6.1.

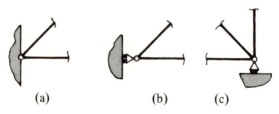

(a)   (b)   (c)

Fig. 6.2

Finally the loading on the framework is specified in terms of the **joint forces**, a joint force being defined as a component of the total direct force acting on a joint. The forces are then specified by a similar data list to that for the joint restraints. Thus each line of the list contains a joint number, a direction number and the direct force component. The single line corresponding to the 10.0 kN force in Fig. 6.1(a), then appears as line 20 in Table 6.1.

6.2.3 Degrees of Freedom
In the calculation of the frameworks in Chapter 5, degrees of freedom were arbitrarily assigned to the unrestrained joints of the frameworks by inspection. However, in the present computer-oriented analysis, a systematic selection of degrees of freedom is necessary. In fact, no distinction is made between the unrestrained and restrained joints at this stage, and two degrees of freedom, in the x and y directions respectively, are assigned to each joint in the order of the joint numbering. Thus for the pin-jointed framework in Fig. 6.1, the eight degrees of freedom are as shown in Fig. 6.3. Note therefore that for joint i, the degrees of freedom are d_{2i-1} in the x direction and d_{2i} in the y direction.

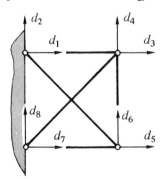

Fig. 6.3

It will be shown below that restraints on particular degrees of freedom at the supported joints can be imposed numerically at a later stage in the analysis.

6.2.4 Member Transformation Matrix
In the examples analysed in Chapter 5, the member transformation matrices were determined by inspection, simply by considering the geometry of the deformation and the relationship between member and global displacements. We recall that this relationship is given by (5.15) as

$$\mathbf{d_m} = \mathbf{t_m d} \, . \qquad\qquad (6.1)$$

In this section we shall describe how $\mathbf{t_m}$ can be generated from the structural data presented in the form discussed in Section 6.2.2.

Suppose a particular member connects joints i and j in a framework as shown in Fig. 6.4(a). We shall assume that i is the lower joint number so that end 1 of the member is connected to i and end 2 to j. The global degrees of freedom at the two ends are respectively d_{2i-1}, d_{2i} and d_{2j-1}, d_{2j}, and the member degrees of freedom are the axial displacements d_{m1} and d_{m2} as shown. If then joint i moves to a new position defined by particular displacements d_{2i-1}, d_{2i}, it is apparent from the geometry of Fig. 6.4(b) that

$$d_{m1} = c_{mx} d_{2i-1} + c_{my} d_{2i} \tag{6.2}$$

where c_{mx} ($= \cos(\alpha_x)$) and c_{my} ($= \cos(\alpha_y)$) are the **direction cosines** of the member. It is convenient in the following discussion to define the relationship (6.2) between the member displacement at one end of a member, and the global displacements at that end, by the matrix $\mathbf{r_m}$. Thus

$$d_{m1} = \mathbf{r_m} \begin{bmatrix} d_{2i-1} \\ d_{2i} \end{bmatrix} \tag{6.3}$$

where

$$\mathbf{r_m} = [c_{mx} \quad c_{my}] \quad . \tag{6.4}$$

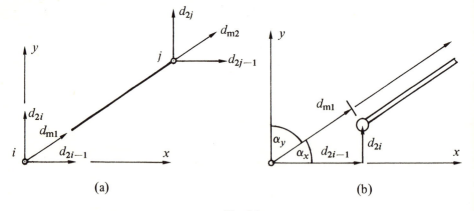

(a)　　　　　　　　　　　　(b)

Fig. 6.4

A similar relationship exists between the member and global displacements at end 2. Thus

$$d_{m2} = \mathbf{r_m} \begin{bmatrix} d_{2j-1} \\ d_{2j} \end{bmatrix} \quad . \tag{6.5}$$

Collecting (6.3) and (6.5) then gives

$$
\mathbf{d}_m = \left[\begin{array}{c|c} \mathbf{r}_m & 0 \\ \hline 0 & \mathbf{r}_m \end{array}\right] \begin{bmatrix} d_{2i-1} \\ d_{2i} \\ d_{2j-1} \\ d_{2j} \end{bmatrix} . \tag{6.6}
$$

(1 × 2 null matrix)

Again, it is convenient to define the relationship (6.6) between the member displacements and the global displacements of the joints to which it is attached, by a further matrix called the **condensed member transformation matrix** $\mathbf{t}'_m$. Thus

$$
\mathbf{d}_m = \mathbf{t}'_m \begin{bmatrix} d_{2i-1} \\ d_{2i} \\ d_{2j-1} \\ d_{2j} \end{bmatrix} \tag{6.7}
$$

where

$$
\mathbf{t}'_m = \left[\begin{array}{c|c} \mathbf{r}_m & 0 \\ \hline 0 & \mathbf{r}_m \end{array}\right] . \tag{6.8}
$$

However, the required transformation matrix $\mathbf{t}_m$ in (6.1), relates the member displacements to *all* the global displacements. In fact, $\mathbf{t}_m$ can be obtained from (6.7) and (6.8) by noting that the displacements of a member are independent of the displacements of the joints to which it is unattached. Thus comparing (6.7) and (6.8) with (6.1) it is apparent that

$$
\text{col. no.} \rightarrow \quad 2i{-}1 \ \ 2i \qquad 2j{-}1 \ \ 2j \qquad (1 \times 2 \text{ null matrix})
$$

$$
\mathbf{t}_m = \left[\begin{array}{cc|c|c|c|c|c|c|c} 0 & 0 & \cdots & \mathbf{r}_m & \cdots & 0 & \cdots & 0 \\ \hline 0 & 0 & \cdots & 0 & \cdots & \mathbf{r}_m & \cdots & 0 \end{array}\right] \tag{6.9}
$$

where the number of columns in $\mathbf{t}_m$ is equal to the total number of global degrees of freedom N.

t_m given by (6.9) can easily be calculated from the available structural data. Thus if l is the length of the member, and l_x, l_y are the projections of this length in the x and y directions respectively, then

$$c_{mx} = l_x/l, \quad c_{my} = l_y/l \qquad (6.10a,b)$$

and

$$\mathbf{r} = \begin{bmatrix} \dfrac{l_x}{l} & \dfrac{l_y}{l} \end{bmatrix} \cdot \qquad (6.11)$$

l_x, l_y and l are calculated from the known joint coordinates as follows:

$$l_x = x_j - x_i, \quad l_y = y_j - y_i, \quad l = \sqrt{l_x^2 + l_y^2} \;. \qquad (6.12a, b, c)$$

Thus for member 3 in the framework in Fig. 6.1, for example,

$$l_x = -3.0 \text{ m}, \quad l_y = -3.0 \text{ m}, \quad l = 4.24 \text{ m}$$

$$\mathbf{r} = [-0.707 \quad -0.707]$$

and

$$t_m = \begin{bmatrix} 0 & 0 & \cdots & -0.707 & -0.707 & \cdots & 0 & 0 & \cdots & 0 \\ 0 & 0 & \cdots & 0 & 0 & \cdots & -0.707 & -0.707 & \cdots & 0 \end{bmatrix}$$

6.2.5 Global Stiffness Matrix

We next consider the problem of assembling the global stiffness matrix, using the assembly equation (5.21),

$$k = \sum_m (t_m^T k_m t_m) \;. \qquad (6.13)$$

We first note that the member stiffness matrices given by (5.12) as

$$k_m = \begin{bmatrix} 1.0 & -1.0 \\ -1.0 & 1.0 \end{bmatrix} \left\{ \dfrac{EA}{l} \right\}_m \qquad (6.14)$$

can easily be evaluated from the member properties provided in the data. The contribution of a particular member to the global stiffness matrix, is then found by inserting (6.9) and (6.14) into (6.13) and carrying out the matrix multiplications. This leads to

$$
t_m^T k_m t_m =
\begin{bmatrix}
0 & 0 & \cdots & 0 & \cdots & 0 & \cdots & 0 \\
0 & 0 & \cdots & 0 & \cdots & 0 & \cdots & 0 \\
\vdots & \vdots & & \vdots & & \vdots & & \vdots \\
0 & 0 & \cdots & r_m^T r_m & \cdots & -r_m^T r_m & \cdots & 0 \\
\vdots & \vdots & & \vdots & & \vdots & & \vdots \\
0 & 0 & \cdots & -r_m^T r_m & \cdots & r_m^T r_m & \cdots & 0 \\
\vdots & \vdots & & \vdots & & \vdots & & \vdots \\
0 & 0 & \cdots & 0 & \cdots & 0 & \cdots & 0
\end{bmatrix}
\tag{6.15}
$$

where column number $\rightarrow$ with columns $2i-1, 2i$ and $2j-1, 2j$; row number $\downarrow$ with rows $2i-1, 2i$ and $2j-1, 2j$; and $(2\times 2$ null matrix$)$.

The contribution is thus evaluated as a sparsely populated matrix of the same size as the global stiffness matrix. In fact using the full member transformation matrix in (6.13) is extremely inefficient, requiring the multiplication of large matrices mainly composed of zero terms. If however, we use the condensed member transformation matrix, t'_m, given by (6.8) then

$$
t_m'^T k_m t_m' =
\begin{bmatrix}
r_m^T r_m & -r_m^T r_m \\
\hline
-r_m^T r_m & r_m^T r_m
\end{bmatrix}
\left\{ \frac{EA}{l} \right\}_m
= k_m^{(g)} .
\tag{6.16}
$$

We thus produce a (4×4) matrix, which contains all the non-zero terms of the contribution of member m to the global stiffness matrix. This matrix is called the **transformed member stiffness matrix** and is denoted by $k_m^{(g)}$. $k_m^{(g)}$ is then the stiffness matrix of the member when its deformation is defined by *four* degrees of freedom at its ends, each of which is parallel to one of the global coordinates.

The assembly of the global stiffness matrix k is carried out by evaluating $k_m^{(g)}$ for each member and augmenting appropriate terms of k with the terms of $k_m^{(g)}$. The appropriate terms of k depend on the connectivity of the member and can be deduced by comparing the right-hand sides of (6.15) and (6.16). Thus for the member connecting joints i and j, particular terms of the global stiffness matrix are augmented as follows:

$$k_{2i-1,\,2i-1} = k_{2i-1,\,2i-1} + k_{m11}^{(g)}$$

$$k_{2i-1,\,2i} = k_{2i-1,\,2i} + k_{m12}^{(g)} \qquad (6.17)$$

$$k_{2i-1,\,2j-1} = k_{2i-1,\,2j-1} + k_{m13}^{(g)} \quad \text{etc.}$$

Thus for member 3 in the framework in Fig. 6.1, for example,

$$\mathbf{k_m} = \begin{bmatrix} 1.0 & -1.0 \\ -1.0 & 1.0 \end{bmatrix} \times 35.36 \text{ MN/m}$$

$$\mathbf{r_m} = \begin{bmatrix} -0.707 & -0.707 \end{bmatrix}$$

$$\mathbf{r_m^T r_m} = \begin{bmatrix} 0.5 & 0.5 \\ 0.5 & 0.5 \end{bmatrix}$$

$$\mathbf{k_m^{(g)}} = \begin{bmatrix} 1.0 & 1.0 & -1.0 & -1.0 \\ 1.0 & 1.0 & -1.0 & -1.0 \\ -1.0 & -1.0 & 1.0 & 1.0 \\ -1.0 & -1.0 & 1.0 & 1.0 \end{bmatrix} \times 17.68 \text{ MN/m}$$

and

$$k_{33} = k_{33} + 17.68 \times 10^6$$

$$k_{34} = k_{34} + 17.68 \times 10^6$$

$$k_{37} = k_{37} - 17.68 \times 10^6 \qquad \text{etc .}$$

6.2.6 Global Force Matrix
In the case of a pin-jointed framework which is loaded by direct forces at the joints, the assembly of the global force matrix is a simple matter of identifying particular joint forces with particular degrees of freedom. A force on joint i in the x direction for example, corresponds to the global degree of freedom d_{2i-1}. Thus, for the framework in Fig. 6.1, the global force matrix contains the single non-zero term $P_4 (= -1.0 \times 10^3 \text{ N})$.

6.2.7 Supports
We next consider the problem of introducing the joint restraints at the supports.

This in fact is a numerical problem, and the several ways of solving it can be discussed with reference to the global equilibrium equations

$$\mathbf{k}\,\mathbf{d} = \mathbf{P}\ . \tag{6.18}$$

In the following discussion it will be helpful to demonstrate the effect of various operations on (6.18) explicitly, so for conciseness, we shall take as an example in this section, a framework with just six degrees of freedom. The equilibrium equations then take the form

$$
\begin{bmatrix}
k_{11} & k_{12} & k_{13} & k_{14} & k_{15} & k_{16} \\
k_{21} & k_{22} & k_{23} & k_{24} & k_{25} & k_{26} \\
k_{31} & k_{32} & k_{33} & k_{34} & k_{35} & k_{36} \\
k_{41} & k_{42} & k_{43} & k_{44} & k_{45} & k_{46} \\
k_{51} & k_{52} & k_{53} & k_{54} & k_{55} & k_{56} \\
k_{61} & k_{62} & k_{63} & k_{64} & k_{65} & k_{66}
\end{bmatrix}
\begin{bmatrix}
d_1 \\ d_2 \\ d_3 \\ d_4 \\ d_5 \\ d_6
\end{bmatrix}
=
\begin{bmatrix}
P_1 \\ P_2 \\ P_3 \\ P_4 \\ P_5 \\ P_6
\end{bmatrix}
. \tag{6.19}
$$

Several of the degrees of freedom in (6.19) are restrained at supports, and it should first be noted, that if an attempt is made to solve (6.19) without applying these restraints, then the numerical procedure becomes unstable. This is because the global stiffness matrix in its unmodified form is **singular**, reflecting the fact that if forces are applied to an unsupported structure, then the structure simply accelerates and the degrees of freedom are not defined. To illustrate the problem of introducing the restraints, we shall assume that two of the degrees of freedom in (6.19) are fixed. Thus suppose that d_3 and d_5 are known and equal to the support settlements d_{s3} and d_{s5}. The corresponding global forces P_3 and P_5 are then *unknown* support reactions which we shall call R_3 and R_5.

The first way of introducing the restraints into (6.19) is simply to rearrange the equations so that the unknown degrees of freedom are separated from the known degrees of freedom as follows:

$$
\left[
\begin{array}{cccc|cc}
k_{11} & k_{12} & k_{14} & k_{16} & k_{13} & k_{15} \\
k_{21} & k_{22} & k_{24} & k_{26} & k_{23} & k_{25} \\
k_{41} & k_{42} & k_{44} & k_{46} & k_{43} & k_{45} \\
k_{61} & k_{62} & k_{64} & k_{66} & k_{63} & k_{65} \\
\hline
k_{31} & k_{32} & k_{34} & k_{36} & k_{33} & k_{35} \\
k_{51} & k_{52} & k_{54} & k_{56} & k_{53} & k_{55}
\end{array}
\right]
\left[
\begin{array}{c}
d_1 \\ d_2 \\ d_4 \\ d_6 \\ \hline d_{s3} \\ d_{s5}
\end{array}
\right]
=
\left[
\begin{array}{c}
P_1 \\ P_2 \\ P_4 \\ P_6 \\ \hline R_3 \\ R_5
\end{array}
\right]
. \tag{6.20}
$$

(6.20) can then be written in symbolic form as

$$
\begin{bmatrix} k_1 & k_2 \\ k_2^T & k_3 \end{bmatrix} \begin{bmatrix} d_u \\ d_s \end{bmatrix} = \begin{bmatrix} P \\ R \end{bmatrix} \qquad (6.21)
$$

Thus

$$
k_1 d_u + k_2 d_s = P \qquad (6.22)
$$

and

$$
k_2^T d_u + k_3 d_s = R \qquad (6.23)
$$

The solution for the unknown displacements d_u is then obtained from (6.22) as

$$
d_u = k_1^{-1}(P - k_2 d_s) \qquad (6.24)
$$

after which the reactions R can be calculated directly from (6.23). It is thus apparent that rearranging the global equilibrium equations enables a solution to be obtained by inverting a stiffness matrix k_1 that is smaller than the original global stiffness matrix k. This is a useful advantage when programming micro-computers. However, generating the various-sized matrices k_1, k_2 and k_3 is quite a complex programming problem, and two simpler alternative methods for introducing restraints will now be discussed.

Suppose the equilibrium equations (6.19) are modified as follows:

$$
\begin{bmatrix} k_{11} & k_{12} & 0 & k_{14} & 0 & k_{16} \\ k_{21} & k_{22} & 0 & k_{24} & 0 & k_{26} \\ 0 & 0 & 1.0 & 0 & 0 & 0 \\ k_{41} & k_{42} & 0 & k_{44} & 0 & k_{46} \\ 0 & 0 & 0 & 0 & 1.0 & 0 \\ k_{61} & k_{62} & 0 & k_{64} & 0 & k_{66} \end{bmatrix} \begin{bmatrix} d_1 \\ d_2 \\ d_3 \\ d_4 \\ d_5 \\ d_6 \end{bmatrix} = \begin{bmatrix} P_1 \\ P_2 \\ d_{s3} \\ P_4 \\ d_{s5} \\ P_6 \end{bmatrix} - \begin{bmatrix} k_{13} & k_{15} \\ k_{23} & k_{25} \\ 0 & 0 \\ k_{43} & k_{45} \\ 0 & 0 \\ k_{63} & k_{65} \end{bmatrix} \begin{bmatrix} d_{s3} \\ d_{s5} \end{bmatrix}
\begin{matrix} (a) \\ (b) \\ (c) \\ (6.25d) \\ (e) \\ (f) \end{matrix}
$$

Equations a, b, d and f are then the same as the corresponding equations in (6.19), with the terms containing the known displacements d_{s3}, d_{s5} transferred to the right-hand sides. Equations c and e are the identities, $d_3 = d_{s3}, d_5 = d_{s5}$. (6.25) can then be written as

$$
k'd = P' \qquad (6.26)
$$

where $\mathbf{k}'$ and $\mathbf{P}'$ are modified global stiffness and force matrices of the same size as the original matrices. The solution for the complete set of displacements $\mathbf{d}$ is then obtained from (6.26) as

$$\mathbf{d} = \mathbf{k}'^{-1} \mathbf{P}' \tag{6.27}$$

and because of the identity equations in (6.25), this automatically contains the required restrained displacements d_{s3} and d_{s5}. Thus the programming problem of introducing a restraint d_{si} say, involves modifying the global stiffness matrix so that its coefficients in the ith row and column are zero except for k_{ii}, which is equal to 1.0. The global force matrix is modified by subtracting from P_j, the term $k_{ji}d_{si}$ $(j \neq i)$ and replacing P_i with d_{si}. If finally the reactions R_3 and R_5 are required, then these can be obtained from (6.19) by finding $\mathbf{P}$ as the matrix product of the *unmodified* matrix $\mathbf{k}$ and the known displacements $\mathbf{d}$. P_3 and P_5 are then the required reactions. This is discussed in greater detail in Section 6.2.10 below.

The simplest way of introducing the support restraints from the programming point of view, is to modify the equilibrium equations (6.19) as follows:

$$
\begin{bmatrix}
k_{11} & k_{12} & k_{13} & k_{14} & k_{15} & k_{16} \\
k_{21} & k_{22} & k_{23} & k_{24} & k_{25} & k_{26} \\
k_{31} & k_{32} & 1.0 \times 10^{30} & k_{34} & k_{35} & k_{36} \\
k_{41} & k_{42} & k_{43} & k_{44} & k_{45} & k_{46} \\
k_{51} & k_{52} & k_{53} & k_{54} & 1.0 \times 10^{30} & k_{56} \\
k_{61} & k_{62} & k_{63} & k_{64} & k_{65} & k_{66}
\end{bmatrix}
\begin{bmatrix}
d_1 \\ d_2 \\ d_3 \\ d_4 \\ d_5 \\ d_6
\end{bmatrix}
=
\begin{bmatrix}
P_1 \\
P_2 \\
(1.0 \times 10^{30})d_{s3} \\
P_4 \\
(1.0 \times 10^{30})d_{s5} \\
P_6
\end{bmatrix}
\begin{matrix}
\text{(a)} \\ \text{(b)} \\ \text{(c)} \\ \text{(6.28d)} \\ \text{(e)} \\ \text{(f)}
\end{matrix}
$$

In this case the arbitrary large numbers 1.0×10^{30} introduced into equations c and e dominate the remaining terms in these equations and they are effectively reduced to

$$(1.0 \times 10^{30})d_3 = (1.0 \times 10^{30})d_{s3} \tag{6.29a}$$

$$(1.0 \times 10^{30})d_5 = (1.0 \times 10^{30})d_{s5} \tag{6.29b}$$

which again are the identities $d_3 = d_{s3}$, and $d_5 = d_{s5}$. Thus the solution

$$\mathbf{d} = \mathbf{k}'^{-1}\mathbf{P}' , \tag{6.30}$$

where $\mathbf{k}'$ and $\mathbf{P}'$ are now the global matrices modified as in (6.28) automatically contains the required restrained displacements d_{s3} and d_{s5}. Further, in solving the simultaneous equations; d_{s3} and d_{s5} are automatically inserted into equations a, b, d and f so that these equations are again the same as the corresponding equations in (6.25). Introducing a restraint d_{si} say, is then seen simply to involve replacing k_{ii} in the global stiffness matrix by 1.0×10^{30}, and P_i in the global force matrix by $1.0 \times 10^{30} \, d_{si}$. The only doubt about this method is that theoretically it can break down if the structural stiffness coefficients are of the same order of magnitude as 10^{30}. In fact in usual SI units, stiffnesses are orders of magnitude less than 10^{10}, and the difficulty does not arise in practical calculations.

6.2.8 Solving the Equilibrium Equations

In the preceding sections, the solution of the equilibrium equations has been presented, for convenience, in terms of the inverted stiffness matrix k^{-1}. In fact, inverting the stiffness matrix is not efficient, unless a structure is to be repeatedly analysed for various loading cases. Where only one loading case is considered, the equilibrium equations are more easily solved as a set of linear simultaneous equations.

The solution of simultaneous equations is a classical problem that has received intensive study over many years. The basic theory of the several methods available has been discussed by many authors, for example by Fox [6.1] and by Wait [6.2]. A discussion of this purely numerical problem is outside the scope of the present text, and we shall simply note here, that the classical Gaussian elimination method is reliable and efficient for problems of moderate size. This method is the basis of the algorithm used in the programs in the appendices to this chapter. A feature of algorithms that have been developed to solve the equilibrium equations in linear structural analysis, is that they utilise the symmetry and banding of the global stiffness matrix to reduce storage and computation time. **Banding** refers to the grouping of the non-zero stiffness coefficients about the leading diagonal of the matrix. This grouping is evident in equation (6.15) in the contribution of a particular member to the matrix. The assembled global stiffness matrix for a small framework will be similar in form to that shown schematically in Fig. 6.5(a), where the non-zero coefficients are indicated by a cross. It is then apparent that since the matrix is symmetrical, it can be completely defined by the coefficients contained within the shaded section shown in the figure. The section includes the leading diagonal, and is of sufficient constant width to enclose half the longest rows of significant coefficients. The width of the section is called the **bandwidth** of the matrix. Referring again to (6.15) it is then apparent that the bandwidth is determined by the member with the greatest difference between the joint numbers at its ends. Suppose this member connects joints i and j. The bandwidth is then equal to half the significance length of row $(2i - 1)$ in (6.15) and it is therefore equal to $(2j - (2i - 1) + 1)$ or $2(j - i + 1)$.

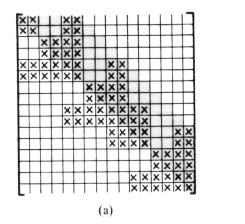

(a)   (b)

Fig. 6.5

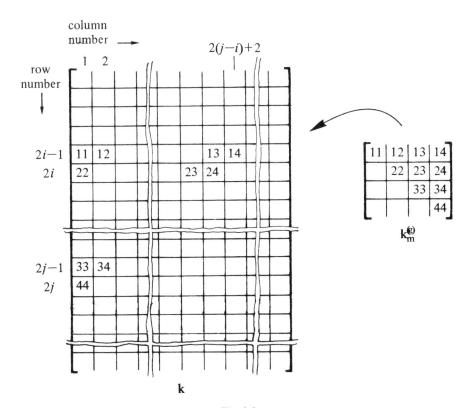

Fig. 6.6

Algorithms for solving banded equations store and operate on the global stiffness matrix in a banded form, that is, on a matrix whose number of rows is the same as that of the global matrix, and whose number of columns is equal to the bandwidth. The banded matrix is then composed of the coefficients of the global stiffness matrix as shown in Fig. 6.5(b). The assembly of this matrix is carried out directly, again by evaluating the transformed stiffness matrix $k_m^{(g)}$ for each member and augmenting the appropriate terms of the banded matrix with the terms of $k_m^{(g)}$, as shown schematically in Fig. 6.6.

Finally, it is clear from the above discussion, that if the banded matrix is to occupy the minimum computer storage for a given framework, it is necessary that the bandwidth should be a minimum. This is achieved by arranging the joint numbering so that the greatest difference between the numbers of connected joints is as small as possible. For simple structures it is usually possible to number the joints satisfactorily by inspection. Thus for the truss bridge in Fig. 6.7, the numbering in Fig. 6.7(a) is satisfactory while that in Fig. 6.7(b) is unsatisfactory. For more complicated structures, computer programs have been developed to select the optimal joint numbering automatically [6.3].

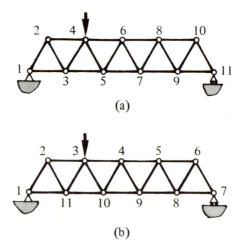

(a)

(b)

Fig. 6.7

6.2.9 Member Forces

Having obtained the solution of the framework in terms of the global displacements, the displacements d_m of a particular member are calculated by (6.7) from the global displacements at its ends, using the condensed transformation matrix t_m'. The member forces are then obtained from the member equilibrium equations (5.13) as follows;

$$P_m = k_m d_m \quad . \tag{6.31}$$

6.2.10 Support Reactions

The final part of the numerical analysis is the calculation of the support reactions. We have briefly mentioned in Section 6.2.7 that when using one of the simpler methods of introducing the support restraints described in that section, the reactions can be generated from the displacement solution **d**, by calculating the global force matrix **P** from

$$\mathbf{P} = \mathbf{k}\,\mathbf{d} \qquad\qquad (6.18)^*$$

where **k** is the unmodified global stiffness matrix. This calculation is equivalent to treating the framework as a free body and obtaining the complete set of forces required to cause the restrained set of displacements. **P** then contains the joint forces on the *unrestrained* joints, and the reactions on the *restrained* joints. The calculation therefore performs two functions: it checks that the correct applied forces can be recovered from the displacement solution and therefore confirms the correct operation of the solving algorithm, and it generates the reactions.

In programming the matrix multiplication in (6.18), account has to be taken of the fact that **k** is stored in its banded form. The procedure is best described schematically. Thus the matrix multiplication using the *normal* form of the matrix is depicted in Fig. 6.8(a). In calculating P_5 for example, we have to obtain the sum of the products of significant terms in row 5 of **k** numbered 1 to 10, with the displacements 1 to 10 as shown. We then note that the terms to the

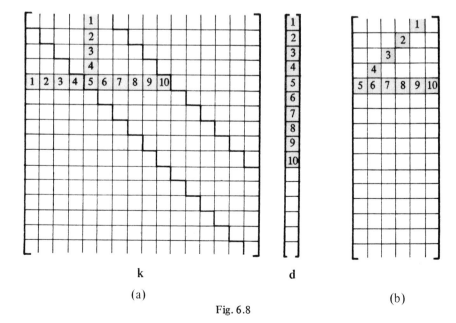

k   d

(a)   (b)

Fig. 6.8

left of the leading diagonal are not directly included in the banded form of the matrix. However, by symmetry they are equal to terms 1 to 4 in column 5 as shown. These terms can therefore be used in the matrix product. The corresponding terms 1 to 4 in the banded matrix, occupy the diagonal shown in Fig. 6.8(b). The programming problem then involves the correct selection of these terms and summing their products with the displacements.

6.2.11 Computer Program

We shall now describe the computer program for solving pin-jointed plane frameworks presented in Appendix A6.1. The programming language is FORTRAN, the widely used high-level language allowing concise clear programming. The language and its use in numerical analysis has been described in many texts, and two that the author has found useful are by Macnab [6.4], and by Monro [6.5].

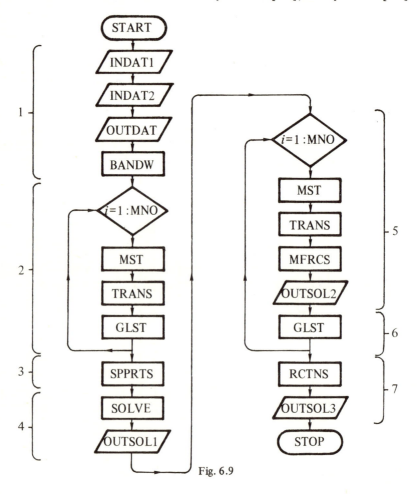

Fig. 6.9

For clarity, the various operations discussed in the previous sections are carried out in separate FORTRAN subroutines. The main program **PJPFA** then uses these subroutines according to the flow diagram in Fig. 6.9. The names of the variables and arrays have been chosen to reflect the notation in the text and their use is summarised in Table 6.2.

Table 6.2

Variables and arrays for **PJPFA**

Type and Name	Use
Integers	
COLNO	Number of columns in banded global stiffness matrix
JFNO	Number of joint forces
JNO	Number of joints
JRNO	Number of joint restraints
MNO	Number of members
RWNO	Number of rows in banded global stiffness matrix
Real arrays	
DG (RWNO)	Global displacements $\mathbf{d}$
DMG (4)	Global displacements at the ends of a member
JCRD (JNO, 2)	Joint coordinates; x, y
JRES (RWNO)	Joint restraints (1000.0 if displacement is unrestrained)
KG (RWNO, COLNO)	Banded global stiffness matrix $\mathbf{k}$ and modified matrix $\mathbf{k}'$
KM (2, 2)	Member stiffness matrix $\mathbf{k}_m$
KMG (4, 4)	Transformed member stiffness matrix $\mathbf{k}_m^{(g)}$
MPRP (MNO, 2)	Member properties: E, A
PG (RWNO)	Global force matrix $\mathbf{P}$ and modified matrix $\mathbf{P}'$
PM (2)	Member force matrix $\mathbf{P}_m$
TM (2, 4)	Condensed member transformation matrix $\mathbf{t}'_m$
WKA (COLNO)	Working area for subroutine **SOLVE**
Integer array	
MCON (MNO, 2)	Member connectivity: i, j

The important features of the main program and subroutines are discussed below.

Program **PJPFA**

PJPFA first declares the variables and arrays to be used in the program†. The coefficients in the arrays JCRD, MPRP, DG, PG, and KG, are initially set to zero in the declaration and those in JRES to 1000.0. The initialisation of JRES enables those coefficients which correspond to joint restraints to be overwritten by small joint settlements, and then to be easily distinguished from the coefficients corresponding to the unrestrained joints.

PJPFA then calls the subroutines according to the flow diagram in Fig. 6.9. In summary: the groups of subroutines indicated by 1 to 7 in the diagram have the following functions:

1. The structural data is read and stored in appropriate arrays.
2. The global stiffness matrix is assembled by cycling through the members.
3. The global stiffness matrix and force matrix are modified for joint restraints.
4. The equilibrium equations are solved and the displacement solution is printed.
5. The member forces are calculated and printed by cycling through the members.
6. In the same cycle, the unmodified global stiffness matrix is reassembled.
7. The joint reactions are calculated and printed.

Subroutine **INDAT1**
Reads and stores the basic structural data discussed in Section 6.2.2, and calculates the number of rows in the stiffness matrix.

Subroutine **INDAT2**
Reads and stores the joint coordinates, member connectivity, member properties, joint restraints and joint forces. The joint restraints are stored in a single-column matrix JRES containing all the degrees of freedom of the structure. The joint forces are stored in the global force matrix PG. (Note that for joint m and direction n, the degree of freedom is d_{2m+n-2}.)

The detailed layout of the data blocks required by **INDAT1** and **INDAT2** is given in Appendix A6.1, Section A6.1.2.

Subroutine **OUTDAT**
Prints the data.

---

† For conciseness, a combined array declaration and data statement has been employed – a statement available only in *extended* versions of FORTRAN.

Subroutine **BANDW**
Calculates the bandwith of the global stiffness matrix, enabling the correct number of columns for the banded global stiffness matrix to be declared in the subsequent subroutines.

Subroutine **MST**
Calculates the member stiffness matrix k_m.

Subroutine **TRANS**
Calculates the condensed member transformation matrix t'_m.

Subroutine **GLST**
Calculates the transformed member stiffness matrix $k_m^{(g)}$. Augments appropriate terms of the banded stiffness matrix with the terms of $k_m^{(g)}$ according to Fig. 6.6.

Subroutine **SPPRTS**
Introduces the support restraints using the simplest method discussed in Section 6.2.7. The particular degrees of freedom that are restrained are determined by the coefficients of JRES that are less than 1000.0. The banded stiffness matrix and force matrix are then modified as described in Section 6.2.7.

Subroutine **SOLVE**
Solves a banded set of N simultaneous equations by Gaussian elimination. The coefficients of the equations are stored in A (N, M), (M < N), the right-hand side in B (N). C (M) is a working area, and the solution is stored in D (N). During the Gaussian elimination, the coefficients in A and B are changed.

Subroutine **OUTSOL1**
Prints the displacement solution.

Subroutine **MFRCS**
Selects the global displacements of the joints at the ends of the member. Calculates the member forces using (6.7) and (6.31).

Subroutine **OUTSOL2**
Prints the member forces.

Subroutine **RCTNS**
Calculates **P** as the matrix product **kd**, according to Fig. 6.8(b).

Subroutine **OUTSOL3**
Prints the global force matrix containing the reactions at the restrained joints.

As an example of the running of the program, the pin-jointed framework in Fig. 6.1 is analysed. The computer output is included in Appendix A6.1. The results are then seen to agree with those produced in Example 5.1.

6.2.12 Pin-jointed Space Frameworks

The numerical analysis of the pin-jointed space framework is similar to that for the plane framework. Here we shall briefly discuss the significant differences.

The joints of the framework are referred to an arbitrarily oriented global x, y and z coordinate system and their positions are therefore specified by three coordinates. Three global degrees of freedom are required at each joint to define the deformation of the framework. Thus the degrees of freedom at joint i are d_{3i-2}, d_{3i-1} and d_{3i} in the x, y and z directions respectively. The joint restraints and joint forces are then specified by displacement and force components in the three directions at the restrained and loaded joints respectively.

The member transformation matrices are obtained from the direction cosines of the member relative to the three global coordinate lines. Thus

$$d_{m1} = c_{mx}d_{3i-2} + c_{my}d_{3i-1} + c_{mz}d_{3i} \tag{6.32a}$$

$$d_{m2} = c_{mx}d_{3j-2} + c_{my}d_{3j-1} + c_{mz}d_{3j} \tag{6.32b}$$

where

$$c_{mx} = l_x/l, \quad c_{my} = l_y/l, \quad c_{mz} = l_z/l \tag{6.33a,b,c}$$

and

$$l = \sqrt{l_x^2 + l_y^2 + l_z^2} \quad . \tag{6.34}$$

Thus

$$\mathbf{r}_m = [c_{mx} \ c_{my} \ c_{mz}] \tag{6.35}$$

and

$$\mathbf{d}_m = \mathbf{t}'_m \begin{bmatrix} d_{3i-2} \\ d_{3i-1} \\ d_{3i} \\ d_{3j-2} \\ d_{3j-1} \\ d_{3j} \end{bmatrix} \tag{6.36}$$

where

$$
t'_m = \left[\begin{array}{c|c} r_m & 0 \\ \hline 0 & r_m \end{array} \right] .
\tag{6.37}
$$

(1 × 3 null matrix)

The transformed member stiffness matrix $k_m^{(g)}$ in (6.16) is therefore a 6 × 6 matrix, and it augments coefficients in the global stiffness matrix occupying rows and columns numbered $3i-2$, $3i-1$, $3i$, and $3j-2$, $3j-1$ and $3j$.

The remaining subroutines are unchanged except for minor modifications to accommodate the increased number of variables. The programming is the same as for the rigid-jointed plane frameworks discussed below.

6.3 RIGID-JOINTED PLANE FRAMEWORKS

6.3.1 Introduction

The structure of the computer program for analysing pin-jointed frameworks is common to programs for analysing all other types of framework. The only major addition necessary for rigid-jointed frameworks is the inclusion of intermediate loading. In this section we shall consider the detailed programming for the analysis of rigid-jointed plane frameworks, referring particularly to the differences between this and the programming discussed in the previous section. We shall use as an example, the pin-footed portal frame discussed in Example 4.9 and reproduced in Fig. 6.10(a). Again all the members are of steel and have the same cross-section for which $A = 4.0 \times 10^3$ mm^2 and $I = 30.0 \times 10^6$ mm^4. The joints and members are numbered as in Fig. 6.10(b).

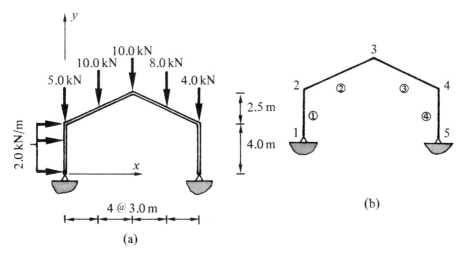

(a)

(b)

Fig. 6.10

6.3.2 Definition of the Framework
A list of data for the pin-footed portal frame is given in Table 6.3.

The basic structural data, comprising numbers of joints, members, joint restraints and joint forces is the same as for pin-jointed frameworks. However, a further parameter is now required to specify the number of intermediate forces. Thus in the case of the portal frame there are three intermediate forces acting on members 1, 2, and 3 respectively, as in Fig. 6.10. The basic structural data is then as shown in line 1 of Table 6.3.

The joint coordinates and member connectivity defined for arbitrarily numbered joints and members are treated in the same way as for the pin-jointed framework, as in lines 2 to 10 in Table 6.3.

Table 6.3
Data for a rigid-jointed plane framework

Line number			Data		
1	5	4	4	3	3
2	1	0.0	0.0		
3	2	0.0	4.0		
4	3	6.0	6.5		
5	4	12.0	4.0		
6	5	12.0	0.0		
7	1	1	2		
8	2	2	3		
9	3	3	4		
10	4	4	5		
11	1	200.0E9	4.0E-3	30.0E-6	
12	2	200.0E9	4.0E-3	30.0E-6	
13	3	200.0E9	4.0E-3	30.0E-6	
14	4	200.0E9	4.0E-3	30.0E-6	
15	1	1	0.0		
16	1	2	0.0		
17	5	1	0.0		
18	5	2	0.0		
19	2	2	-5.0E3		
20	3	2	-10.0E3		
21	4	2	-4.0E3		
22	1	1	2	2.0E3	
23	2	2	1	-10.0E3	0.5
24	3	2	1	-8.0E3	0.5

The physical properties of the members needed for calculating the member stiffness matrices given by (5.47) are E, A, I. These are provided by a data list as in lines 11 to 14 in Table 6.3.

A support restraint in a rigid-jointed plane framework, fixes one of the three displacement components at a joint, the components being the linear displacements in the x and y directions and the rotation about the z axis. Pinned and roller supports restrain one or more of the linear displacements and an encastered support in addition restrains the rotation. The direction of a joint restraint can be defined numerically by the integers 1, 2 or 3; 1 or 2 referring to the x or y directions, and 3 referring to the rotation. The joint restraint data for the pin-footed portal frame is then as shown in lines 15 to 18 of Table 6.3.

The loading on a joint is similarly defined in terms of the joint forces, being either a direct force component in the x or y directions or a couple, again distinguished by 1, 2 or 3. Thus the joint force data for the pin-footed portal frame is shown in lines 19 to 21 of Table 6.3.

Finally we specify the intermediate loading on the framework, in terms of intermediate forces. For simplicity we shall again restrict the intermediate forces to two types, a concentrated force component, and a uniformly distributed force component acting on the whole member. For each intermediate force the following data is required:

  (i) the member on which it acts,
 (ii) the direction in which it acts $(1 \equiv x, 2 \equiv y)$,
(iii) the type of force ($1 \equiv$ concentrated, $2 \equiv$ uniformly distributed),
 (iv) the magnitude of the force (N or N/m), and,
  (v) in the case of the concentrated force, its position from the end of the member attached to the joint with the lower number, expressed as a proportion of the length of the member.

The intermediate force data for the pin-footed portal frame is then as shown in lines 22 to 24 of Table 6.3.

6.3.3 Degrees of Freedom

Three degrees of freedom at each joint completely define the deformation of the framework; they are the linear displacements in the x and y directions and the rotation about the z axis. Again they are assigned to each joint in the order of the joint numbering. Thus for the portal frame in Fig. 6.10, the 15 degrees of freedom are as shown in Fig. 6.11. At joint i, therefore, the degrees of freedom are d_{3i-2}, d_{3i-1}, in the x and y directions respectively, and d_{3i} about the z axis.

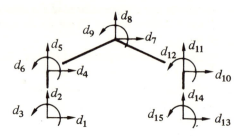

Fig. 6.11

6.3.4 Member Transformation Matrix

The deformation of a member in a rigid-jointed plane framework is defined by three member degrees of freedom at each end, being the linear displacements in the x_m and y_m directions and the rotation about the z_m axis. Consider the member connecting joints i and j of the framework as in Fig. 6.12(a), with $i < j$. At end i, the global degrees of freedom $d_{3i-2}, d_{3i-1}, d_{3i}$ and local degrees of freedom d_{m1}, d_{m2}, d_{m3} are in the directions shown. If then joint i moves to a new position defined by particular displacements $d_{3i-2}, d_{3i-1}, d_{3i}$, it is apparent from the geometry of Fig. 6.12(b) that

$$d_{m1} = d_{3i-2} \cos(\alpha_x) + d_{3i-1} \cos(\alpha_y)$$

$$d_{m2} = -d_{3i-2} \cos(\alpha_y) + d_{3i-1} \cos(\alpha_x) \qquad (6.38)$$

$$d_{m3} = d_{3i} \quad .$$

Thus

$$\begin{bmatrix} d_{m1} \\ d_{m2} \\ d_{m3} \end{bmatrix} = r_m \begin{bmatrix} d_{3i-2} \\ d_{3i-1} \\ d_{3i} \end{bmatrix} \qquad (6.39)$$

where r_m again is the matrix relating the member displacements at one end of a member to the global displacements at that end.

Thus

$$r_m = \begin{bmatrix} c_{mx} & c_{my} & 0 \\ -c_{my} & c_{mx} & 0 \\ 0 & 0 & 1.0 \end{bmatrix} \quad . \qquad (6.40)$$

A similar relationship exists between the member and global displacements at end 2. Thus

$$\mathbf{d_m} = \mathbf{t'_m} \begin{bmatrix} d_{3i-2} \\ d_{3i-1} \\ d_{3i} \\ d_{3j-2} \\ d_{3j-1} \\ d_{3j} \end{bmatrix} \tag{6.41}$$

where $\mathbf{t'_m}$ given by

$$\mathbf{t'_m} = \begin{bmatrix} \mathbf{r_m} & \mathbf{0} \\ \hline \mathbf{0} & \mathbf{r_m} \end{bmatrix} \quad \text{(3 × 3 null matrix)} \tag{6.42}$$

is again the condensed member transformation matrix.

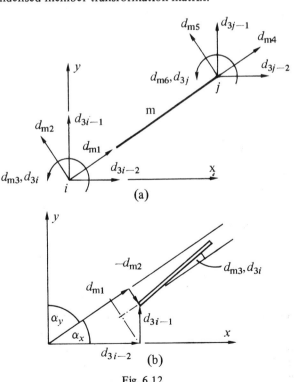

Fig. 6.12

6.3.5 Global Stiffness Matrix

The stiffness matrix for a member in a rigid-jointed plane framework is given by (5.47). For convenience in the following discussion, this can be expressed in terms of the submatrices k_{m1}, k_{m2} and k_{m3} as

$$k_m = \left[\begin{array}{c|c} k_{m1} & k_{m2} \\ \hline k_{m2}^T & k_{m3} \end{array} \right] \tag{6.43}$$

where

$$k_{m1} = \left[\begin{array}{ccc} EA/l & 0 & 0 \\ 0 & 12.0EI/l^3 & 6.0EI/l^2 \\ 0 & 6.0EI/l^2 & 4.0EI/l \end{array} \right]_m , \tag{6.44a}$$

$$k_{m2} = \left[\begin{array}{ccc} -EA/l & 0 & 0 \\ 0 & -12.0EI/l^3 & 6.0EI/l^2 \\ 0 & -6.0EI/l^2 & 2.0EI/l \end{array} \right]_m , \tag{6.44b}$$

$$k_{m3} = \left[\begin{array}{ccc} EA/l & 0 & 0 \\ 0 & 12.0EI/l^3 & -6.0EI/l^2 \\ 0 & -6.0EI/l^2 & 4.0EI/l \end{array} \right] . \tag{6.44c}$$

Again, the submatrices can easily be evaluated from the member properties provided in the data.

The transformed member stiffness matrix $k_m^{(g)}$ is a six by six matrix, composed of products of r_m, and k_{m1}, k_{m2} and k_{m3} as follows;

$$k_m^{(g)} = t_m'^T k_m t_m' = \left[\begin{array}{c|c} r_m^T k_{m1} r_m & r_m^T k_{m2} r_m \\ \hline r_m^T k_{m2}^T r_m & r_m^T k_{m3} r_m \end{array} \right] \tag{6.45}$$

and the rows and columns numbered $3i-2$, $3i-1$, $3i$, and $3j-2$, $3j-1$ and $3j$ of the global stiffness matrix are augmented with the sub-matrices of $k_m^{(g)}$ as shown in Fig. 6.13. It is then apparent that the bandwidth of the global stiffness matrix is equal to $3(j-i+1)$ where $(j-i)$ is the maximum difference between the numbers of the connected joints in the framework.

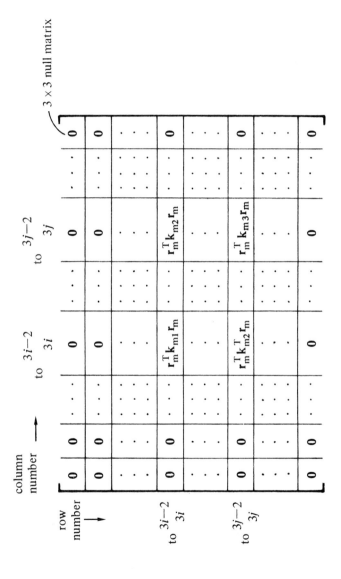

Fig. 6.13

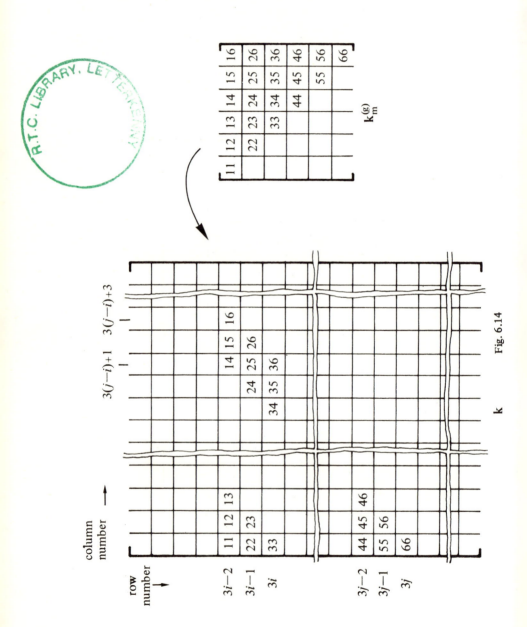

Fig. 6.14

The direct assembly of the banded global stiffness matrix is carried out by augmenting appropriate terms of the matrix with terms from the transformed member stiffness matrices, as shown in Fig. 6.14.

6.3.6 Global Force Matrix

The direct forces and couples acting on the joints of the framework can be placed immediately in the global force matrix, in the same way as the forces on the pin-jointed framework. However, it is also necessary to calculate the global forces P_Q that are equivalent to the intermediate forces specified in the structural data. P_Q are assembled from the member fixing forces using (5.61, b and c) as follows:

$$P_F = \sum_m (t_m^T P_{Fm}) \tag{6.46}$$

$$P_Q = -P_F . \tag{6.47}$$

We therefore need to obtain the member fixing forces P_{Fm}.

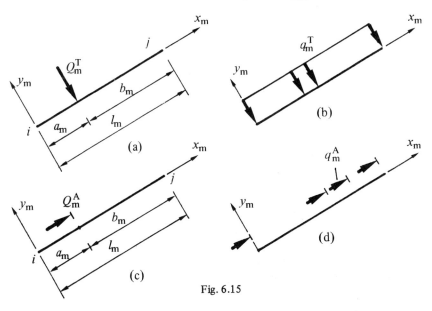

Fig. 6.15

The member fixing forces for *transverse* concentrated and distributed forces are given by (5.45) and (5.46), and for *axial* concentrated and distribured forces by (5.49) and (5.50). Here for clarity we shall call the transverse and axial intermediate forces acting on a member m, Q_m^T, q_m^T and Q_m^A, q_m^A respectively. They are assumed to act in the directions shown in Figs. 6.15(a) to (d). Substituting into the above equations and collecting terms, we then obtain:

$$
\mathbf{P}_{\mathrm{Fm}} = \begin{bmatrix} -Q^A(bl) \\[6pt] Q^T \dfrac{b^2}{l}(l+2a) \\[6pt] Q^T ab^2 \\[6pt] -Q^A(al) \\[6pt] Q^T \dfrac{a^2}{l}(l+2b) \\[6pt] -Q^T a^2 b \end{bmatrix}_{\mathrm{m}} (1/l_{\mathrm{m}}^2) \quad \text{or} = \begin{bmatrix} -q^A \\[6pt] q^T \\[6pt] q^T(l/6) \\[6pt] -q^A \\[6pt] q^T \\[6pt] -q^T(l/6) \end{bmatrix}_{\mathrm{m}} (l_{\mathrm{m}}/2) \quad . \quad (6.48a, b)
$$

The intermediate forces Q_{m}, q_{m} say, specified in the data are assumed to act in the global coordinate directions. It therefore remains to determine the corresponding transverse and axial components required in (6.48). These can be readily deduced from Fig. 6.16. Thus, if a concentrated force Q_{m} is directed in the x-direction, as in Fig. 6.16(a), then

$$Q_{\mathrm{m}}^A = Q_{\mathrm{m}} c_{\mathrm{m}x}, \quad Q_{\mathrm{m}}^T = Q_{\mathrm{m}} c_{\mathrm{m}y} \qquad (6.49a, b)$$

while if it is directed in the y direction as in Fig. 6.16(b), then

$$Q_{\mathrm{m}}^A = Q_{\mathrm{m}} c_{\mathrm{m}y}, \quad Q_{\mathrm{m}}^T = -Q_{\mathrm{m}} c_{\mathrm{m}x} \qquad . \qquad (6.50a, b)$$

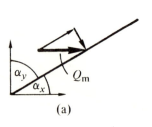
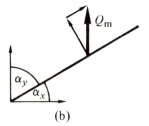

(a)   Fig. 6.16   (b)

Finally, we note that when assembling the global fixing-force matrix as in (6.46), it is again inefficient to use the total member transformation matrix $\mathbf{t}_{\mathrm{m}}$. If rather, we determine

$$\mathbf{P}_{\mathrm{Fm}}^{(g)} = \mathbf{t}_{\mathrm{m}}'^{T} \mathbf{P}_{\mathrm{Fm}} \qquad (6.51)$$

then $\mathbf{P}_{\mathrm{Fm}}^{(g)}$ is the **transformed member fixing-force matrix**, giving the fixing forces at the joints to which the member is attached, in the global coordinate directions.

The appropriate terms of the global fixing force matrix are then augmented by $\mathbf{P}_{Fm}^{(g)}$.

6.3.7 Member Forces
The support restraints are introduced and the solution for the global displacements is obtained, in the same way as for pin-jointed frameworks. The displacements $\mathbf{d}_m$ of a particular member are then calculated by (6.41) from the global displacements at the ends, using the condensed member transformation matrix $\mathbf{t}'_m$. The member forces are then obtained from the member equilibrium equations (5.34) as

$$\mathbf{P}_m = \mathbf{k}_m \mathbf{d}_m + \mathbf{P}_{Fm} \quad . \tag{6.52}$$

6.3.8 Support Reactions
As in Section 6.2.10, the support reactions are found by calculating the global force matrix from the displacement solution $\mathbf{d}$ using the unmodified global stiffness matrix. However, the intermediate forces represented by $\mathbf{P}_Q$, must also be included in the global equilibrium equations. Thus in the presence of intermediate forces, (6.18) is modified to

$$\mathbf{k}\,\mathbf{d} = \mathbf{P} + \mathbf{P}_Q \tag{6.53}$$

and the required global force matrix is given by

$$\mathbf{P} = \mathbf{k}\,\mathbf{d} - \mathbf{P}_Q \quad . \tag{6.54}$$

Again, $\mathbf{P}$ then contains the joint forces on the unrestrained joints, and the reactions on the restrained joints.

6.3.9 Computer Program
We shall now describe the computer program for solving rigid-jointed plane frameworks presented in Appendix A6.2. The main program **RJPFA** uses variables, arrays and subroutines with the same names and functions as those used by **PJPFA** and they are summarised in Table 6.2, in Section 6.2.11, and in the flow diagram in Fig. 6.9. However, in order to store the matrices of increased size discussed in this section, the dimensions of some of the arrays are altered to the following:

DMG(6), KM(6,6), KMG(6,6), MPRP(MNO,3), PM(6), TM(6,6)

RJPFA also accounts for the intermediate loading by calling three additional subroutines, **MFX**, **GLFQ** and **GLF**, as in an extended section of the flow diagram shown in Fig. 6.17. These use the additional variables listed in Table 6.4.

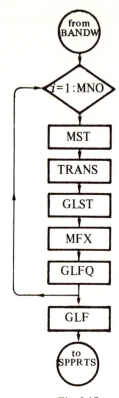

Fig. 6.17

Table 6.4
Additional variables and arrays for **RJPFA**

Type	Use
Integer	
IFNO	Number of intermediate forces
Real arrays	
INTFR (IFNO, 2)	Magnitude of intermediate force (N or N/m), position of concentrated force (a/l)
PGQ (RWNO)	Equivalent global force matrix $\mathbf{P}_Q$
PFM (MNO, 6)	Member fixing forces P_{Fm1} to P_{Fm6}
PFMG (6)	Transformed member fixing force matrix $\mathbf{P}_{Fm}^{(g)}$
Integer arrays	
INTFI(IFNO, 3)	Member on which intermediate force acts, direction, type

The important features of the additional subroutines and a significant change in **INDAT2** are discussed below.

Subroutine **INDAT2**
Reads and stores additional information concerning the intermediate forces, using the two new arrays INTFI and INTFR described in Table 6.4, INTFI is an integer array that stores the part of the information presented as integers for a particular sequence of intermediate forces. INTFR is a real array that stores the remaining part of the information presented as real variables, for the same sequence of forces.
        The detailed layout of the data blocks required by **INDAT1** and **INDAT2** is given in Appendix A6.2, Section A6.2.2.

Subroutine **MFX**
Calculates the member fixing force matrix $\mathbf{P}_{Fm}$, and stores it as a row in PFM. At the end of the cycle through the members, PFM contains all the fixing force matrices.

Subroutine **GLFQ**
Calculates the transformed member fixing force matrix $\mathbf{P}_{Fm}^{(g)}$, and augments appropriate terms of the global equivalent force matrix $\mathbf{P}_Q$.

Subroutine **GLF**
Adds $\mathbf{P}_Q$ to the global force matrix **P**.

As an example of the running of the program, the pin-footed portal frame shown in Fig. 6.10 is analysed. The computer output is included in Appendix A6.2. The results can be compared with the solution of the same problem in Example 4.9 by the flexibility method. The slight differences in the bending moments shown in Fig. 4.23 and the member couples in the output are because axial effects were ignored in Example 4.9. If the areas of members are increased to say 4.0 m^2, then the results produced by **RJPFA** are identical to those produced in that example.

6.4 RIGID-JOINTED SPACE FRAMEWORKS

6.4.1 Introduction
The computer program described in the previous section is restricted to analysing plane frameworks. In this section we shall describe the significant differences between this and a program for analysing space frameworks. We shall use as an example the framework considered in Examples 4.14 and 5.10 and reproduced in Fig. 6.18(a). All the members are of steel and are of the same cross-section

for which $A = 4.0 \times 10^3$ mm^2, $I_y = 15.0 \times 10^6$ mm^4, $I_z = 30.0 \times 10^6$ mm^4 and $J = 20.0 \times 10^6$ mm^4. The joint and member notation and member coordinate systems are shown in Fig. 6.18(b).

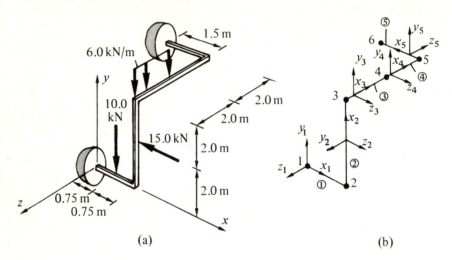

(a)   (b)

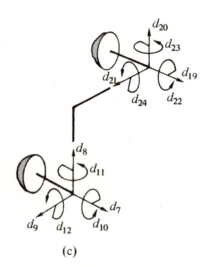

(c)

Fig. 6.18

6.4.2 Definition of the Framework

A list of data for the space framework in Fig. 6.18 is given in Table 6.5.

The parameters for the basic structural data are the same as for rigid-jointed plane frameworks, being the numbers of joints, members, joint restraints, joint forces and intermediate forces. In the case of the framework in Fig. 6.18(a), an extra joint is inserted in the top beam at the end of the distributed force as shown in Fig. 6.18(b). This device is necessary to enable the intermediate loading to be described by a uniformly distributed force over the whole of member 3. The basic structural data is then given by the first line in Table 6.5.

Table 6.5
Data for a rigid-jointed space framework

Line number	Data						
1	6	5	12	0	3		
2	1	0.0	0.0	0.0			
3	2	1.5	0.0	0.0			
4	3	1.5	4.0	0.0			
5	4	1.5	4.0	-2.0			
6	5	1.5	4.0	-4.0			
7	6	0.0	4.0	-4.0			
8	1	1	2	0			
9	2	2	3	-1.5708			
10	3	3	4	0			
11	4	4	5	0			
12	5	5	6	0			
13	1	200.0E9	76.9E9	4.0E-3	15.0E-6	30.0E-6	20.0E-6
14	2	200.0E9	76.9E9	4.0E-3	15.0E-6	30.0E-6	20.0E-6
15	3	200.0E9	76.9E9	4.0E-3	15.0E-6	30.0E-6	20.0E-6
16	4	200.0E9	76.9E9	4.0E-3	15.0E-6	30.0E-6	20.0E-6
17	5	200.0E9	76.9E9	4.0E-3	15.0E-6	30.0E-6	20.0E-6
18	1	1	0.0				
19	1	2	0.0				
20	1	3	0.0				
21	1	4	0.0				
22	1	5	0.0				
23	1	6	0.0				
24	6	1	0.0				
25	6	2	0.0				
26	6	3	0.0				
27	6	4	0.0				
28	6	5	0.0				
29	6	6	0.0				
30	1	2	1	-10.0E3	0.5		
31	2	1	1	-15.0E3	0.5		
32	3	2	2	-6.0E3			

The joint coordinates relative to the global coordinate system, and the member connectivity are defined in data lists in the same way as for rigid-jointed plane frameworks. However, for general space frameworks, this information is insufficient to completely define the structural geometry. Thus, whereas in a plane frameworks, each member is oriented so that its x_m–y_m plane coincides with the global x–y plane, in a space framework this is not always so. Thus in a framework in the form of a dome for example, the member x_m–y_m planes will be perpendicular to the surface of the dome. The orientations of the member planes therefore have to be defined.

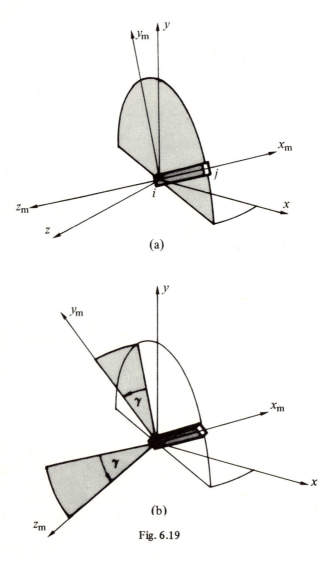

(a)

(b)

Fig. 6.19

Suppose a member connects joints i and j in the structure as in Fig. 6.19(a), where i and j are arbitrarily positioned in space. Assuming $i < j$, then the member x_m coordinate is defined, and runs from i to j as shown. A reference position for defining the orientation of the member plane is then chosen, this being the position in which the $x_m - y_m$ plane shown shaded in the figure contains the global y axis. Any other orientation of the $x_m - y_m$ plane can be defined by assuming that the change in orientation is achieved by a clockwise **rotation** γ of the member about its own x_m axis as shown in Fig. 6.19(b). Thus specifying γ in addition to the connectivity, completely defines the position and orientation of the member in space. There is however, one case for which this notation breaks down: when the member is vertical and the x_m and y axes coincide. A suitable reference position for such a member is the position in which the $x_m - y_m$ plane contains the global x axis. The actual orientation of the $x_m - y_m$ plane is then defined by a clockwise rotation γ about the x_m axis from this position as shown in Fig. 6.20. γ can be included as an extra parameter in the member connectivity data list. The joint coordinate and connectivity data lists for the space framework example are then as shown in lines 2 to 12 of Table 6.5.

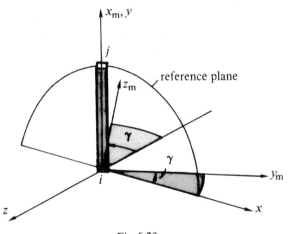

Fig. 6.20

The physical properties of the members needed for calculating the member stiffness matrices given in Table 5.7 are E, G, A, I_y, I_z and J. These are provided by a data list as in lines 13 to 17 of Table 6.5.

A support restraint for a rigid-jointed space framework fixes one of the 6 displacement components at a joint; the components being the linear displacements in the x, y and z directions and the rotations about the x, y and z axes. The direction of the support restraint is specified in the data list by an integer from 1 to 6. Thus the joint restraint data for the space framework example is as shown in lines 18 to 29 of Table 6.5.

The loading on a joint is similarly defined in terms of the joint forces, being the direct forces in the x, y and z directions and the couples about the x, y and z axes, again distinguished by integers 1 to 6.

Finally we specify the intermediate loading on the framework. The specification is identical to that for plane frameworks except that for space frameworks it is possible for an intermediate force to be directed in the z direction, indicated by the direction integer 3. Thus the intermediate force data for the framework in Fig. 6.18 is as shown in lines 30 to 32 of Table 6.5.

6.4.3 Degrees of Freedom

Six degrees of freedom at each joint completely define the deformation of the space framework, being the three linear displacements in the x, y and z directions, and the three rotations about the x, y and z axes. 36 degrees of freedom are therefore needed to describe the space framework in Fig. 6.18(a). Some of these are shown in Fig. 6.18(c). At joint i therefore, the degrees of freedom are d_{6i-5}, d_{6i-4} to d_{6i}.

6.4.4 Member Transformation Matrix

In the previous sections, we have derived transformation matrices for members simply by considering the geometrical relationship between the displacements at their ends. However, for arbitrarily oriented members in a space framework, the relationship is complicated and it is much simpler to derive the member transformation matrices using a classical result from the theory of differential geometry.

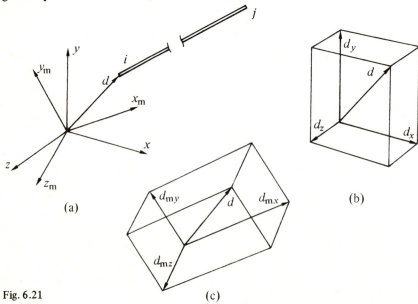

Fig. 6.21

Consider a member connecting joints i and j in a framework as in Fig. 6.21(a). The member coordinates x_m, y_m and z_m are defined by the positions of i and j and by γ. Suppose joint i and end 1 of the member move to a new position defined by the displacement vector d as in the figure. d can then be represented by its components d_x, d_y, d_z in the global coordinate system, and d_{mx} d_{my} d_{mz} in the member coordinate system, being the sides of the parallelepipeds containing d as a diagonal as in Fig. 6.21(b) and (c). d_x, d_y, d_z are equal to the linear global displacements d_{6i-5}, d_{6i-4}, d_{6i-3} of joint i and d_{mx} d_{my} d_{mz} are equal to the linear member displacements d_1, d_2, d_3 of end 1 of member m.

We next invoke the result from the theory of differential geometry [6.6]. This states that if any vector, such as d above, is defined by its components d_{mx}, d_{my}, d_{mz} and d_x, d_y, d_z in two cartesian coordinate systems, then the components are linearly related by the cosines of the angles between the coordinate lines of the two coordinate systems as follows:

$$\begin{bmatrix} d_{mx} \\ d_{my} \\ d_{mz} \end{bmatrix} = \begin{bmatrix} c_{xmx} & c_{xmy} & c_{xmz} \\ c_{ymx} & c_{ymy} & c_{ymz} \\ c_{zmx} & c_{zmy} & c_{zmz} \end{bmatrix} \begin{bmatrix} d_x \\ d_y \\ d_z \end{bmatrix} . \tag{6.55}$$

c_{xmx}, c_{xmy}, etc. in (6.55) are the **direction cosines** of the member coordinate lines with respect to the global coordinate lines, and the subscripts indicate which lines are involved. Thus c_{ymz} for example, is the cosine of the angle between the y_m and z coordinate lines.

(6.55) is a relationship between the member and global linear displacements at end 1 of the member. It can therefore be rewritten in the form

$$\begin{bmatrix} d_{m1} \\ d_{m2} \\ d_{m3} \end{bmatrix} = c_m \begin{bmatrix} d_{6i-5} \\ d_{6i-4} \\ d_{6i-3} \end{bmatrix} \tag{6.56}$$

where c_m is the **direction cosine matrix** given by

$$c_m = \begin{bmatrix} c_{xmx} & c_{xmy} & c_{xmz} \\ c_{ymx} & c_{ymy} & c_{ymz} \\ c_{zmx} & c_{zmy} & c_{zmz} \end{bmatrix} . \tag{6.57}$$

Provided the rotations are small, they too are the components of a single vector, the **rotation vector**, at the end of the member [6.7]. Thus the components in the two coordinate systems, d_{m4}, d_{m5}, d_{m6} and $d_{6i-2}, d_{6i-1}, d_{6i}$ are also related by (6.55). Thus

$$\begin{bmatrix} d_{m4} \\ d_{m5} \\ d_{m6} \end{bmatrix} = c_m \begin{bmatrix} d_{6i-2} \\ d_{6i-1} \\ d_{6i} \end{bmatrix} . \tag{6.58}$$

Collecting (6.56) and (6.58), it is apparent that the matrix r_m, relating the displacements at end 1 is now given by

$$r_m = \left[\begin{array}{c|c} c_m & 0 \\ \hline 0 & c_m \end{array} \right] \overset{(3 \times 3 \text{ null matrix})}{.} \tag{6.59}$$

Again, the same relationship exists at both ends of the member, and the condensed member transformation matrix is thus

$$t'_m = \left[\begin{array}{c|c} r_m & 0 \\ \hline 0 & r_m \end{array} \right] \overset{(6 \times 6 \text{ null matrix})}{.} \tag{6.60}$$

The problem of obtaining t'_m for a member in a space framework is therefore reduced to obtaining the direction cosine matrix given by (6.57).

Calculating the direction cosines of the x_m coordinate line is as before, a simple problem since

$$c_{xmx} = l_x/l, \; c_{xmy} = l_y/l, \; c_{xmz} = l_z/l . \tag{6.61a,b,c}$$

However, the direction cosines of the y_m and z_m coordinate lines for a member rotated through γ, are much more difficult to calculate directly. They are however, obtainable by considering relationships between the components of d in a sequence of simpler coordinate systems.

Suppose that the final orientation of the member coordinate system in Fig. 6.19(b) is obtained by starting from the position where the member system coincides with the global coordinate system, and then carrying out three rotations about the member axes in turn as follows:

(i) an anticlockwise rotation α about the y_m axis as in Fig. 6.22(a),
(ii) a clockwise rotation β about the z_m axis as in Fig. 6.22(b) and
(iii) a clockwise rotation γ about the x_m axis as in Fig. 6.22(c).

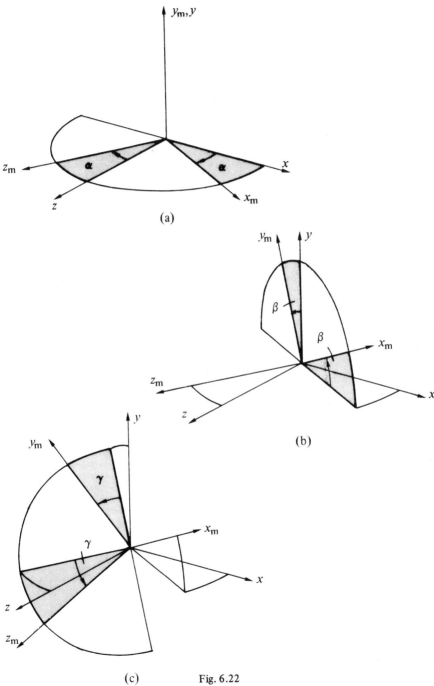

(a)

(b)

(c)   Fig. 6.22

We then introduce the following notation; $d_m^{(\alpha)}$, $d_m^{(\beta)}$, $d_m^{(\gamma)}$ are the linear member displacement components relative to the coordinate systems at the end of each rotation while $c_m^{(\alpha)}$, $c_m^{(\beta)}$ and $c_m^{(\gamma)}$ are the direction cosine matrices between the coordinate systems at the end and beginning of each rotation. Since then the general relationship (6.55) exists between the vector components in any two of the coordinate systems, we have:

$$d_m^{(\alpha)} = c_m^{(\alpha)} d \tag{6.62}$$

$$d_m^{(\beta)} = c_m^{(\beta)} d_m^{(\alpha)} = c_m^{(\beta)} c_m^{(\alpha)} d \tag{6.63}$$

$$d_m^{(\gamma)} = c_m^{(\gamma)} d_m^{(\beta)} = c_m^{(\gamma)} c_m^{(\beta)} c_m^{(\alpha)} d \quad . \tag{6.64}$$

But $d_m^{(\gamma)}$ are the member displacements in the final member coordinate system. Therefore, comparing (6.56) and (6.64) we have

$$c_m = c_m^{(\gamma)} c_m^{(\beta)} c_m^{(\alpha)} \quad . \tag{6.65}$$

The direction cosines corresponding to the individual rotations are easy to obtain from geometry. Thus, from Fig. 6.22(a),

$$c_m^{(\alpha)} = \begin{bmatrix} \cos(\alpha) & 0 & \sin(\alpha) \\ 0 & 1.0 & 0 \\ -\sin(\alpha) & 0 & \cos(\alpha) \end{bmatrix} \tag{6.66}$$

(using $\cos(\pi/2 - \alpha) = \sin(\alpha)$, $\cos(\pi/2 + \alpha) = -\sin(\alpha)$).
From Fig. 6.22(b)

$$c_m^{(\beta)} = \begin{bmatrix} \cos(\beta) & \sin(\beta) & 0 \\ -\sin(\beta) & \cos(\beta) & 0 \\ 0 & 0 & 1.0 \end{bmatrix} \tag{6.67}$$

and from Fig. 6.22(c)

$$c_m^{(\gamma)} = \begin{bmatrix} 1.0 & 0 & 0 \\ 0 & \cos(\gamma) & \sin(\gamma) \\ 0 & -\sin(\gamma) & \cos(\gamma) \end{bmatrix} \quad . \tag{6.68}$$

Whence carrying out the matrix product in (6.65) we obtain

$$c_m = \begin{bmatrix} \cos(\beta)\cos(\alpha) & \sin(\beta) & \cos(\beta)\sin(\alpha) \\[2mm] \begin{pmatrix} -\cos(\gamma)\sin(\beta)\cos(\alpha) \\ -\sin(\gamma)\sin(\alpha) \end{pmatrix} & \cos(\gamma)\cos(\beta) & \begin{pmatrix} -\cos(\gamma)\sin(\beta)\sin(\alpha) \\ +\sin(\gamma)\cos(\alpha) \end{pmatrix} \\[2mm] \begin{pmatrix} \sin(\gamma)\sin(\beta)\cos(\alpha) \\ -\cos(\gamma)\sin(\alpha) \end{pmatrix} & -\sin(\gamma)\cos(\beta) & \begin{pmatrix} \sin(\gamma)\sin(\beta)\sin(\alpha) \\ +\cos(\gamma)\cos(\alpha) \end{pmatrix} \end{bmatrix} \quad (6.69)$$

It then remains to obtain c_m from the structural data provided. First, we note that γ, the final rotation of the member, is supplied directly as a data parameter. Thus only α and β need be obtained from the geometry. Consider the member in Fig. 6.23 where α and β are the angles shown. l_x, l_y, l_z and l are immediately calculable from the coordinates of i and j, and calling the distance between i and A, l_{xz} say,

$$l_{xz} = (l_x^2 + l_z^2)^{\frac{1}{2}} \quad . \tag{6.70}$$

Thus

$$\cos(\alpha) = l_x/l_{xz}, \quad \sin(\alpha) = l_z/l_{xz} \tag{6.71a,b}$$

$$\cos(\beta) = l_{xz}/l, \quad \sin(\beta) = l_y/l \quad . \tag{6.71c,d}$$

Therefore, by substituting into (6.69), c_m can be evaluated from the structural data.

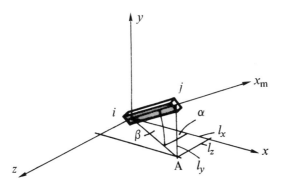

Fig. 6.23

Finally we consider the special case of the vertical member shown in Fig. 6.20. The member coordinate system can be obtained from a starting position coinciding with the global coordinate system, by carrying out the rotations

$$\alpha = \pi, \quad \beta = \pi/2, \quad \gamma = \gamma \quad . \tag{6.72}$$

Thus

$$
c_m^{(\alpha)} = \begin{bmatrix} -1.0 & 0 & 0 \\ 0 & 1.0 & 0 \\ 0 & 0 & -1.0 \end{bmatrix}, \quad c_m^{(\beta)} = \begin{bmatrix} 0 & 1.0 & 0 \\ -1.0 & 0 & 0 \\ 0 & 0 & 1.0 \end{bmatrix},
$$

$$
c_m^{(\gamma)} = \begin{bmatrix} 1.0 & 0 & 0 \\ 0 & \cos(\gamma) & \sin(\gamma) \\ 0 & -\sin(\gamma) & \cos(\gamma) \end{bmatrix}. \qquad (6.73a, b, c)
$$

and c_m is given by

$$
c_m = \begin{bmatrix} 0 & 1.0 & 0 \\ \cos(\gamma) & 0 & -\sin(\gamma) \\ -\sin(\gamma) & 0 & -\cos(\gamma) \end{bmatrix}. \qquad (6.74)
$$

6.4.5 Global Stiffness Matrix

The member stiffness matrix for a member in a rigid-jointed space framework is given in Table 5.7. This again can be expressed in terms of the submatrices k_{m1}, k_{m2} and k_{m3} as

$$
k_m = \left[\begin{array}{c|c} k_{m1} & k_{m2} \\ \hline k_{m2}^T & k_{m3} \end{array} \right] \qquad (6.43)*
$$

where k_{m1}, k_{m2} and k_{m3} are now appropriate 6×6 matrices selected from the table. The transformed member stiffness matrix $k_m^{(g)}$ is a 12×12 matrix, given by (6.45), and rows and columns numbered $6i-5$ to $6i$, and $6j-5$ to $6j$ of the global stiffness matrix are augmented with submatrices of $k_m^{(g)}$ in a similar manner as in Fig. 6.13 for the plane framework. It is then apparent that the bandwidth of the global stiffness matrix is equal to $6(j - i + 1)$ where $(j - i)$ is again the maximum difference between the numbers of the connected joints in the framework.

6.4.6 Global Force Matrix

In order to make use of (6.46) and (6.47) we must again calculate the member fixing forces $\mathbf{P}_{Fm}$. For this purpose, the intermediate loading on a member, presented in the data as force components in the global x, y or z directions, has to be expressed as two transverse components in the y_m and z_m directions together with an axial component in the x_m direction. Again we can invoke the vector transformation (6.55), which converts the intermediate force components Q_{mx}, Q_{my} and Q_{mz} into three components in the x_m, y_m and z_m directions. The member fixed-end forces are then calculable from (6.48) but care has to be taken over their signs. Thus in deriving (6.48) for the member in the plane framework, the transverse force shown in Fig. 6.15 is taken to act in the *opposite* direction to the y_m axis. The y_m component generated above has therefore to be reversed in sign. We then note that the corresponding fixing forces at end 1 of the member are positive, as shown in Fig. 6.24(a). However, if we consider the transverse force in the z_m direction as in Fig. 6.24(b), then a *positive* transverse component produces a *negative* linear fixing force at end 1, and a *positive* couple.

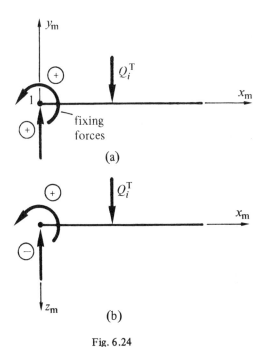

Fig. 6.24

6.4.7 Solution

The introduction of restraints, the solution in terms of the global displacements and the calculation of the member forces and the reactions are all undertaken in the same way as for the rigid-jointed plane framework.

6.4.8 Computer Program

The computer program for solving rigid-jointed space frameworks, **RJSFA†** is organised in exactly the same way as **RJPFA**, and uses the same variables, arrays and subroutines, the arrays being appropriately increased in size. Differences in detail reflect the special points discussed in this section. As an example of the running of the program, the space framework in Fig. 6.18 has been analysed and and the computer output is given in Appendix A6.3. Allowing for axial effects, the reactions at joint 1 are then seen to agree with those produced in Example 4.14 by the flexibility method.

6.5 CONCLUSION

The purpose of this chapter has been to introduce the reader to the problem of programming a computer to solve linear frameworks. The programs that have been described contain the subroutines that are the core of any numerical analysis based on the stiffness method. The method is used in a vast body of research and commercial software which is now available for solving all types of structures. We shall conclude this chapter by briefly reviewing some of the developments that are possible from the starting point of the programs presented here. For clarity, the developments are discussed under three headings.

1. *Linear elastic analysis*
    (i) Include members that are pin-ended at one or both ends.
    (ii) Include flexible supports and inclined roller supports as in Fig. 6.25(a). Note that these supports can be modelled by suitable pin-ended members as shown in Fig. 6.25(b).

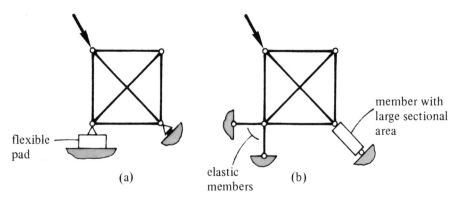

flexible pad

(a)

elastic members

(b)

member with large sectional area

Fig. 6.25

† This computer program is available on tape from Ellis Horwood Limited, Market Cross House, Cooper Street, Chichester, West Sussex, PO19 1EB.

(iii) Enhance the input facilities to
    (a) generate the joint coordinates of regular frameworks automatically,
    (b) assign properties to groups of identical members,
    (c) fix groups of supported joints, and
    (d) generate dead loading and standard combinations of live loading.
(iv) Allow single members to be modified without recalculating the whole of the global stiffness matrix [6.8].
(v) Include tapered or curved members.
(vi) Solve as special cases, continuous beams and grillages, and for bridge analysis, include member shear flexibility and prestress loading [6.9].

2. *Non-linear elastic analysis*
(vii) Account for changes in the stiffness of compression members [6.10].
(viii) Account for changes in the geometry of tension structures such as cable nets [6.11].

3. *Elasto-plastic analysis*
(ix) Include plastic yielding of the material of the member and predict ultimate limit states [6.12].

The stiffness method is also the basis of the well-known finite-element method for solving plate and shell structures [6.13]. The finite element is a two-dimensional shell element whose deformation is completely defined by element degrees of freedom and associated displacement functions, the displacement functions being chosen so that a compatible displacement field is generated in the deformed structure. The element stiffness matrix is calculated from the virtual work equation, and the transformation matrix, from the orientation of the element defined by the coordinates of its corners. All the necessary information is then available to carry out the stiffness method in the manner described in this chapter.

6.6 PROGRAMMING PROBLEMS

6.1 Using any convenient programming language, convert **PJPFA** and **RJPFA** to run on an available mini- computer or mainframe computer.

6.2 Add subroutines to **PJPFA** and **RJPFA** for reading in the properties of different materials and different member cross-sections, and then assigning these properties to appropriate groups of members in a framework.

6.3 Add subroutines to **RJPFA** to account for members with one or both ends pinned.

6.4 Write a program for a micro-computer, for solving a continuous beam by the stiffness method. (Use the simple direct methods for assembling the global stiffness matrix and equivalent global force matrix suggested by (5.28) and (5.33).)

6.5 Write a program for solving a continuous beam by the flexibility method. (Note the following points:

(i) The releases should be taken as internal moment releases at the supports as in Example 4.11, and as shown in Fig. 6.26(a). The bending moments m_i and m_j due to unit release forces at supports i and j are then as shown in Figs. 6.26(b) and (c).

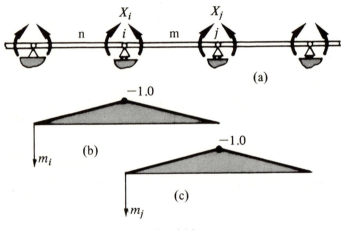

Fig. 6.26

(ii) When assembling the flexibility matrix use (4.33). It is then immediately apparent that $f_{ij} = 0$ if $|i{-}j|>1$, simply because the moments m_i only occur in the two spans m and n adjacent to release i. In calculating f_{ii} and f_{ij}, the product integral $\oint\left(\dfrac{m_i m_j}{EI}\right)$ dx taken over spans n and m leads to

$$f_{ii} = \left\{\frac{l}{3EI}\right\}_n + \left\{\frac{l}{3EI}\right\}_m , \quad f_{ij} = \left\{\frac{l}{6EI}\right\}_m . \qquad (6.75a, b)$$

(iii) The initial relative displacements at the releases u_P are calculated using (4.32). It is then apparent that intermediate forces only contribute to u_P at the releases at the ends of the spans on which they act. The concentrated force Q_m acting on span m as shown in Fig. 6.27(a) causes the initial relative displacements

$$u_{Pi} = -Q_m \left\{ \frac{ab}{6EIl} \right\}_m (l+b), \quad u_{Pj} = -Q_m \left\{ \frac{ab}{6EIl} \right\}_m (l+a) \ .$$

$$(6.76a, b)$$

The uniformly distributed force q_m acting on the whole of span m as shown in Fig. 6.27(b), causes the initial relative displacements

$$u_{Pi} = u_{Pj} = -q_m \left\{ \frac{l^3}{24EI} \right\}_m \ . \qquad\qquad (6.77a, b)$$

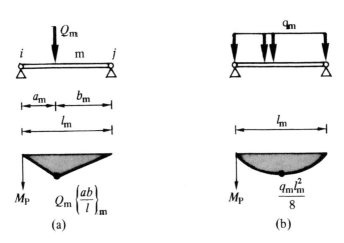

Fig. 6.27

(iv) The solution is given in terms of the release forces **X**, by (4.23).)

6.6 Write a program for solving pin-jointed space frameworks.

6.7 Write an independent version of the program **RJSFA**, for solving rigid-jointed space frameworks.

APPENDIX A6.1 PIN-JOINTED PLANE FRAMEWORK ANALYSIS COMPUTER PROGRAM

A6.1.1 PJPFA FORTRAN Listing

```
1          PROGRAM      PJPFA
2          INTEGER      JNO,MNO,JRNO,JFNO,RWNO,COLNO
3          REAL         JCRD(10,2)/20*0.0/,MPRP(10,2)/20*0.0/,
4        &      JRES(20)/20*1000.0/,DG(20)/20*0.0/,PG(20)/20*0.0/,
5        &      KG(20,10)/200*0.0/,KM(2,2),TM(2,4),PM(2),
6        &      WKA(10)
7          INTEGER      MCON(10,2)
8   C
9   C
10         CALL INDAT1(JNO,MNO,JRNO,JFNO,RWNO)
11         CALL INDAT2(JNO,MNO,JRNO,JFNO,RWNO,
12       &             JCRD,MPRP,JRES,PG,MCON)
13         CALL OUTDAT(JNO,MNO,JRNO,JFNO,RWNO,
14       &             JCRD,MPRP,JRES,PG,MCON)
15         CALL BANDW(MNO,MCON,COLNO)
16         DO 10 I= 1,MNO
17         CALL MST(I,JNO,MNO,JCRD,MPRP,MCON,KM)
18         CALL TRANS(I,JNO,MNO,JCRD,MCON,TM)
19         CALL GLST(I,MNO,RWNO,COLNO,MCON,KM,TM,KG)
20      10 CONTINUE
21         CALL SPPRTS(RWNO,COLNO,JRES,PG,KG)
22         CALL SOLVE(RWNO,COLNO,KG,PG,WKA,DG)
23         CALL OUTSOL1(JNO,RWNO,DG)
24         DO 20 I= 1,MNO
25         CALL MST(I,JNO,MNO,JCRD,MPRP,MCON,KM)
26         CALL TRANS(I,JNO,MNO,JCRD,MCON,TM)
27         CALL MFRCS(I,MNO,RWNO,DG,KM,TM,MCON,PM)
28         CALL OUTSOL2(I,MNO,PM,MCON)
29         CALL GLST(I,MNO,RWNO,COLNO,MCON,KM,TM,KG)
30      20 CONTINUE
31         CALL RCTNS(RWNO,COLNO,KG,DG,PG)
32         CALL OUTSOL3(JNO,RWNO,PG)
33  C
34         STOP
35         END
36  C
37  C
38  C
39         SUBROUTINE   INDAT1(JNO,MNO,JRNO,JFNO,RWNO)
40         INTEGER      JNO,MNO,JRNO,JFNO,RWNO
41  C
42  C
43         READ(5,500)  JNO,MNO,JRNO,JFNO
44         RWNO= 2*JNO
45  C
46     500 FORMAT(4I10)
47  C
48         RETURN
49         END
50  C
51  C
52  C
```

```
53              SUBROUTINE  INDAT2(JNO,MNO,JRNO,JFNO,RWNO,
54          &                   JCRD,MPRP,JRES,PG,MCON)
55              INTEGER     JNO,MNO,JRNO,JFNO,RWNO
56              REAL        JCRD(JNO,2),MPRP(MNO,2),JRES(RWNO),PG(RWNO)
57    C         INTEGER     MCON(MNO,2)
58    C
59    C
60              READ(5,500)  (M,(JCRD(M,N),N=1,2),I=1,JNO)
61              READ(5,510)  (M,(MCON(M,N),N=1,2),I=1,MNO)
62              READ(5,520)  (M,(MPRP(M,N),N=1,2),I=1,MNO)
63              READ(5,530)  (M,N,JRES(2*M+N-2),I=1,JRNO)
64              READ(5,540)  (M,N,PG(2*M+N-2),I=1,JFNO)
65    C
66        500 FORMAT(I10,2F10.0)
67        510 FORMAT(3I10)
68        520 FORMAT(I10,2E10.0)
69        530 FORMAT(2I10,F10.0)
70        540 FORMAT(2I10,E10.0)
71    C
72              RETURN
73              END
74    C
75    C
76    C
77              SUBROUTINE  OUTDAT(JNO,MNO,JRNO,JFNO,RWNO,
78          &                   JCRD,MPRP,JRES,PG,MCON)
79              INTEGER     JNO,MNO,JRNO,JFNO,RWNO
80              REAL        JCRD(JNO,2),MPRP(MNO,2),JRES(RWNO),PG(RWNO)
81              INTEGER     MCON(MNO,2)
82    C
83    C
84              WRITE(6,500) JNO,MNO,JRNO,JFNO
85              WRITE(6,510)
86              WRITE(6,520) (M,(JCRD(M,N),N=1,2),M=1,JNO)
87              WRITE(6,530)
88              WRITE(6,540) (M,(MCON(M,N),N=1,2),M=1,MNO)
89              WRITE(6,550)
90              WRITE(6,560) (M,(MPRP(M,N),N=1,2),M=1,MNO)
91              WRITE(6,570)
92              WRITE(6,520) (M,(JRES(N),N=(2*M-1),(2*M)),M=1,JNO)
93              WRITE(6,580)
94              WRITE(6,590) (M,(PG(N),N=(2*M-1),(2*M)),M=1,JNO)
95    C
96        500 FORMAT(1H1,'PIN-JOINTED PLANE FRAMEWORK ANALYSIS',
97          &        /////1H ,'DATA SUPPLIED',
98          &        /////1H ,'NUMBER OF JOINTS      ',I5,
99          &        /1H ,'NUMBER OF MEMBERS     ',I5,
100         &        //1H ,'NUMBER OF JOINT RESTRAINTS',I5,
101         &        /1H ,'NUMBER OF JOINT FORCES   ',I6)
102       510 FORMAT(///1H ,'JOINT COORDINATES',
103         &        //1H ,10X,'JOINT',11X,'X(M)',11X,'Y(M)')
104       520 FORMAT(1H ,I13,F17.3,F15.3)
105       530 FORMAT(///1H ,'MEMBER CONNECTIVITY',
106         &        //1H ,10X,'MEMBER',8X,'JOINT1',9X,'JOINT2')
107       540 FORMAT(1H ,I13,I15,I15)
108       550 FORMAT(///1H ,'MEMBER PROPERTIES',
109         &        //1H ,10X,'MEMBER',7X,'E(N/M2)',7X,'AREA(M2)')
```

```
110    560 FORMAT(1H ,I13,E17.3,E15.3)
111    570 FORMAT(///1H ,'JOINT RESTRAINTS',
112        &        /1H ,5X,'(LX=LINEAR RESTRAINT IN X DIRECTION,',
113        &        /1H ,6X,'1000.000=FREE)',
114        &        //1H ,10X,'JOINT',10X,'LX(M)',10X,'LY(M)')
115    580 FORMAT(///1H ,'JOINT FORCES',
116        &        /1H ,5X,'(PX=FORCE IN X DIRECTION)',
117        &        //1H ,10X,'JOINT',10X,'PX(N)',10X,'PY(N)')
118    590 FORMAT(1H ,I13,F17.2,F15.2)
119  C
120        RETURN
121        END
122  C
123  C
124  C
125        SUBROUTINE    BANDW(MNO,MCON,COLNO)
126        INTEGER       MNO,COLNO
127        INTEGER       MCON(MNO,2)
128  C
129  C
130        K1=0
131        DO 10 N= 1,MNO
132        I= MCON(N,1)
133        J= MCON(N,2)
134        K= (J-I)
135        IF(K.GT.K1) K1= K
136     10 CONTINUE
137        COLNO= 2*(K1+1)
138  C
139        RETURN
140        END
141  C
142  C
143  C
144        SUBROUTINE    MST(IMEM,JNO,MNO,JCRD,MPRP,MCON,KM)
145        INTEGER       IMEM,JNO,MNO
146        REAL          JCRD(JNO,2),MPRP(MNO,2),KM(2,2)
147        INTEGER       MCON(MNO,2)
148  C
149  C
150        I= MCON(IMEM,1)
151        J= MCON(IMEM,2)
152        RLX= JCRD(J,1)-JCRD(I,1)
153        RLY= JCRD(J,2)-JCRD(I,2)
154        RL= SQRT(RLX**2+RLY**2)
155        A= MPRP(IMEM,1)*MPRP(IMEM,2)/RL
156        KM(1,1)= A
157        KM(1,2)= -A
158        KM(2,1)= -A
159        KM(2,2)= A
160  C
161        RETURN
162        END
163  C
164  C
165  C
```

```
166          SUBROUTINE   TRANS(IMEM,JNO,MNO,JCRD,MCON,TM)
167          INTEGER      IMEM,JNO,MNO
168          REAL         JCRD(JNO,2),TM(2,4)
169          INTEGER      MCON(MNO,2)
170   C
171   C
172          I= MCON(IMEM,1)
173          J= MCON(IMEM,2)
174          RLX= JCRD(J,1)-JCRD(I,1)
175          RLY= JCRD(J,2)-JCRD(I,2)
176          RL= SQRT(RLX**2+RLY**2)
177          RCX= RLX/RL
178          RCY= RLY/RL
179          TM(1,1)= RCX
180          TM(1,2)= RCY
181          TM(1,3)= 0.0
182          TM(1,4)= 0.0
183          TM(2,1)= 0.0
184          TM(2,2)= 0.0
185          TM(2,3)= RCX
186          TM(2,4)= RCY
187   C
188          RETURN
189          END
190   C
191   C
192   C
193          SUBROUTINE   GLST(IMEM,MNO,RWNO,COLNO,MCON,KM,TM,KG)
194          INTEGER      IMEM,MNO,RWNO,COLNO
195          REAL         KM(2,2),TM(2,4),KG(RWNO,COLNO),KMG(4,4)
196          INTEGER      MCON(MNO,2)
197          INTEGER      P,Q,R
198   C
199   C
200          DO 10 I= 1,4
201          DO 10 J= 1,4
202          KMG(I,J)= 0.0
203          DO 10 M= 1,2
204          DO 10 N= 1,2
205     10 KMG(I,J)= KMG(I,J)+TM(M,I)*KM(M,N)*TM(N,J)
206   C
207          IF(IMEM.NE.1) GOTO 20
208          DO 30 M= 1,RWNO
209          DO 30 N= 1,COLNO
210     30 KG(M,N)= 0.0
211     20 CONTINUE
212          I= MCON(IMEM,1)
213          J= MCON(IMEM,2)
214          K= 2*(J-I)
215          DO 40 M= 1,2
216          DO 40 N= 1,2
217          P= (2*I-1)+M-1
218          Q= (2*J-1)+M-1
219          R= 1-M+N
220          IF(R.LT.1) GOTO 40
221          KG(P,R)= KG(P,R)+KMG(M,N)
```

```
222              KG(Q,R)= KG(Q,R)+KMG(M+2,N+2)
223          40 KG(P,K+R)= KG(P,K+R)+KMG(M,N+2)
224    C
225              RETURN
226              END
227    C
228    C
229    C
230              SUBROUTINE    SPPRTS(RWNO,COLNO,JRES,PG,KG)
231              INTEGER       RWNO,COLNO
232              REAL          JRES(RWNO),PG(RWNO),KG(RWNO,COLNO)
233    C
234    C
235              DO 10 I= 1,RWNO
236              IF(JRES(I).GT.999.9) GOTO 10
237              KG(I,1)= 1.0E30
238              PG(I)= JRES(I)*1.0E30
239          10 CONTINUE
240    C
241              RETURN
242              END
243    C
244    C
245    C
246              SUBROUTINE    SOLVE(N,M,A,B,C,D)
247              REAL          A(N,M),B(N),C(M),D(N)
248    C
249    C
250              DO 10 K= 1,N
251              B(K)= B(K)/A(K,1)
252              IF(K.EQ.N) GOTO 100
253              DO 20 J= 2,M
254              C(J)= A(K,J)
255              A(K,J)= A(K,J)/A(K,1)
256          20 CONTINUE
257              DO 30 L= 2,M
258              I= K+L-1
259              IF(N.LT.I) GOTO 30
260              J= 0
261              DO 40 L1= L,M
262              J= J+1
263              A(I,J)= A(I,J)-C(L)*A(K,L1)
264          40 CONTINUE
265              B(I)= B(I)-C(L)*B(K)
266          30 CONTINUE
267          10 CONTINUE
268         100 D(K)= B(K)
269              K= K-1
270              IF(K.EQ.0) GOTO 200
271              DO 50 J= 2,M
272              L= K+J-1
273              IF(N.LT.L) GOTO 50
274              B(K)= B(K)-A(K,J)*B(L)
275          50 CONTINUE
276              GOTO 100
277         200 CONTINUE
278    C
```

```
279          RETURN
280          END
281   C
282   C
283   C
284          SUBROUTINE    OUTSOL1(JNO,RWNO,DG)
285          INTEGER       RWNO
286          REAL          DG(RWNO)
287   C
288   C
289          WRITE(6,500)
290          WRITE(6,510) (M,(DG(N),N=(2*M-1),(2*M)),M= 1,JNO)
291   C
292     500 FORMAT(/////1H ,'RESULTS',
293        &        /////1H ,'JOINT DISPLACEMENTS',
294        &        //1H ,10X,'JOINT',10X,'LX(M)',10X,'LY(M)')
295     510 FORMAT(1H ,I13,1F17.6,1F15.6)
296   C
297          RETURN
298          END
299   C
300   C
301   C
302          SUBROUTINE    MFRCS(IMEM,MNO,RWNO,DG,KM,TM,MCON,PM)
303          INTEGER       IMEM,MNO,RWNO
304          REAL          DG(RWNO),KM(2,2),TM(2,4),PM(2),DMG(4)
305          INTEGER       MCON(MNO,2)
306   C
307   C
308          I= MCON(IMEM,1)
309          J= MCON(IMEM,2)
310          DMG(1)= DG(2*I-1)
311          DMG(2)= DG(2*I)
312          DMG(3)= DG(2*J-1)
313          DMG(4)= DG(2*J)
314          DO 10 I= 1,2
315          PM(I)= 0.0
316          DO 10 J= 1,2
317          DO 10 K= 1,4
318       10 PM(I)= PM(I)+KM(I,J)*TM(J,K)*DMG(K)
319   C
320          RETURN
321          END
322   C
323   C
324   C
325          SUBROUTINE    OUTSOL2(IMEM,MNO,PM,MCON)
326          INTEGER       IMEM,MNO
327          REAL          PM(2)
328          INTEGER       MCON(MNO,2)
329   C
330   C
331          IF(IMEM.GT.1) GOTO 10
332          WRITE(6,500)
333       10 WRITE(6,510) (IMEM,MCON(IMEM,M),PM(M),M=1,2)
334   C
```

```
335      500 FORMAT(///1H ,'LOCAL MEMBER FORCES',
336          &       //1H ,3X,'MEMBER/JOINT',9X,'PMX(N)')
337      510 FORMAT(/1H ,I12,'/',I2,F15.2,/1H ,I12,'/',I2,F15.2)
338 C
339          RETURN
340          END
341 C
342 C
343 C
344          SUBROUTINE   RCTNS(RWNO,COLNO,KG,DG,PG)
345          INTEGER      RWNO,COLNO
346          REAL         KG(RWNO,COLNO),DG(RWNO),PG(RWNO)
347 C
348 C
349          DO 10 I= 1,RWNO
350          PG(I)= 0.0
351          DO 20 J= 1,COLNO
352          N= I+J-1
353          IF(N.GT.RWNO) GOTO 20
354          PG(I)= PG(I)+KG(I,J)*DG(N)
355       20 CONTINUE
356          DO 30 J= 2,COLNO
357          N= I-J+1
358          IF(N.LT.1) GOTO 30
359          PG(I)= PG(I)+KG(N,J)*DG(N)
360       30 CONTINUE
361       10 CONTINUE
362 C
363          RETURN
364          END
365 C
366 C
367 C
368          SUBROUTINE   OUTSOL3(JNO,RWNO,PG)
369          INTEGER      JNO,RWNO
370          REAL         PG(RWNO)
371 C
372 C
373          WRITE(6,500)
374          WRITE(6,510) (M,(PG(N),N=(2*M-1),(2*M)),M=1,JNO)
375      500 FORMAT(///1H ,'GLOBAL FORCES AND REACTIONS',
376          &       //1H ,10X,'JOINT',10X,'PX(N)',10X,'PY(N)')
377      510 FORMAT(1H ,I13,F17.2,F15.2)
378 C
379          RETURN
380          END
```

A6.1.2 PJPFA Data Presentation

Data block	Number of rows	Description of data	FORTRAN format
1	1	number of joints; number of members; number of joint restraints; number of joint forces	4I10
2	number of joints	joint number; joint coordinates (m)	I10, 2F10.0
3	number of members	member number; member connectivity (lower joint number first)	3I10
4	number of members	member number; Young's modulus, E (N/m^2); sectional area, A (m^2)	I10, 2E10.0
5	number of joint restraints	joint number; direction number; displacement (m)	2I10, F10.0
6	number of joint forces	joint number; direction number; force (N)	2I10, E10.0

A6.1.3 PJPFA Example

```
PIN-JOINTED PLANE FRAMEWORK ANALYSIS

DATA SUPPLIED

NUMBER OF JOINTS        4
NUMBER OF MEMBERS       5

NUMBER OF JOINT RESTRAINTS    4
NUMBER OF JOINT FORCES        1
```

```
JOINT COORDINATES

        JOINT          X(M)            Y(M)
          1            0.000           3.000
          2            3.000           3.000
          3            3.000           0.000
          4            0.000           0.000



MEMBER CONNECTIVITY

        MEMBER         JOINT1          JOINT2
          1              1               2
          2              2               3
          3              2               4
          4              3               4
          5              1               3



MEMBER PROPERTIES

        MEMBER         E(N/M2)         AREA(M2)
          1            0.200E+12       0.500E-03
          2            0.200E+12       0.500E-03
          3            0.200E+12       0.750E-03
          4            0.200E+12       0.750E-03
          5            0.200E+12       0.500E-03



JOINT RESTRAINTS
      (LX=LINEAR RESTRAINT IN X DIRECTION,
       1000.000=FREE)

        JOINT          LX(M)           LY(M)
          1            0.000           0.000
          2            1000.000        1000.000
          3            1000.000        1000.000
          4            0.000           0.000



JOINT FORCES
      (PX=FORCE IN X DIRECTION)

        JOINT          PX(N)           PY(N)
          1            0.00            0.00
          2            0.00           -10000.00
          3            0.00            0.00
          4            0.00            0.00
```

RESULTS

JOINT DISPLACEMENTS

JOINT	LX(M)	LY(M)
1	0.000000	-0.000000
2	0.000183	-0.000527
3	-0.000078	-0.000410
4	-0.000000	-0.000000

LOCAL MEMBER FORCES

MEMBER/JOINT	PMX(N)
1/ 1	-6090.33
1/ 2	6090.33
2/ 2	3909.67
2/ 3	-3909.67
3/ 2	8613.02
3/ 4	-8613.02
4/ 3	3909.67
4/ 4	-3909.67
5/ 1	-5529.11
5/ 3	5529.11

GLOBAL FORCES AND REACTIONS

JOINT	PX(N)	PY(N)
1	-10000.00	3909.67
2	-0.00	-10000.00
3	0.00	0.00
4	10000.00	6090.33

APPENDIX A6.2 RIGID-JOINTED PLANE FRAMEWORK ANALYSIS COMPUTER PROGRAM

A6.2.1 RJPFA FORTRAN Listing

```
 1          PROGRAM      RJPFA
 2          INTEGER      JNO,MNO,JRNO,JFNO,IFNO,RWNO,COLNO
 3          REAL         JCRD(20,2)/40*0.0/,MPRP(20,3)/60*0.0/,
 4       &  JRES(60)/60*1000.0/,INTFR(20,2)/40*0.0/,
 5       &  DG(60)/60*0.0/,PG(60)/60*0.0/,
 6       &  KG(60,15)/900*0.0/,KM(6,6)/36*0.0/,TM(6,6)/36*0.0/,
 7       &  PM(6),PFM(20,6)/120*0.0/,PGQ(60)/60*0.0/,WKA(15)
 8          INTEGER      MCON(20,2),INTFI(20,3)
 9  C
10  C
11          CALL INDAT1(JNO,MNO,JRNO,JFNO,IFNO,RWNO)
12          CALL INDAT2(JNO,MNO,JRNO,JFNO,IFNO,RWNO,
13       &         JCRD,MPRP,JRES,PG,INTFR,MCON,INTFI)
14          CALL OUTDAT(JNO,MNO,JRNO,JFNO,IFNO,RWNO,
15       &         JCRD,MPRP,JRES,PG,INTFR,MCON,INTFI)
16          CALL BANDW(MNO,MCON,COLNO)
17          DO 10 I= 1,MNO
18          CALL MST(I,JNO,MNO,JCRD,MPRP,MCON,KM)
19          CALL TRANS(I,JNO,MNO,JCRD,MCON,TM)
20          CALL GLST(I,MNO,RWNO,COLNO,MCON,KM,TM,KG)
21          CALL MFX(I,JNO,MNO,IFNO,
22       &         JCRD,INTFR,MCON,INTFI,PFM)
23          CALL GLFQ(I,MNO,RWNO,MCON,PFM,TM,PGQ)
24   10     CONTINUE
25          CALL GLF(RWNO,PG,PGQ)
26          CALL SPPRTS(RWNO,COLNO,JRES,PG,KG)
27          CALL SOLVE(RWNO,COLNO,KG,PG,WKA,DG)
28          CALL OUTSOL1(JNO,RWNO,DG)
29          DO 20 I= 1,MNO
30          CALL MST(I,JNO,MNO,JCRD,MPRP,MCON,KM)
31          CALL TRANS(I,JNO,MNO,JCRD,MCON,TM)
32          CALL MFRCS(I,MNO,RWNO,DG,KM,TM,MCON,PFM,PM)
33          CALL OUTSOL2(I,MNO,PM,MCON)
34          CALL GLST(I,MNO,RWNO,COLNO,MCON,KM,TM,KG)
35   20     CONTINUE
36          CALL RCTNS(RWNO,COLNO,KG,DG,PGQ,PG)
37          CALL OUTSOL3(JNO,RWNO,PG)
38  C
39          STOP
40          END
41  C
42  C
43  C
44          SUBROUTINE   INDAT1(JNO,MNO,JRNO,JFNO,IFNO,RWNO)
45          INTEGER      JNO,MNO,JRNO,JFNO,IFNO,RWNO
46  C
47  C
48          READ(5,500)  JNO,MNO,JRNO,JFNO,IFNO
49          RWNO= 3*JNO
50  C
```

```
51      500 FORMAT(5I10)
52   C
53          RETURN
54          END
55   C
56   C
57   C
58          SUBROUTINE    INDAT2(JNO,MNO,JRNO,JFNO,IFNO,RWNO,
59         &                     JCRD,MPRP,JRES,PG,INTFR,MCON,INTFI)
60          INTEGER       JNO,MNO,JRNO,JFNO,IFNO,RWNO
61          REAL          JCRD(JNO,2),MPRP(MNO,3),JRES(RWNO),PG(RWNO),
62         &              INTFR(IFNO,2)
63          INTEGER       MCON(MNO,2),INTFI(IFNO,3)
64   C
65   C
66          READ(5,500)   (M,(JCRD(M,N),N=1,2),I=1,JNO)
67          READ(5,510)   (M,(MCON(M,N),N=1,2),I=1,MNO)
68          READ(5,520)   (M,(MPRP(M,N),N=1,3),I=1,MNO)
69          READ(5,530)   (M,N,JRES(3*M+N-3),I=1,JRNO)
70          IF(JFNO.EQ.0) GOTO 10
71          READ(5,540)   (M,N,PG(3*M+N-3),I=1,JFNO)
72       10 IF(IFNO.EQ.0) GOTO 20
73          READ(5,550)   ((INTFI(M,N),N=1,3),(INTFR(M,N),N=1,2),M=1,IFNO)
74       20 CONTINUE
75   C
76      500 FORMAT(I10,2F10.0)
77      510 FORMAT(3I10)
78      520 FORMAT(I10,3E10.0)
79      530 FORMAT(2I10,F10.0)
80      540 FORMAT(2I10,E10.0)
81      550 FORMAT(3I10,E10.0,F10.0)
82   C
83          RETURN
84          END
85   C
86   C
87   C
88          SUBROUTINE    OUTDAT(JNO,MNO,JRNO,JFNO,IFNO,RWNO,
89         &                     JCRD,MPRP,JRES,PG,INTFR,MCON,INTFI)
90          INTEGER       JNO,MNO,JRNO,JFNO,IFNO,RWNO
91          REAL          JCRD(JNO,2),MPRP(MNO,3),JRES(RWNO),PG(RWNO),
92         &              INTFR(IFNO,2)
93          INTEGER       MCON(MNO,2),INTFI(IFNO,3)
94   C
95   C
96          WRITE(6,500) JNO,MNO,JRNO,JFNO,IFNO
97          WRITE(6,510)
98          WRITE(6,520) (M,(JCRD(M,N),N=1,2),M=1,JNO)
99          WRITE(6,530)
100         WRITE(6,540) (M,(MCON(M,N),N=1,2),M=1,MNO)
101         WRITE(6,550)
102         WRITE(6,560) (M,(MPRP(M,N),N=1,3),M=1,MNO)
103         WRITE(6,570)
104         WRITE(6,575) (M,(JRES(N),N=(3*M-2),(3*M)),M=1,JNO)
105         IF(JFNO.EQ.0) GOTO 10
106         WRITE(6,580)
107         WRITE(6,590) (M,(PG(N),N=(3*M-2),(3*M)),M=1,JNO)
```

```
108     10 IF(IFNO.EQ.0) GOTO 20
109        WRITE(6,600)
110        WRITE(6,610) (M,(INTFI(M,N),N=1,3),M=1,IFNO)
111        WRITE(6,620)
112        WRITE(6,630) (M,(INTFR(M,N),N=1,2),M=1,IFNO)
113     20 CONTINUE
114   C
115    500 FORMAT(1H1,'RIGID-JOINTED PLANE FRAMEWORK ANALYSIS',
116        &       /////1H ,'DATA SUPPLIED',
117        &       /////1H ,'NUMBER OF JOINTS     ',I5,
118        &       /1H ,'NUMBER OF MEMBERS    ',I5,
119        &       //1H ,'NUMBER OF JOINT RESTRAINTS',I5,
120        &       /1H ,'NUMBER OF JOINT FORCES   ',I6,
121        &       //1H ,'NUMBER OF INTERMEDIATE FORCES',I2)
122    510 FORMAT(///1H ,'JOINT COORDINATES',
123        &       //1H ,10X,'JOINT',11X,'X(M)',11X,'Y(M)')
124    520 FORMAT(1H ,I13,F17.3,F15.3)
125    530 FORMAT(///1H ,'MEMBER CONNECTIVITY',
126        &       //1H ,10X,'MEMBER',8X,'JOINT1',9X,'JOINT2')
127    540 FORMAT(1H ,I13,I15,I15)           \
128    550 FORMAT(///1H ,'MEMBER PROPERTIES',
129        &       //1H ,10X,'MEMBER',7X,'E(N/M2)',7X,'AREA(M2)',
130        &       10X,'I(M4)')
131    560 FORMAT(1H ,I13,E17.3,2E15.3)
132    570 FORMAT(///1H ,'JOINT RESTRAINTS',
133        &       /1H ,5X,'(LX=LINEAR RESTRAINT IN X DIRECTION,',
134        &       /1H ,6X,'RZ=ROTATIONAL RESTRAINT ABOUT Z AXIS,',
135        &       /1H ,6X,'1000.000=FREE)',
136        &       //1H ,10X,'JOINT',10X,'LX(M)',10X,'LY(M)',
137        &       8X,'RZ(RAD)')
138    575 FORMAT(1H ,I13,F17.3,2F15.3)
139    580 FORMAT(///1H ,'JOINT FORCES',
140        &       /1H ,5X,'(PX=FORCE IN X DIRECTION,',
141        &       /1H ,6X,'CZ=COUPLE ABOUT Z AXIS)',
142        &       //1H ,10X,'JOINT',10X,'PX(N)',10X,'PY(N)',9X,'CZ(NM)')
143    590 FORMAT(1H ,I13,F17.2,2F15.2)
144    600 FORMAT(///1H ,'INTERMEDIATE FORCES',
145        &       //1H ,4X,'FORCE NUMBER',8X,'MEMBER',6X,'DIRECTION',11X,
146        &       'TYPE')
147    610 FORMAT(1H ,I13,3I15)
148    620 FORMAT(//1H ,4X,'FORCE NUMBER',3X,'FORCE(N OR N/M)',3X,
149        &       'POSITION')
150    630 FORMAT(1H ,I13,F17.2,F15.2)
151   C
152        RETURN
153        END
154   C
155   C
156   C
157        SUBROUTINE   BANDW(MNO,MCON,COLNO)
158        INTEGER      MNO,COLNO
159        INTEGER      MCON(MNO,2)
160   C
161   C
```

```
162          K1=0
163          DO 10 N= 1,MNO
164          I= MCON(N,1)
165          J= MCON(N,2)
166          K= (J-I)
167          IF(K.GT.K1) K1= K
168       10 CONTINUE
169          COLNO= 3*(K1+1)
170   C
171          RETURN
172          END
173   C
174   C
175   C
176          SUBROUTINE   MST(IMEM,JNO,MNO,JCRD,MPRP,MCON,KM)
177          INTEGER      IMEM,JNO,MNO
178          REAL         JCRD(JNO,2),MPRP(MNO,3),KM(6,6)
179          INTEGER      MCON(MNO,2)
180   C
181   C
182          I= MCON(IMEM,1)
183          J= MCON(IMEM,2)
184          RLX= JCRD(J,1)-JCRD(I,1)
185          RLY= JCRD(J,2)-JCRD(I,2)
186          RL= SQRT(RLX**2+RLY**2)
187          A= MPRP(IMEM,1)*MPRP(IMEM,2)/RL
188          KM(1,1)= A
189          KM(1,4)= -A
190          KM(4,4)= A
191          A= 12.0*MPRP(IMEM,1)*MPRP(IMEM,3)/RL**3
192          KM(2,2)= A
193          KM(2,5)= -A
194          KM(5,5)= A
195          A= 6.0*MPRP(IMEM,1)*MPRP(IMEM,3)/RL**2
196          KM(2,3)= A
197          KM(2,6)= A
198          KM(3,5)= -A
199          KM(5,6)= -A
200          A= 4.0*MPRP(IMEM,1)*MPRP(IMEM,3)/RL
201          KM(3,3)= A
202          KM(3,6)= A/2.0
203          KM(6,6)= A
204          DO 10 M= 1,6
205          DO 10 N= 1,6
206       10 IF(M.GT.N) KM(M,N)= KM(N,M)
207   C
208          RETURN
209          END
210   C
211   C
212   C
213          SUBROUTINE   TRANS(IMEM,JNO,MNO,JCRD,MCON,TM)
214          INTEGER      IMEM,JNO,MNO
215          REAL         JCRD(JNO,2),TM(6,6),RM(3,3)/9*0.0/
216          INTEGER      MCON(MNO,2)
217   C
218   C
```

```
219          I= MCON(IMEM,1)
220          J= MCON(IMEM,2)
221          RLX= JCRD(J,1)-JCRD(I,1)
222          RLY= JCRD(J,2)-JCRD(I,2)
223          RL= SQRT(RLX**2+RLY**2)
224          RCX= RLX/RL
225          RCY= RLY/RL
226          RM(1,1)= RCX
227          RM(1,2)= RCY
228          RM(2,1)= -RCY
229          RM(2,2)= RCX
230          RM(3,3)= 1.0
231          DO 20 M= 1,3
232          DO 20 N= 1,3
233          TM(M,N)= RM(M,N)
234       20 TM(M+3,N+3)= RM(M,N)
235  C
236          RETURN
237          END
238  C
239  C
240  C
241          SUBROUTINE  GLST(IMEM,MNO,RWNO,COLNO,MCON,KM,TM,KG)
242          INTEGER     IMEM,MNO,RWNO,COLNO
243          REAL        KM(6,6),TM(6,6),KG(RWNO,COLNO),KMG(6,6)
244          INTEGER     MCON(MNO,2)
245          INTEGER     P,Q,R
246  C
247  C
248          DO 10 I= 1,6
249          DO 10 J= 1,6
250          KMG(I,J)= 0.0
251          DO 10 M= 1,6
252          DO 10 N= 1,6
253       10 KMG(I,J)= KMG(I,J)+TM(M,I)*KM(M,N)*TM(N,J)
254  C
255          IF(IMEM.NE.1) GOTO 20
256          DO 30 M= 1,RWNO
257          DO 30 N= 1,COLNO
258       30 KG(M,N)= 0.0
259       20 CONTINUE
260          I= MCON(IMEM,1)
261          J= MCON(IMEM,2)
262          K= 3*(J-I)
263          DO 40 M= 1,3
264          DO 40 N= 1,3
265          P= (3*I-2)+M-1
266          Q= (3*J-2)+M-1
267          R= 1-M+N
268          IF(R.LT.1) GOTO 40
269          KG(P,R)= KG(P,R)+KMG(M,N)
270          KG(Q,R)= KG(Q,R)+KMG(M+3,N+3)
271       40 KG(P,K+R)= KG(P,K+R)+KMG(M,N+3)
272  C
273          RETURN
274          END
275  C
```

```
276  C
277  C
278        SUBROUTINE MFX(IMEM,JNO,MNO,IFNO,
279     &                JCRD,INTFR,MCON,INTFI,PFM)
280        INTEGER    IMEM,JNO,MNO,IFNO
281        REAL       JCRD(JNO,2),INTFR(IFNO,2),PFM(MNO,6),TMP(6)
282        INTEGER    MCON(MNO,2),INTFI(IFNO,3)
283  C
284  C
285        IF(IFNO.EQ.0) GOTO 10
286        I= MCON(IMEM,1)
287        J= MCON(IMEM,2)
288        RLX= JCRD(J,1)-JCRD(I,1)
289        RLY= JCRD(J,2)-JCRD(I,2)
290        RL= SQRT(RLX**2+RLY**2)
291        CMX= RLX/RL
292        CMY= RLY/RL
293        DO 10 I= 1,IFNO
294        IF(INTFI(I,1).NE.IMEM) GOTO 10
295        RQ= INTFR(I,1)
296        IF(INTFI(I,2).EQ.2) GOTO 20
297        RQA= RQ*CMX
298        RQT= RQ*CMY
299        GOTO 30
300     20 RQA= RQ*CMY
301        RQT= -RQ*CMX
302     30 IF(INTFI(I,3).EQ.2) GOTO 40
303        A= INTFR(I,2)*RL
304        B= (RL-A)
305        C= 1/(RL**2)
306        TMP(1)= -RQA*B*RL*C
307        TMP(2)= (RQT*B*B*(RL+2.0*A)/RL)*C
308        TMP(3)= RQT*A*B*B*C
309        TMP(4)= -RQA*A*RL*C
310        TMP(5)= (RQT*A*A*(RL+2.0*B)/RL)*C
311        TMP(6)= -RQT*A*A*B*C
312        GOTO 50
313     40 C= RL/2
314        TMP(1)= -RQA*C
315        TMP(2)= RQT*C
316        TMP(3)= (RQT*RL/6.0)*C
317        TMP(4)= -RQA*C
318        TMP(5)= RQT*C
319        TMP(6)= (-RQT*RL/6.0)*C
320     50 DO 60 J= 1,6
321        PFM(IMEM,J)= PFM(IMEM,J)+TMP(J)
322     60 CONTINUE
323     10 CONTINUE
324  C
325        RETURN
326        END
327  C
328  C
329  C
```

```
330        SUBROUTINE GLFQ(IMEM,MNO,RWNO,MCON,PFM,TM,PGQ)
331        INTEGER     IMEM,MNO,RWNO
332        REAL        PFM(MNO,6),TM(6,6),PGQ(RWNO),PFMG(6)
333        INTEGER     MCON(MNO,2)
334        INTEGER     P,Q
335   C
336   C
337        DO 10 I= 1,6
338        PFMG(I)= 0.0
339        DO 10 M= 1,6
340    10 PFMG(I)= PFMG(I)+TM(M,I)*PFM(IMEM,M)
341        I= MCON(IMEM,1)
342        J= MCON(IMEM,2)
343        DO 20 M= 1,3
344        P= (3*I-2)+M-1
345        Q= (3*J-2)+M-1
346        PGQ(P)= PGQ(P)-PFMG(M)
347    20 PGQ(Q)= PGQ(Q)-PFMG(M+3)
348   C
349        RETURN
350        END
351   C
352   C
353   C
354        SUBROUTINE   GLF(RWNO,PG,PGQ)
355        INTEGER     RWNO
356        REAL        PG(RWNO),PGQ(RWNO)
357   C
358   C
359        DO 10 I= 1,RWNO
360    10 PG(I)= PG(I)+PGQ(I)
361   C
362        RETURN
363        END
364   C
365   C
366   C
367        SUBROUTINE   SPPRTS(RWNO,COLNO,JRES,PG,KG)
368        INTEGER     RWNO,COLNO
369        REAL        JRES(RWNO),PG(RWNO),KG(RWNO,COLNO)
370   C
371   C
372        DO 10 I= 1,RWNO
373        IF(JRES(I).GT.999.9) GOTO 10
374        KG(I,1)= 1.0E30
375        PG(I)= JRES(I)*1.0E30
376    10 CONTINUE
377   C
378        RETURN
379        END
380   C
381   C
382   C
383        SUBROUTINE   SOLVE(N,M,A,B,C,D)
384        REAL        A(N,M),B(N),C(M),D(N)
385   C
386   C
```

```
387            DO 10 K= 1,N
388            B(K)= B(K)/A(K,1)
389            IF(K.EQ.N) GOTO 100
390            DO 20 J= 2,M
391            C(J)= A(K,J)
392            A(K,J)= A(K,J)/A(K,1)
393        20 CONTINUE
394            DO 30 L= 2,M
395            I= K+L-1
396            IF(N.LT.I) GOTO 30
397            J= 0
398            DO 40 L1= L,M
399            J= J+1
400            A(I,J)= A(I,J)-C(L)*A(K,L1)
401        40 CONTINUE
402            B(I)= B(I)-C(L)*B(K)
403        30 CONTINUE
404        10 CONTINUE
405       100 D(K)= B(K)
406            K= K-1
407            IF(K.EQ.0) GOTO 200
408            DO 50 J= 2,M
409            L= K+J-1
410            IF(N.LT.L) GOTO 50
411            B(K)= B(K)-A(K,J)*B(L)
412        50 CONTINUE
413            GOTO 100
414       200 CONTINUE
415 C
416            RETURN
417            END
418 C
419 C
420 C
421            SUBROUTINE    OUTSOL1(JNO,RWNO,DG)
422            INTEGER       RWNO
423            REAL          DG(RWNO)
424 C
425 C
426            WRITE(6,500)
427            WRITE(6,510) (M,(DG(N),N=(3*M-2),(3*M)),M= 1,JNO)
428 C
429       500 FORMAT(//////1H ,'RESULTS',
430         &        //////1H ,'JOINT DISPLACEMENTS',
431         &        //1H ,10X,'JOINT',10X,'LX(M)',10X,'LY(M)',
432         &        8X,'RZ(RAD)')
433       510 FORMAT(1H ,I13,1F17.6,2F15.6)
434 C
435            RETURN
436            END
437 C
438 C
439 C
```

```
440          SUBROUTINE    MFRCS(IMEM,MNO,RWNO,DG,KM,TM,MCON,PFM,PM)
441          INTEGER       IMEM,MNO,RWNO
442          REAL          DG(RWNO),KM(6,6),TM(6,6),PM(6),PFM(MNO,6),DMG(6)
443          INTEGER       MCON(MNO,2)
444   C
445   C
446          I= MCON(IMEM,1)
447          J= MCON(IMEM,2)
448          DMG(1)= DG(3*I-2)
449          DMG(2)= DG(3*I-1)
450          DMG(3)= DG(3*I)
451          DMG(4)= DG(3*J-2)
452          DMG(5)= DG(3*J-1)
453          DMG(6)= DG(3*J)
454          DO 10 I= 1,6
455          PM(I)= 0.0
456          DO 20 J= 1,6
457          DO 20 K= 1,6
458       20 PM(I)= PM(I)+KM(I,J)*TM(J,K)*DMG(K)
459       10 PM(I)= PM(I)+PFM(IMEM,I)
460   C
461          RETURN
462          END
463   C
464   C
465   C
466          SUBROUTINE    OUTSOL2(IMEM,MNO,PM,MCON)
467          INTEGER       IMEM,MNO
468          REAL          PM(6)
469          INTEGER       MCON(MNO,2)
470   C
471   C
472          IF(IMEM.GT.1) GOTO 10
473          WRITE(6,500)
474       10 WRITE(6,510) (IMEM,MCON(IMEM,M),(PM(3*(M-1)+N),N=1,3),M=1,2)
475   C
476      500 FORMAT(///1H ,'LOCAL MEMBER FORCES',
477          &        //1H ,3X,'MEMBER/JOINT',9X,'PMX(N)',
478          &         9X,'PMY(N)',8X,'CMZ(NM)')
479      510 FORMAT(/1H ,I12,'/',I2,3F15.2,/1H ,I12,'/',I2,3F15.2)
480   C
481          RETURN
482          END
483   C
484   C
485   C
486          SUBROUTINE    RCTNS(RWNO,COLNO,KG,DG,PGQ,PG)
487          INTEGER       RWNO,COLNO
488          REAL          KG(RWNO,COLNO),DG(RWNO),PGQ(RWNO),PG(RWNO)
489   C
490   C
```

```
491          DO 10 I= 1,RWNO
492          PG(I)= 0.0
493          DO 20 J= 1,COLNO
494          N= I+J-1
495          IF(N.GT.RWNO) GOTO 20
496          PG(I)= PG(I)+KG(I,J)*DG(N)
497       20 CONTINUE
498          DO 30 J= 2,COLNO
499          N= I-J+1
500          IF(N.LT.1) GOTO 30
501          PG(I)= PG(I)+KG(N,J)*DG(N)
502       30 CONTINUE
503          PG(I)= PG(I)-PGQ(I)
504       10 CONTINUE
505    C
506          RETURN
507          END
508    C
509    C
510    C
511          SUBROUTINE     OUTSOL3(JNO,RWNO,PG)
512          INTEGER        JNO,RWNO
513          REAL           PG(RWNO)
514    C
515    C
516          WRITE(6,500)
517          WRITE(6,510) (M,(PG(N),N=(3*M-2),(3*M)),M=1,JNO)
518      500 FORMAT(///1H ,'GLOBAL FORCES AND REACTIONS',
519          &        //1H ,10X,'JOINT',10X,'PX(N)',10X,'PY(N)',
520          &          9X,'CZ(NM)')
521      510 FORMAT(1H ,I13,F17.2,2F15.2)
522    C
523          RETURN
524          END
```

A6.2.2 RJPFA Data Presentation

Data block	Number of rows	Description of data	FORTRAN format
1	1	number of joints; number of members; number of joint restraints; number of joint forces; number of intermediate forces	5I10
2	number of joints	joint number; joint coordinates (m)	I10, 2F10.0
3	number of members	member number; member connectivity	3I10
4	number of members	member number; Young's modulus, E (N/m^2); sectional area, A (m^2); second moment of area, I (m^4)	I10, 3E10.0
5	number of joint restraints	joint number; direction number; displacement (m)	2I10, F10.0
6	number of joint forces	joint number; direction number; force (N)	2I10, E10.0
7	number of intermediate forces	member number; direction number; type number; force (N or N/m); position of concentrated force (a/l)	3I10, E10.0, F10.0

A6.2.3 RJPFA Portal Frame Example

```
RIGID-JOINTED PLANE FRAMEWORK ANALYSIS

DATA SUPPLIED
NUMBER OF JOINTS          5
NUMBER OF MEMBERS         4

NUMBER OF JOINT RESTRAINTS     4
NUMBER OF JOINT FORCES         3
NUMBER OF INTERMEDIATE FORCES  3
```

JOINT COORDINATES

JOINT	X(M)	Y(M)
1	0.000	0.000
2	0.000	4.000
3	6.000	6.500
4	12.000	4.000
5	12.000	0.000

MEMBER CONNECTIVITY

MEMBER	JOINT1	JOINT2
1	1	2
2	2	3
3	3	4
4	4	5

MEMBER PROPERTIES

MEMBER	E(N/M2)	AREA(M2)	I(M4)
1	0.200E+12	0.400E-02	0.300E-04
2	0.200E+12	0.400E-02	0.300E-04
3	0.200E+12	0.400E-02	0.300E-04
4	0.200E+12	0.400E-02	0.300E-04

JOINT RESTRAINTS
     (LX=LINEAR RESTRAINT IN X DIRECTION,
     RZ=ROTATIONAL RESTRAINT ABOUT Z AXIS,
     1000.000=FREE)

JOINT	LX(M)	LY(M)	RZ(RAD)
1	0.000	0.000	1000.000
2	1000.000	1000.000	1000.000
3	1000.000	1000.000	1000.000
4	1000.000	1000.000	1000.000
5	0.000	0.000	1000.000

JOINT FORCES
     (PX=FORCE IN X DIRECTION,
     CZ=COUPLE ABOUT Z AXIS)

JOINT	PX(N)	PY(N)	CZ(NM)
1	0.00	0.00	0.00
2	0.00	-5000.00	0.00
3	0.00	-10000.00	0.00
4	0.00	-4000.00	0.00
5	0.00	0.00	0.00

```
INTERMEDIATE FORCES

    FORCE NUMBER        MEMBER      DIRECTION        TYPE
          1               1             1              2
          2               2             2              1
          3               3             2              1


    FORCE NUMBER    FORCE(N OR N/M)    POSITION
          1             2000.00          0.00
          2           -10000.00          0.50
          3            -8000.00          0.50



RESULTS


JOINT DISPLACEMENTS

        JOINT        LX(M)           LY(M)          RZ(RAD)
          1        -0.000000       -0.000000       -0.001957
          2         0.011167       -0.000091       -0.005349
          3         0.022077       -0.026495        0.001850
          4         0.032976       -0.000094       -0.001242
          5         0.000000       -0.000000       -0.011745


LOCAL MEMBER FORCES

    MEMBER/JOINT        PMX(N)          PMY(N)         CMZ(NM)

        1/ 1          18166.67         122.94          0.00
        1/ 2         -18166.67        7877.06      -15508.24

        2/ 2          12335.23        9124.21       15508.24
        2/ 3          -8489.08         106.56       13799.11

        3/ 3           9899.34       -3278.05      -13799.11
        3/ 4         -12976.26       10662.67      -31508.24

        4/ 4          18833.33        7877.06       31508.24
        4/ 5         -18833.33       -7877.06          0.00


GLOBAL FORCES AND REACTIONS

        JOINT          PX(N)           PY(N)          CZ(NM)
          1          -122.94         18166.67         0.00
          2            -0.00         -5000.00         0.00
          3            -0.00        -10000.00        -0.00
          4            -0.00         -4000.00         0.00
          5         -7877.06         18833.33         0.00
```

APPENDIX A6.3 RIGID-JOINTED SPACE FRAMEWORK ANALYSIS COMPUTER PROGRAM

A6.3.1 RJSFA Example

RIGID-JOINTED SPACE FRAMEWORK ANALYSIS

DATA SUPPLIED

NUMBER OF JOINTS   6
NUMBER OF MEMBERS   5

NUMBER OF JOINT RESTRAINTS   12
NUMBER OF JOINT FORCES   0
NUMBER OF INTERMEDIATE FORCES 3

JOINT COORDINATES

JOINT	X(M)	Y(M)	Z(M)
1	0.000	0.000	0.000
2	1.500	0.000	0.000
3	1.500	4.000	0.000
4	1.500	4.000	-2.000
5	1.500	4.000	-4.000
6	0.000	4.000	-4.000

MEMBER CONNECTIVITY AND ROTATION

MEMBER	JOINT1	JOINT2	GAMMA(RAD)
1	1	2	0.000
2	2	3	-1.571
3	3	4	0.000
4	4	5	0.000
5	5	6	0.000

MEMBER PROPERTIES

MEMBER	E(N/M2)	G(N/M2)	AREA(M2)
1	0.200E+12	0.769E+11	0.400E-02
2	0.200E+12	0.769E+11	0.400E-02
3	0.200E+12	0.769E+11	0.400E-02
4	0.200E+12	0.769E+11	0.400E-02
5	0.200E+12	0.769E+11	0.400E-02

MEMBER	IY(M4)	IZ(M4)	J(M4)
1	0.150E-04	0.300E-04	0.200E-04
2	0.150E-04	0.300E-04	0.200E-04
3	0.150E-04	0.300E-04	0.200E-04
4	0.150E-04	0.300E-04	0.200E-04
5	0.150E-04	0.300E-04	0.200E-04

```
JOINT RESTRAINTS
    (LX= LINEAR RESTRAINT IN X DIRECTION,
     RX= ROTATIONAL RESTRAINT ABOUT X AXIS,
     1000.000= FREE)

        JOINT        LX(M)            LY(M)            LZ(M)
                     RX(RAD)          RY(RAD)          RZ(RAD)

          1          0.000            0.000            0.000
                     0.000            0.000            0.000

          2          1000.000         1000.000         1000.000
                     1000.000         1000.000         1000.000

          3          1000.000         1000.000         1000.000
                     1000.000         1000.000         1000.000

          4          1000.000         1000.000         1000.000
                     1000.000         1000.000         1000.000

          5          1000.000         1000.000         1000.000
                     1000.000         1000.000         1000.000

          6          0.000            0.000            0.000
                     0.000            0.000            0.000


INTERMEDIATE FORCES

    FORCE NUMBER      MEMBER      DIRECTION       TYPE
         1              1             2             1
         2              2             1             1
         3              3             2             2


    FORCE NUMBER   FORCE(N OR N/M)   POSITION
         1           -10000.00         0.50
         2           -15000.00         0.50
         3           -6000.00          0.00
```

RESULTS

JOINT DISPLACEMENTS

JOINT	LX(M) RX(RAD)	LY(M) RY(RAD)	LZ(M) RZ(RAD)
1	0.000000 0.000000	0.000000 0.000000	0.000000 0.000000
2	-0.000024 0.000609	0.001158 -0.001454	0.001214 0.002487
3	-0.022971 -0.000727	0.001108 -0.006454	0.002146 0.004315
4	-0.009777 -0.000243	-0.000317 -0.006240	0.002149 0.002183
5	-0.000004 0.000295	-0.000052 -0.003034	0.002152 0.000050
6	0.000000 0.000000	0.000000 0.000000	0.000000 0.000000

LOCAL MEMBER FORCES

MEMBER/JOINT	PMX(N) CMX(NM)	PMY(N) CMY(NM)	PMZ(N) CMZ(NM)
1/ 1	12755.81 -624.45	20082.35 3894.23	-1314.08 3240.53
1/ 2	-12755.81 624.45	-10082.35 -1923.11	1314.08 19383.00
2/ 2	10082.35 1923.11	1314.04 19383.00	-12755.81 624.38
2/ 3	-10082.35 -1923.11	-1314.09 1640.24	-2244.19 4631.88
3/ 3	1314.08 -1640.22	10082.35 1923.11	-2244.19 4631.88
3/ 4	-1314.08 1640.22	1917.65 2565.28	2244.19 3532.82
4/ 4	1314.08 -1640.22	-1917.65 -2565.28	-2244.19 -3532.82
4/ 5	-1314.08 1640.22	1917.65 7053.67	2244.19 -302.47
5/ 5	2244.19 -302.47	-1917.65 -7053.67	1314.08 -1640.22
5/ 6	-2244.19 302.47	1917.65 5082.54	-1314.08 -1236.25

GLOBAL FORCES AND REACTIONS

JOINT	PX(N) CX(NM)	PY(N) CY(NM)	PZ(N) CZ(NM)
1	12755.81 -624.45	20082.35 3894.23	-1314.08 3240.53
2	0.00 0.00	0.00 0.00	0.00 0.00
3	0.00 0.00	0.00 0.00	0.00 0.00
4	0.00 0.00	0.00 0.00	0.00 0.00
5	0.00 0.00	0.00 0.00	0.00 0.00
6	2244.19 -302.47	1917.65 5082.54	1314.08 1236.25

REFERENCES

[6.1] Fox, L. (1964), *An Introduction to Numerical Linear Algebra,* Oxford University Press.

[6.2] Wait, R. (1979), *The Numerical Solution of Algebraic Equations,* Wiley Interscience, New York.

[6.3] Collins, R. J. (1973), Bandwidth reduction by automatic renumbering, *Int. J. Num. Methods in Engng.,* **16,** 345–356.

[6.4] Macnab, C. S. (1974), *Fortran,* Blackie, Glasgow.

[6.5] Monro, D. M. (1977), *Computing with Fortran IV,* Edward Arnold, London.

[6.6] Bickley, W. G., and Gibson, R. E. (1962), *Via Vector to Tensor,* English Universities Press, London, Chapter 5.

[6.7] Fung, Y. C. (1965), *Foundations of Solid Mechanics,* Prentice-Hall, Englewood Cliffs, N.J., Section 4.4.

[6.8] Majid, K. I. (1974), *Optimum Design of Structures,* Newnes–Butterworth, London, Chapter 6.

[6.9] Hambly, E. C. (1976), *Bridge Deck Behaviour,* Chapman & Hall, London.

[6.10] Allen, H. G., and Bulson, P. S. (1980), *Background to Buckling,* McGraw-Hill, London.

[6.11] Møllmann, H. (1974), *Analysis of Hanging Rooves by Means of the Displacement Method,* Polyteknisk Forlag, Lyngby, Denmark.

[6.12] Jennings, A., and Majid, K. I. (1965), An elasto-plastic analysis by computer for framed structures loaded up to collapse, *Structural Engineer*, 43, 407–412.

[6.13] Zienkiewicz, O. C. (1977), *The Finite Element Method*, 2nd edn., McGraw-Hill, London.

Solutions to Problems and Hints

CHAPTER 1

1.1 12.5×10^{-3} m², 38.60×10^{-6} m⁴, 168.23×10^{-6} m⁴; 101.2 MN/m² at the bottom, -77.2 MN/m² at the top. **1.2** 125.6 kNm (compression governing). **1.3** 11.75 kNm (compression governing) at $y = 150.0$ mm, $z = -75.0$ mm. **1.4** 230.0×10^6 $(\pi r^2 t)$ Nm (r and t in m); $\stackrel{\triangle}{=} 3.0$ mm. (Divide the collapse bending moment by the partial strength factor. Multiply the wind bending moment by the partial load factor.)

1.5 $\stackrel{\triangle}{=} 20.0$ mm. (In order to linearise the equation for determining the thickness t, neglect second or higher order terms in t when calculating the second moment of area of the composite section.) **1.6** 0.345 m; 4.76×10^{-3} m⁴; 68.81 MN/m², -36.23 MN/m².

1.7 96.7 MNm. (Compression of the top flange plate governs.)

1.8 $\tau_{xy} = (20.0 - 3.56 \times 10^3 y^2)$ MN/m² (y in m); 20.0 MN/m².

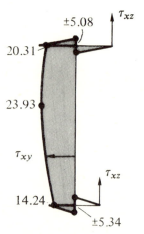

(all stresses are in MN/m²)

Fig. S1

1.9 23.93 MN/m². The distribution of shear stresses is shown in Fig. S1. (Define variable positions in the web and flanges by coordinates y and z respectively, measured from the centroid of the section. Show that $d\tau_{xy}/dy = 0$ at $y = 0$.) **1.12** 69.0 MN/m², 657.1 Nm.

CHAPTER 2

2.1 The non-zero reactions from left to right are as follows: (a) $-P/2$, $P/2$; (b) $-P$; (c) P, P; (d) P/l, $-P/l$; (e) $3P/2$, $3P/2$; (f) $Pl/3$; (g) 2.47, 0.99 kN; (h) 20.0, 12.0 kN; (i) -0.75, 0.75 kN. The shear force and bending moment diagrams are shown in Figs. S2(a) to (i). **2.2** (a) 2.06, 6.44 kN, 2.06 kNm; (b) 2.0P, 2.0P, -5.0 Pa; (c) 2.0P, 2.0P; (d) -5.0, 15.0, 5.0 kN, -5.0 kNm; (e) 6.25, 13.75 kN, 6.25, -13.75 kNm. The shear force and bending moment diagrams are shown in Figs. S3 (a) to (e). **2.3** 2.62, 7.38 kN; 4.24, 0.76 kN; 0.54, 7.46 kN; 8.23, 1.14 kN. The shear force and bending moment diagrams are shown in Fig. S4. **2.4** (a) -10.0 kN, 25.0 kNm; (b) -7.07, 27.07 kN, 48.28 kNm; (c) 10.0 kN, 10.0 kNm; (d) 10.0 kN, -10.0 kNm; (e) 10.0 kNm; (f) 2.5, 7.5, -2.5, 2.5 kN; (g) 2.5, 7.5 kN; (h) -10.0, -2.5, 7.5 kN. The stress resultant diagrams are shown in Figs. S5 (a) to (h). **2.5** $-(8.0 + 28.0 \sin(\theta))$ kNm. **2.6** 1250.0, 500.0, -1250.0, 1250.0 kN; $(16.67x^2 - 500.0x)$, $-(8.33x^2 - 750.0x + 15{,}000.0)$ kNm. **2.7** -7.81, -2.5, -25.0, 10.0, 2.5, 7.81 kN. The stress resultant diagrams are shown in Fig. S6. **2.8** $-pr^2 (1.0 - \cos(\theta))$, $pr^2 (\theta - \sin(\theta))$. **2.9**† (a) $-0.625P$, $-0.625P$, 0.884P, 0, $-0.625P$, $-1.25P$, 0.884P, 0.625P, $-1.0P$, $-1.25P$, 0.530P, 0.875P, $-0.875P$, $-0.875P$, 1.237P, 0, $-0.875P$; (b) 0, 0, 5.77, -2.89, -5.77, 5.77, -11.55 kN; (c) 1.46P, $-3.97P$, 3.34P, $-2.72P$, 1.0P, 0, 1.46P, $-3.97P$, 3.34P; (d) -75.4, 59.6, -80.0, 29.8, -37.7 kN; (e) -2.5, 7.07, -1.86, -4.71, 3.73, -5.0, -5.0 kN; (f) -77.1, 9.1, 51.7, -46.8, -44.0, 49.2 kN; (g) -1.45, 4.71, -2.59 kN; (h) -15.26, 0, 22.30, -18.18, 15.28, -0.67 kN.

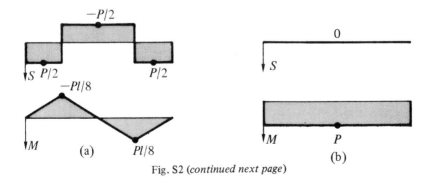

Fig. S2 (*continued next page*)

---

† For those structures where members are not individually identified by letters, member forces are quoted starting with members on the left-hand sides of the structures in the figures, and moving through the structures from left to right.

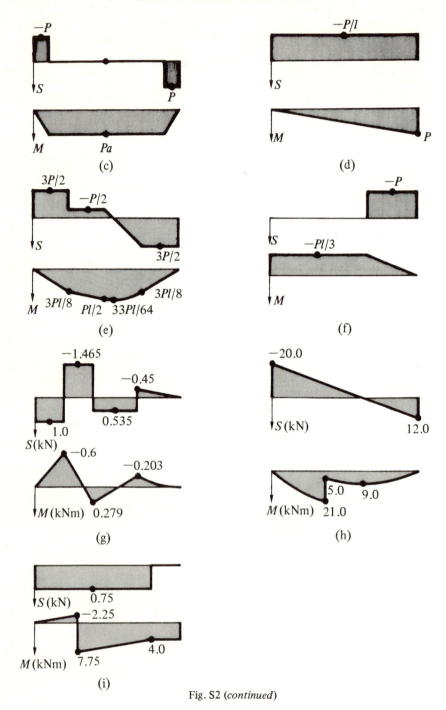

Fig. S2 (*continued*)

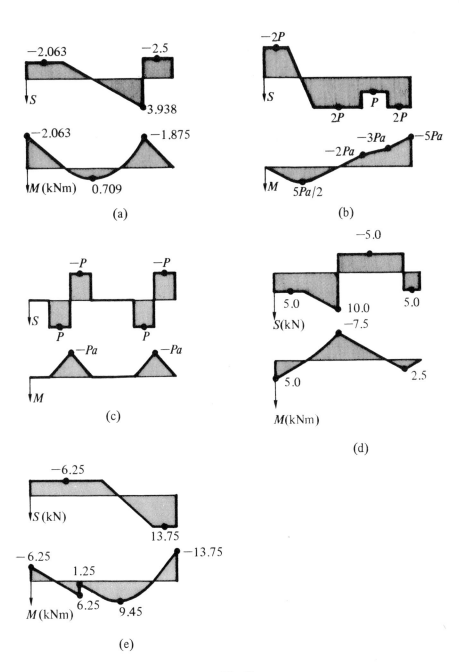

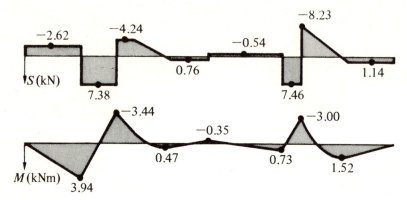

Fig. S4

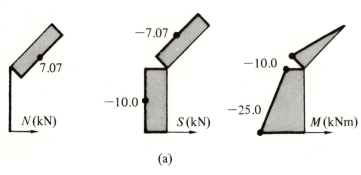

(a)

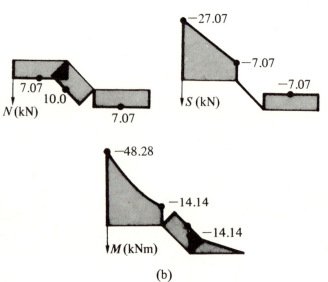

(b)

Fig. S5 (*continued next page*)

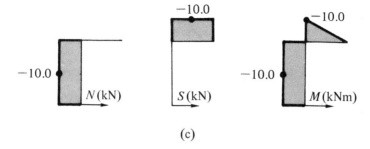

(c)

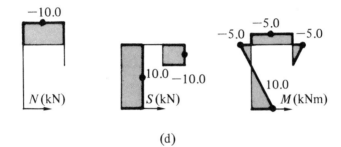

(d)

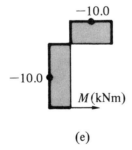

(e)

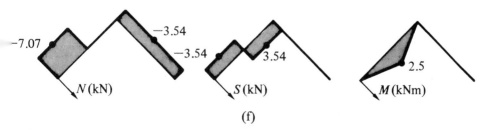

(f)

Fig. S5 (*continued next page*)

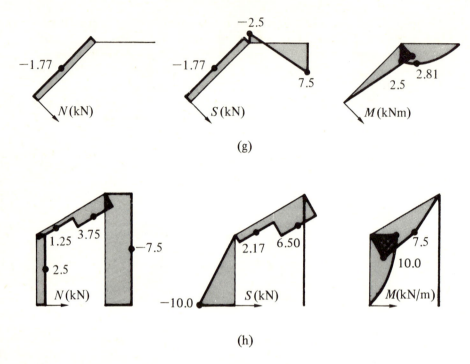

(g)

(h)

Fig. S5 (*continued*)

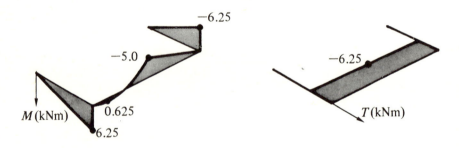

Fig. S6

2.10 The Pratt truss. (Calculate the support reactions. Then take vertical sections and consider vertical equilibrium to show that the longer diagonal members are in tension in the Pratt truss and in compression in the Howe truss.) The upper chord members of the trusses are in compression. The central vertical member of the Pratt truss halves the length of the central upper chord member and restricts its tendency to buckle. **2.11** $- 0.407P$. (Take sections cutting each internal member in turn, and consider rotational equilibrium about an axis through the tip of the cantilever to obtain the forces in the members.) **2.12** $N_a = N_c = N_e = -0.577P$, $N_b = N_d = 0.577P$, $N_f = 0$, $N_g = -0.577P$, $N_h = -1.155P$, $N_i = -1.44P$. The vertical members either restrict the buckling of the upper chord members or provide intermediate suspension points for alternative load positions. (Use the method of joints to determine the forces in the vertical members. Take vertical sections and consider vertical and rotational equilibrium to obtain the forces in the diagonals and upper chord members.) **2.13** $N_a = -0.75P$, $N_b = 0.75P$. (Take a curved section cutting members a and b and the two vertical members on the left-hand side of the panel containing a and b. Then consider horizontal and rotational equilibrium.) **2.14** $N_e = P$. (Take sections cutting the 5 members in turn and consider rotational equilibrium about an axis through the left-hand support for members a, b, c and d, and about an axis through the right-hand support for member e.) **2.15** $N_a = 2.50$, $N_b = 2.0$, $N_c = 1.0$, $N_d = -4.04$ kN. (Take a section through the apex of the truss, cutting member a and consider rotational equilibrium about an axis through the apex. Then take a section cutting members a, b, c and d and consider rotational equilibrium about convenient axes.) **2.16** $N_a = 26.67$, $N_b = 40.0$, $N_c = 26.67$ kN. **2.17** $N_a = -11.80$ kN. (Take sections through the pylon, and consider rotational equilibrium about axes through the top to show that the 5.0 kN force only generates forces in the main legs of the pylon. Consider the side of the pylon containing member a as a plane pin-jointed framework. The 10.0 kN force acts in this plane. Obtain the coordinates of the joints relative to a coordinate system lying in this plane. Thence use the method of tension coefficients to calculate the force in member a). **2.18** $M = 37, R = 3, J = 20$; $M = 17, R = 3, J = 10$; $M = 21, R = 9, J = 10$.

CHAPTER 3

3.1 3.27, 5.86 mm. **3.2** 0.639. **3.3** $9.74°$, $1.394\ Pa/EA$. **3.4** 5.0 mm; 29.3 mm. **3.5** 12.40 mm. (Note that $\Delta = Nl/EA = \sigma_{xx}l/E$.) **3.6** $0.55, 2.65$ mm. **3.7** $0.0355 \times Pl^3/EI$; $5\ Pl^3/(384\ EI)$; $0.0347\ Pl^2/EI$ (upwards). **3.9** $0.121\ pa^4/EI$. **3.10** $y = 10.0 - 49.1 \times 10^{-6}\ (22.5\ x^2 + x^3/3)$ m. x is measured from the centre-line of the column, y is measured from the ground. Both dimensions are in metres. The column is subjected to a constant axial force at all load positions, so that the axial deflection of the column does not affect the trajectory. The beam is not subjected to axial forces. ($g = 9.81$ m/sec^2.) **3.11** 0.126 m. **3.13** 6.6 mm. **3.15** $(pr^4/EI)\ (2.0 + \pi^2 EI/(2.0GJ))$; $(\pi pr^3/(2.0EI))\ (-1.0 + EI/GJ)$. **3.16** 6.93 mm.

3.17
$$
\begin{bmatrix}
6.48 & 4.14 & -1.38 \\
4.14 & 6.48 & -0.35 \\
-1.38 & -0.35 & 1.98
\end{bmatrix}
\left(\frac{a}{EA}\right)
$$

CHAPTER 4

4.1 $1.06P$. **4.2** -37.71, 26.67, 3.33, 30.0, -4.71, 4.71, -30.0, -3.33, 33.33, -47.14 kN. **4.3** $\begin{bmatrix} 0.15 & 0.0167 \\ 0.0167 & 0.15 \end{bmatrix} \times 10^{-6}$ m/N; -6.27, 3.73, -5.27, 1.8, -1.27, -3.12, 3.15, 2.62, -4.45, -1.85, -1.85 kN. **4.4** The force in the central vertical member is 7.04 kN, the forces in the remaining horizontal and vertical members are 3.52 kN, the forces in the diagonals are -4.98 kN. (Include the temperature change in the expressions for Δ_P given in (4.8) and (4.9).) **4.5** 1.26 mm. (Include Δ_C in Δ_P.) **4.6** -19.43, 0.91, -1.22, -14.0, -0.57 kN. **4.7** 2.25 kN. **4.8** 0, 0.62, -1.35, 1.04, -1.21, -7.47, 6.01 kN. **4.9** 8.16, 6.19 kN. The bending moment diagram is shown in Fig. S7. **4.10** 158.18 kN. **4.11** (In (3.22) consider $\Delta\theta$ to be an appropriate linear function of y.) **4.12** Pab^2/l^2, $-Pa^2b/l^2$. **4.13** $\begin{bmatrix} 10.20 & 1.92 \\ 1.92 & 10.20 \end{bmatrix} (1/I_{\text{beam}}) \times 10^{-9}$ m/N; 5.33, 11.02 kN (both downwards); -37.28, -43.93 kNm (on either side of joint 1), -139.82, -77.12 kNm (on either side of joint 2). **4.14** -43.56, -33.23 kNm. (Treat the culvert as being simply supported on its two lower corners. Make a cut in the centre of the top of the culvert and solve for the two unknown release forces.)

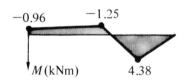

-0.96 -1.25

M (kNm)   4.38

Fig. S7

4.15 $N = 6$;

$$\begin{bmatrix} 1.67a^2 & 1.0a^2 & -2.0a & 0.33a^2 & -0.5a^2 & -0.5a \\ 1.0a^2 & 1.67a^2 & -2.0a & 0.5a^2 & -1.0a^2 & -1.0a \\ -2.0a & -2.0a & 4.0 & -0.5a & 1.0a & 1.0 \\ 0.33a^2 & 0.5a^2 & -0.5a & 1.67a^2 & -1.0a^2 & -2.0a \\ -0.5a^2 & -1.0a^2 & 1.0a & -1.0a^2 & 1.67a^2 & 2.0a \\ -0.5a & -1.0a & 1.0 & -2.0a & 2.0a & 4.0 \end{bmatrix} (a/EI)$$

In determining the flexibility matrix, the directions of the unit release forces are taken such that the forces on the *columns* at the releases are in the x and y directions and clockwise about the z axis. **4.16** (i) $N = 3I$; (ii) $N = 3I$; (iii) $N = 6I$. **4.17** 1.42, 2.12 kN. The bending moment diagram is shown in Fig. S8. **4.18** 0; $(4.0pr/\pi)$ $(1.0 + \pi^2 EI/(4.0GJ))/(1.0 + 3.0EI/GJ)$. The linear reaction in the y direction and the couple reactions about the x and z axes would be non-zero. (Consider the ring beam described by a local coordinate system x_m, y_m and z_m with x_m tangential to the axis of the beam at any point and y_m vertical. The uniformly distributed force gives rise only to the stress resultants M_z, S_y and T. The linear reactions in the x and z directions and the couple reaction about the y axis give rise only to N, S_z and M_y. The flexibility method would therefore demonstrate these reactions to be zero.) **4.19** $2.12Pa/EA$; $11.72 \times 10^3/(EA_a)$; 32.3 mm. **4.20** 11.6 mm. The bending moments in the central column are twice those in the left-hand column. (Neglecting axial effects, the displacements of the central column are the same as those in the left-hand column, and would be calculated using the same procedure as for the left-hand column.) **4.21** $N_e = 2.04$ kN, $N_f = -17.29$ kN. (Add $-\mathbf{r}d_s$ to the expression for $\mathbf{u}_P$ as in (4.43).) **4.22** Both couples are positive, equal to $6EId_s/l^2$, where (EI) is the bending stiffness of the beam.

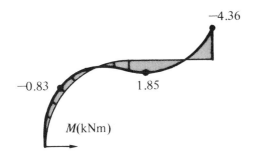

Fig. S8

CHAPTER 5

5.1 $\begin{bmatrix} 400.0 & 0 \\ 0 & 400.0 \end{bmatrix}$ MN/m; 0.625, 1.083 mm; 62.5, 120.7, -32.3, 108.2, -108.2, 32.3, -120.7, -62.5 MN.

5.2

(a)
$$\begin{bmatrix} 5.0 & 8.66 & -5.0 & -8.66 \\ 8.66 & 15.0 & -8.66 & -15.0 \\ -5.0 & -8.66 & 5.0 & 8.66 \\ -8.66 & -15.0 & 8.66 & 15.0 \end{bmatrix} \text{MN/m} ;$$

(b)
$$\begin{bmatrix} 0 & 0 & 0 & 0 \\ 0 & 0 & 0 & 0 \\ 0 & 0 & 11.93 & 5.96 \\ 0 & 0 & 5.96 & 2.98 \end{bmatrix} \text{MN/m},$$

5.3
$$\begin{bmatrix} (2+\sqrt{2}) & 0 & -\sqrt{2} & 0 & 1.0 \\ 0 & 2.0 & 0 & 0 & -1.0 \\ -\sqrt{2} & 0 & (2+\sqrt{2}) & 0 & -1.0 \\ 0 & 0 & 0 & 2.0 & -1.0 \\ 1.0 & -1.0 & -1.0 & -1.0 & (2+\sqrt{2}) \end{bmatrix} (EA/2\sqrt{2}a) .$$

k is calculated assuming joint 3 is restrained horizontally. (One of the free joints has to be restrained horizontally, otherwise rotational equilibrium about an axis through the point of suspension cannot be maintained. If this restraint were not present, the global stiffness matrix would be singular and k^{-1} indeterminate.)

5.4
$$\begin{bmatrix} 4.0 & 2.0 & 0 & 0 \\ 2.0 & 8.0 & 2.0 & 0 \\ 0 & 2.0 & 8.0 & 2.0 \\ 0 & 0 & 2.0 & 4.0 \end{bmatrix} (EI/a) .$$

(Solve as a continuous beam subject to concentrated couples at joints 2 and 3.)

5.5 -26.56 kNm; -29.86 kNm; -112.5 kNm.

5.6

$$\begin{bmatrix} 3.0 & 1.5 & 0 & 0 \\ 1.5 & 5.5 & 1.25 & 0 \\ 0 & 1.25 & 5.5 & 1.5 \\ 0 & 0 & 1.5 & 3.0 \end{bmatrix} \times 10^9 \text{ Nm/rad;} \qquad \begin{bmatrix} -416.7 \\ -638.6 \\ 501.7 \\ 237.4 \end{bmatrix} \times 10^3 \text{ Nm.}$$

The global force at joint 3 is reduced by about 3.5% — a rough estimate of the error involved in the approximation.

5.7 61.52, 47.85 kNm (both positive). **5.8** 0.0259 × 10⁻³, 0.0056 × 10⁻³ rad; −1.29 kNm. **5.9** −43.56, −33.23 kNm. **5.10** 0.351 × 10⁻³, 2.469 × 10⁻³ rad, 22.96 mm; 28.40, −47.37 kNm.

5.11

$$\begin{bmatrix} 57.4 & 20.0 & 8.7 & 0 & 11.3 & -11.3 \\ 20.0 & 57.4 & 0 & 8.7 & 11.3 & -11.3 \\ 8.7 & 0 & 70.7 & 20.0 & 11.3 & -4.7 \\ 0 & 8.7 & 20.0 & 70.7 & 11.3 & -4.7 \\ 11.3 & 11.3 & 11.3 & 11.3 & 19.8 & -19.8 \\ -11.3 & -11.3 & -4.7 & -4.7 & -19.8 & 28.6 \end{bmatrix} \begin{matrix} \text{MNm/rad,} \\ \text{MN/m, etc.,} \end{matrix} \begin{bmatrix} 1.76 \\ 0 \\ -0.26 \\ 0 \\ 4.6 \\ 7.6 \end{bmatrix} \begin{matrix} \text{kNm} \\ \\ \text{kNm} \\ \\ \text{kN} \\ \text{kN} \end{matrix}$$

5.12

$$\begin{bmatrix} 40.0 & 10.0 & 6.3 \\ 10.0 & 37.3 & 4.3 \\ 6.3 & 4.3 & 33.0 \end{bmatrix} \begin{matrix} \text{MNm/rad,} \\ \text{MN/m, etc.,} \end{matrix} \begin{bmatrix} 0 \\ -1.78 \\ -5.33 \end{bmatrix} \begin{matrix} \\ \text{kNm} \\ \text{kN} \end{matrix}$$

(Note that when the framework is subjected to a sidesway d_3, joint 2 moves at right angles to the axis of the inclined column, the horizontal component of the displacement being equal to d_3. This inclined displacement of joint 2 should be taken into account when calculating the transformation matrices of the beam and the column.)
5.13 The member end couples are given by: 1.60, −1.29; 1.29, −0.54; 0.09, 0.04; 0.45, −0.25 kNm. **5.14** The *bending moments* at the ends of the spans of the beam are given by: −70.24, −76.83 kNm; −15.37, −15.37 kNm; −76.83 −70.24 kNm. **5.15** −33.32, −53.51 kNm (on either side of joint 2), −33.05, −20.58 kNm (on either side of joint 4).

5.16

$$\begin{bmatrix} 299.8 & 0 & 3.2 & -298.5 & 0 & 0 \\ 0 & 267.2 & 8.0 & 0 & -4.8 & 8.0 \\ 3.2 & 8.0 & 28.6 & 0 & -8.0 & 9.0 \\ -298.5 & 0 & 0 & 299.8 & 0 & 3.2 \\ 0 & -4.8 & -8.0 & 0 & 267.2 & -8.0 \\ 0 & 8.0 & 9.0 & 3.2 & -8.0 & 28.6 \end{bmatrix} \begin{matrix} \\ \\ \\ \text{MNm/rad,} \\ \text{MN/m, etc.,} \\ \\ \\ \end{matrix} \begin{bmatrix} 50.0 \\ 0 \\ 41.67 \\ 0 \\ 0 \\ 0 \end{bmatrix} \begin{matrix} \text{kN} \\ \\ \text{kNm} \\ \\ \\ \\ \end{matrix}$$

5.17

$$\begin{bmatrix} 2515.0 & 0 & 15.0 & -2500.0 & 0 & 0 \\ 0 & 2515.0 & 15.0 & 0 & -15.0 & 15.0 \\ 15.0 & 15.0 & 40.0 & 0 & -15.0 & 10.0 \\ -2500.0 & 0 & 0 & 3048.6 & -933.2 & 9.7 \\ 0 & -15.0 & -15.0 & -933.2 & 1641.0 & -9.4 \\ 0 & 15.0 & 10.0 & 9.7 & -9.4 & 37.3 \end{bmatrix} \begin{matrix} \\ \\ \\ \text{MNm/rad,} \\ \text{MN/m, etc.,} \\ \\ \end{matrix} \begin{bmatrix} 0 \\ 0 \\ 0 \\ -4.0 \\ -2.31 \\ -1.78 \end{bmatrix} \begin{matrix} \\ \\ \\ \text{kN} \\ \text{kN} \\ \text{kNm} \end{matrix} \;.$$

5.18 $$\begin{bmatrix} -12.5 \\ -21.7 \\ 12.5 \\ 21.7 \end{bmatrix} \text{kN}, \quad \begin{bmatrix} 0 \\ 0 \\ 22.4 \\ 11.2 \end{bmatrix} \text{kN} \;.$$

5.19 $-Pa^3/(48.0E_bI_b + \sqrt{2}a^2 E_tA_t), \; Pa^2/(64.0E_bI_b).$

5.20

$$\begin{bmatrix} 300.0 & -90.0 & 150.0 & 0 & 0 & 0 \\ -90.0 & 141.3 & 0 & 90.0 & 0 & -69.3 \\ 150.0 & 0 & 600.0 & 150.0 & 0 & 0 \\ 0 & 90.0 & 150.0 & 300.0 & 0 & 0 \\ 0 & 0 & 0 & 0 & 52.0 & 0 \\ 0 & -69.3 & 0 & 0 & 0 & 86.6 \end{bmatrix} \begin{matrix} \\ \\ \text{MNm/rad} \\ \text{MN/m, etc.,} \\ \\ \\ \end{matrix} \begin{bmatrix} -31.25 \\ -37.25 \\ 31.25 \\ 0 \\ 0 \\ 0 \end{bmatrix} \begin{matrix} \text{kNm} \\ \text{kN} \\ \text{kNm} \\ \\ \\ \\ \end{matrix}$$

Solutions to Problems and Hints

The degrees of freedom are the rotational displacement of joint 1, the vertical and rotational displacements of joint 2, the rotational displacement of joint 3, and the horizontal and vertical displacements of joint 4.

Index